ASTRONOMY AND
ASTROPHYSICS LIBRARY

Eugene F. Milone
and William J. F. Wilson

Solar System Astrophysics

Planetary Atmospheres and
the Outer Solar System

 Springer

Eugene F. Milone
Dept. Physics & Astronomy
University of Calgary, Calgary
2500 University Drive NW
Calgary, Alberta T2N 1N4
Canada
milone@ucalgary.ca

Willam J. F. Wilson
Dept. Physics & Astronomy
University of Calgary, Calgary
2500 University Drive NW
Calgary, Alberta T2N 1N4
Canada
wjfwilso@ucalgary.ca

Cover Illustration: "Mysterium Cosmographicum" by David Mouritsen

Library of Congress Control Number: 2007933183

Astronomy and Astrophysics Library Series ISSN 0941-7834

ISBN: 978-0-387-73156-8   e-ISBN: 978-0-387-73157-5
ISBN: 978-0-387-73153-7 (set)

Printed on acid-free paper.

Observing the Sun, along with a few other aspects of astronomy, can be dangerous. Neither the publisher nor the author accepts any legal responsibility or liability for personal loss or injury caused, or alleged to have been caused, by any information or recommendation contained in this book.

9 8 7 6 5 4 3 2 1

springer.com

Preface

This work is appearing in two parts because its mass is the result of combining detailed exposition and recent scholarship. Book I, dealing mainly with the inner solar system, and Book II, mainly on the outer solar system, represent the combined, annually updated, course notes of E. F. Milone and W. J. F. Wilson for the undergraduate course in solar system astrophysics that has been taught as part of the Astrophysics Program at the University of Calgary since the 1970s. The course, and so the book, assumes an initial course in astronomy and first-year courses in mathematics and physics. The relevant concepts of mathematics, geology, and chemistry that are required for the course are introduced within the text itself.

Solar System Astrophysics is intended for use by second- and third-year astrophysics majors, but other science students have also found the course notes rewarding. We therefore expect that students and instructors from other disciplines will also find the text a useful treatment. Finally, we think the work will be a suitable resource for amateurs with some background in science or mathematics. Most of the mathematical formulae presented in the text are derived in logical sequences. This makes for large numbers of equations, but it also makes for relatively clear derivations. The derivations are found mainly in Chapters 2–6 in Book I, subtitled, *Background Science and the Inner Solar System*, and in Chapters 10 and 11 in Book II subtitled, *Planetary Atmospheres and the Outer Solar System*. Equations are found in the other chapters as well but these contain more expository material and recent scholarship than some of the earlier chapters. Thus, Chapters 8 and 9, and 12–16 contain some useful derivations, but also much imagery and results of modern studies.

The first volume starts with a description of historical perceptions of the solar system and universe, in narrowing perspective over the centuries, reflecting the history (until the present century, when extra-solar planets again have begun to broaden our focus). The second chapter treats the basic concepts in the geometry of the circle and of the sphere, reviewing and extending material from introductory astronomy courses, such as spherical coordinate transformations. The third chapter then reviews basic mechanics and two-body systems, orbital description, and the computations of ephemerides, then progresses to the restricted three-body and n-body cases, and concludes with a discussion of perturbations. The fourth chapter treats the core of the solar system, the Sun, and is not a bad introduction to solar or stellar astrophysics; the place of the Sun in the galaxy and in the context of other stars is described, and radiative transport, optical depth, and limb-darkening are introduced. In Chapter 5, the structure and composition of the Earth are discussed, the Adams–Williamson equation is derived, and its use for determining the march of pressure and density with radius described. In Chapter 6, the thermal structure and energy transport through the Earth are treated, and in this chapter the basic ideas of thermodynamics are put to use. Extending the discussion of the Earth's interior, Chapter 7 describes the rocks and minerals in the Earth and their crystalline structure. Chapter 8 treats the Moon, its structures, and its origins, making use of the developments of the preceding chapters. In Chapter 9, the surfaces of the other terrestrial planets are described, beginning with Mercury. In each of the three sections of this chapter, a brief historical discussion is followed by descriptions of modern ground-based and space mission results, with some of the spectacular imagery of Venus and Mars. The chapter concludes with a description of the evidence for water and surface modification on Mars. This concludes the discussion of the inner solar system.

The second volume begins in Chapter 10 with an extensive treatment of the physics and chemistry of the atmosphere and ionosphere of the Earth and an introduction to meteorology, and this discussion is extended to the atmospheres of Venus and Mars. Chapter 11 treats the magnetospheres of the inner planets, after a brief exposition of electromagnetic theory. In Chapter 12, we begin to treat the outer solar system, beginning with the gas giants. The structure, composition, and particle environments around these planets are discussed, and this is continued in Chapter 13, where the natural satellites and rings of these objects are treated in detail, with abundant use made of the missions to the outer planets. In Chapter 14, we discuss comets, beginning with an historical introduction that highlights the importance of comet studies to the development of modern astronomy. It summarizes the ground- and space-based imagery and discoveries, but makes use of earlier derivations to discuss cometary orbits. This chapter ends with the demise of comets and the physics of meteors. Chapter 15 treats the study of meteorites and the remaining small bodies of the solar system, the asteroids (*aka* minor planets, planetoids), and the outer solar system "Kuiper Belt" objects, and the closely related objects known as centaurs, plutinos, cubewanos, and others, all of which are numbered as asteroids. The chapter ends with discussions of the origin of the solar system and of debris disks around other stars, which point to widespread evidence of the birth of other planetary systems. Finally, in Chapter 16, we discuss the methods and results of extra-solar planet searches, the distinctions among stars, brown dwarfs, and planets, and we explore the origins of planetary systems in this wider context.

At the end of each chapter we have a series of challenges. Instructors may use these as homework assignments, each due two weeks after the material from that chapter were discussed in class; *we* did! The general reader may find them helpful as focusing aids.

January 22, 2008   *E. F. Milone & W. J. F. Wilson*

Acknowledgments

These volumes owe their origin to more than 30 years of solar system classes in the Astrophysics Program at the University of Calgary, called, at various times, Geophysics 375, Astrophysics 301, 309, and 409. Therefore, we acknowledge, first, the students who took these courses and provided feedback. It is also a pleasure to thank the following people for their contributions:

David Mouritsen, formerly of Calgary and now in Toronto, provided for Chapter 1 and our covers an image of his original work of art, an interpretation of Kepler's *Mysterium Cosmographicum*, in which the orbits of the planets are inscribed within solid geometric figures.

In Chapter 3, the latest version of David Bradstreet's software package, *Binary Maker 3* was used to create an image to illustrate restricted three-body solutions.

University of Calgary Professor Emeritus Alan Clark gave us an image of an active region and detailed comments on the solar physics material of Chapter 4; Dr Rouppe van der Voort of the University of Oslo provided high-quality images of two other active region figures, for Chapter 4; the late Dr Richard Tousey of the US Naval Research Laboratory provided slides of some of the images, subsequently scanned for Chapter 4; limb-darkened spectral distribution plots were provided by Dr Robert L. Kurucz, of the Harvard-Smithsonian Center for Astrophysics; Dr. Charles Wolff, of Goddard Space Flight Center, NASA, reviewed the solar oscillations sections and provided helpful suggestions.

Dr. D. J. Stevenson provided helpful criticism of our lunar origins figures, and Dr. Robin Canup kindly prepared panels of her lunar simulations for our Fig. 8.10.

Dr Andrew Yau provided excellent notes as a guest lecturer in Asph 409 on the Martian atmosphere and its evolution, which contributed to our knowledge of the material presented in Chapters 9, 10, and 11; similarly, lectures by Professor J. S. Murphree of the University of Calgary illuminated the magnetospheric material described in Chapter 11. Dr. H. Nair was kind enough to provide both permission and data for Fig. 11.3.

NASA's online photo gallery provided many of the images in Chapters 8, 9, 12, 13, 14, and some of those in Chapter 15; additional images were provided by the Naval Research Laboratory (of both the Sun and the

Moon). Some of these and other images involved work by other institutions, such as the U.S. Geological Survey, the Jet Propulsion Laboratory, Arizona State Univ., Cornell, the European Space Agency, the Italian Space Agency (ASI), CalTech, Univ. of Arizona, Space Science Institute, Boulder, the German Air and Space Center (DLR), Brown University, the Voyagers and the Cassini Imaging Teams, CICLOPS, the Hubble Space Telescope, University of Maryland, the Minor Planet Center, Applied Physics Laboratory of the Johns Hopkins University, and the many individual sources, whether cited in captions or not, who contributed their talents to producing these images.

Dr. John Trauger provided a high resolution UV image of Saturn and its auroras for Chapter 12.

Dr William Reach, Caltech, provided an infrared mosaic image of Comet Schwassmann–Wachmann 3, and Mr John Mirtle of Calgary provided many of the comet images for Chapter 14, including those of Comets 109P/Swift–Tuttle, C/1995 O1 (Hale–Bopp), C/Hyakutake, Lee, C/Ikeya–Zhang, Brorsen–Metcalfe, and Machholz; Dr. Michael J. Mumma of NASA's GSFC provided an important correction, and Professor Michael F. A'Hearn of the University of Maryland critiqued the comet content of Chapter 14.

Dr Allan Treiman, of the Lunar and Planetary Institute, Houston, was kind enough to provide background material on the debate over the ALH84001 organic life question, mentioned in Chapter 15. Mr Matthias Busch made available his *Easy-Sky* images of asteroid family distributions, and the Minor Planet Center provided the high-resolution figures of the distributions of the minor planets in the inner and outer solar system.

Dr Charles Lineweaver, University of New South Wales, provided a convincing illustration for the brown dwarf desert, illustrated in Chapter 16; University of Calgary graduate student Michael Williams provided several figures from his MSc thesis for Chapter 16.

Mr Alexander Jack assisted in updating and improving the readability of equations and text in some of the early chapters, and he and Ms Veronica Jack assisted in developing the tables of the extra-solar planets and their host stars for Chapter 16.

In addition, we thank the many authors, journals, and publishers who have given us permission to use their figures and tabular material or adaptations thereof, freely. Finally, it is also a pleasure to thank Springer editors Dr Hans Koelsch, Dr Harry Blom, and their associate, Christopher Coughlin, for their support for this project.

Contents

Planetary Atmospheres and the Outer Solar System

10. Planetary Atmospheres 1
- 10.1. Atmospheric Constituents 1
- 10.2. Atmospheric Structure 4
  - 10.2.1. Pressure Variation with Height 4
  - 10.2.2. Temperature Variation with Height 7
- 10.3. Circulation in the Atmosphere 11
  - 10.3.1. Centrifugal and Coriolis Forces 11
  - 10.3.2. Physical Effects of the Centrifugal and Coriolis Forces 13
    - 10.3.2.1. The Centrifugal Force 13
    - 10.3.2.2. The Coriolis Force 15
  - 10.3.3. Pressure Gradient Force 16
  - 10.3.4. Friction 17
  - 10.3.5. Geostrophic Balance and Geostrophic Winds 18
  - 10.3.6. Thermal Effects 19
    - 10.3.6.1. Thermal Circulation 19
    - 10.3.6.2. The Thermal Wind 20
  - 10.3.7. Global Circulation 22
    - 10.3.7.1. The Observed Surface Pattern 22
    - 10.3.7.2. The Hadley Cell 24
    - 10.3.7.3. The Ferrel and Polar Cells 25
    - 10.3.7.4. Eddie Motions in the Westerlies 26
    - 10.3.7.5. Air Masses and Fronts 27
    - 10.3.7.6. Jet Streams 27
- 10.4. Atmospheric Effects on the Heat Budget 30
  - 10.4.1. The Earth 30
    - 10.4.1.1. Troposphere of the Earth 31
    - 10.4.1.2. Stratosphere and Mesosphere 32
    - 10.4.1.3. Thermosphere 33
    - 10.4.1.4. Exosphere 34
  - 10.4.2. Mars 34
    - 10.4.2.1. Troposphere 36

10.4.2.2.	Stratomesosphere	36
10.4.2.3.	Thermosphere	36

10.4.3. Venus... 36
    10.4.3.1. Troposphere 37
    10.4.3.2. Stratomesosphere 37
    10.4.3.3. Thermosphere 37
10.5. Planetary Circulation Effects 38
  10.5.1. Circulation and the Coriolis Force.................. 38
  10.5.2. Meridional (N–S) Circulation...................... 39
  10.5.3. Zonal (E–W) Circulation.......................... 39
    10.5.3.1. Mars 40
    10.5.3.2. Venus 41
      10.5.3.2.1. Atmospheric Superrotation 41
      10.5.3.2.2. Cyclostrophic Balance 41
      10.5.3.2.3. Atmospheric Angular Momentum.................. 42
      10.5.3.2.4. Superrotation vs. Other Circulation Patterns......... 43
  10.5.4. Other Considerations 44
    10.5.4.1. Latent Heat 44
    10.5.4.2. Thermal Inertia 44
    10.5.4.3. Brunt-Väisälä Frequency:................ 44
    10.5.4.4. Diffusion and Mixing in Planetary Atmospheres 45
      10.5.4.4.1. Diffusion 45
      10.5.4.4.2. Diffusion vs. Mixing 47
      10.5.4.4.3. The Homopause............. 48
  10.5.5. Chemical Cycles 48
    10.5.5.1. Carbon Cycle (Earth) 48
    10.5.5.2. Oxygen Cycle (Earth) 48
    10.5.5.3. Nitrogen Cycle (Earth) 49
    10.5.5.4. Sulfur Cycle (Earth).................... 51
    10.5.5.5. Sulfur Cycle (Venus).................... 52
    10.5.5.6. Thermospheric Chemistry of Neutrals (Earth)................................. 53
  10.5.6. Excess Radiation................................. 54

11. Planetary Ionospheres and Magnetospheres 57
11.1. Earth: Ionospheric Layers 57
  11.1.1. The F Layer 57
    11.1.1.1. Atoms and Ions in the F Layer 57
    11.1.1.2. Production Mechanisms 58

		11.1.1.3.	Loss Mechanisms	58
		11.1.1.4.	Ion Concentration vs. Altitude	59
		11.1.1.5.	Charge Separation	60
	11.1.2.	The E Layer		60
		11.1.2.1.	Atoms and Ions in the E Layer	60
		11.1.2.2.	Production Mechanisms	60
		11.1.2.3.	Loss Mechanisms	61
	11.1.3.	The D Layer		61
		11.1.3.1.	Dominant Ions	61
		11.1.3.2.	Production Mechanisms	62
		11.1.3.3.	Loss Mechanisms	63
	11.1.4.	Reflection of Radio Waves		63
11.2.	Atmospheric and Ionospheric Chemistry on Mars and Venus			64
	11.2.1.	Neutral Atmosphere of Mars		64
	11.2.2.	Neutral Atmosphere of Venus		65
	11.2.3.	Ionosphere of Mars		66
		11.2.3.1.	Dominant Ions	66
		11.2.3.2.	Production Mechanisms	66
		11.2.3.3.	Loss Mechanisms	67
	11.2.4.	Ionosphere of Venus		67
		11.2.4.1.	Dominant Ions	67
		11.2.4.2.	Production and Loss Mechanisms	68
	11.2.5.	Atmospheric Escape Mechanisms		68
		11.2.5.1.	Jeans Escape	68
		11.2.5.2.	Suprathermal Atoms and Ions	69
11.3.	Solar Wind			69
11.4.	Maxwell's Equations and the Plasma Frequency			70
	11.4.1.	Maxwell's Equations		70
	11.4.2.	Application to a Polarized Wave		72
11.5.	The Earth's Magnetosphere			75
	11.5.1.	Forces Acting on Charged Particles		79
		11.5.1.1.	The Lorentz Force	79
		11.5.1.2.	The Gravitational Force	79
	11.5.2.	$\vec{E}$ Uniform and Time-Independent; $\vec{B} = 0$		79
	11.5.3.	$\vec{B}$ Uniform and Time-Independent; $\vec{E} = 0$		80
	11.5.4.	Guiding Center		82
	11.5.5.	Diamagnetism		83
	11.5.6.	$\vec{E} \times \vec{B}$ Drift and Field-Aligned Currents		84
	11.5.7.	$\vec{E} \times \vec{B}$ Drift with Collisions		86
	11.5.8.	Polarization Drift		87
	11.5.9.	Gradient and Curvature Drift		88
		11.5.9.1.	Gradient Drift	88
		11.5.9.2.	Curvature Drift	89

11.6.   Electric Currents in the Ionosphere and Magnetosphere....   90
        11.6.1. The Ionospheric Dynamo.........................   90
                11.6.1.1.   The S_q ("Solar Quiet") Current System..   90
                11.6.1.2.   The L_q ("Lunar Quiet") Current
                            System..............................   90
        11.6.2. Boundary Current   91
        11.6.3. Ring Current   92
                11.6.3.1.   Magnetic Mirrors.....................   92
                11.6.3.2.   Characteristics of the Motion...........   93
                11.6.3.3.   Trapping and Precipitation.............   95
                11.6.3.4.   The Ring Current   98
        11.6.4. Magnetic Storms.................................   99
        11.6.5. Magnetospheric Convection   100
        11.6.6. The Magnetotail Current Sheet   101
        11.6.7. Magnetospheric Substorms.......................   102
        11.6.8. Coupling Between the Magnetosphere
                and the Ionosphere..............................   103
11.7.   Magnetospheres of Mercury, Venus, and Mars.............   107
        11.7.1. Mercury   107
        11.7.2. Venus...   110
        11.7.3. Mars..   113

12. The Giant Planets **119**
12.1.   Jupiter...   119
        12.1.1. Visible Phenomena...............................   121
        12.1.2. Jovian Atmospheric Structure....................   126
12.2.   Saturn   128
12.3.   Uranus...   131
12.4.   Neptune..   133
12.5.   Internal Pressures   136
12.6.   Excess Radiation   137
12.7.   Ionospheres of the Giant Planets   139
12.8.   The Jovian Magnetosphere   140
        12.8.1. Inner Magnetosphere of Jupiter   141
        12.8.2. Middle Magnetosphere of Jupiter.................   141
        12.8.3. Outer Magnetosphere of Jupiter..................   142
        12.8.4. Interaction with Io   143
                12.8.4.1.   DAM.................................   145
                12.8.4.2.   Neutral and Ionized Population of the
                            Jovian Magnetosphere..................   145
        12.8.5. Io as a Source of Particles......................   146
                12.8.5.1.   Rate of Supply   146
                12.8.5.2.   Loss Mechanisms from Io...............   146

13. Satellite and Ring Systems **151**

13.1. Satellites 151

13.1.1. The Moons of Mars 161

13.1.2. The Moons of Jupiter 163

13.1.2.1. Io 164

13.1.2.2. Europa 166

13.1.2.3. Ganymede 168

13.1.2.4. Callisto 170

13.1.3. The Moons of Saturn 173

13.1.3.1. Titan 175

13.1.4. Uranian Moons 183

13.1.5. Neptunian Moons 184

13.1.5.1. Triton 185

13.1.6. Pluto–Charon 186

13.2. Origins of Ring systems 189

13.3. Ring Structures 192

13.3.1. Jovian Rings 192

13.3.2. Saturnian Rings 192

13.3.3. Uranian Rings 198

13.3.4. Neptunian Rings 200

13.3.5. Nature and Possible Origins of the
Ring Structures 201

13.4. Orbital Stability of the Moons and the Case of Pluto 203

13.4.1. Satellite Stability 203

13.4.2. Conjectures about Pluto 205

13.5. Origins of the Moons 207

14. Comets and Meteors **213**

14.1. Comets in History 213

14.1.1. Early History 213

14.1.2. Tycho Brahe and the Comet of 1577 215

14.1.3. Later Historical Studies 216

14.2. Comet Designations 218

14.3. Cometary Orbits 220

14.4. Typical and Historically Important Comets 225

14.5. Cometary Structure 229

14.6. Cometary Composition 231

14.7. Origins of Comets 237

14.8. Cometary Demise 240

14.9. Meteor Showers 242

14.10. Meteors 243

14.10.1. Basic Meteor Phenomena and Circumstances 243

14.10.2. Meteor Heating and Incandescence 243

14.11. Micrometeorites 249
14.12. Dust Destinies 251
        14.12.1. Radiation Pressure 251

15. Meteorites, Asteroids and the Age and Origin
     of the Solar System **257**
     15.1.   Stones from Heaven 257
             15.1.1. Categories and Nomenclature of Meteorites 258
                     15.1.1.1.   Broad Categories 258
                     15.1.1.2.   Another Distinction: *Falls* and *Finds* 259
                     15.1.1.3.   Nomenclature 259
             15.1.2. Petrographic Categories 259
             15.1.3. Meteorite Groupings and Subgroupings 260
                     15.1.3.1.   Undifferentiated Meteorites 260
                     15.1.3.2.   Differentiated Meteorites 262
     15.2.   Undifferentiated Meteorites: the Chondrites 265
             15.2.1. Defining the Chondrites 265
             15.2.2. Carbonaceous Chondrites 269
             15.2.3. Ordinary Chondrites 270
             15.2.4. Enstatites 270
             15.2.5. The R Group 271
             15.2.6. Former Members, from the IAB Clan 271
             15.2.7. Origins of the Chondrites 271
     15.3.   DSR Meteorites 273
             15.3.1. The Igneous Clan 273
             15.3.2. Other DSR Meteorites 274
     15.4.   Iron Meteorites 277
     15.5.   Ages and Origins of Meteorites 279
             15.5.1. Radiogenic Ages 279
             15.5.2. Gas Retention Ages 282
             15.5.3. Cosmic Ray Exposure Ages 283
             15.5.4. Case Study: The Zagami SNC Basaltic
                     Shergottite 283
     15.6.   Other Sources of Evidence for Meteoritic Origins 284
     15.7.   Parent Bodies and the Asteroids 286
             15.7.1. The Discovery of Ceres 286
             15.7.2. Nomenclature 287
             15.7.3. Families of Orbits 290
             15.7.4. Dimensions and Masses of Asteroids 293
                     15.7.4.1.   Asteroid Dimensions and Albedo 293
                     15.7.4.2.   Asteroid Masses and Densities 297
             15.7.5. Asteroids and Meteorites 298

15.8.   Implications for the Origin of the Solar System 302
15.9.   The Solar Nebula 303
15.10. The Proto-Planetary Disk 305

16. Extra-Solar Planetary Systems 313
16.1.   Historical Perspective 313
16.2.   Methods to Find "Small"-Mass Companions 338
        16.2.1. Radial Velocity Variations of the Visible
                Component 338
        16.2.2. Transit eclipses 342
        16.2.3. Astrometric Variations 347
        16.2.4. Gravitational Lensing 348
        16.2.5. Direct Imaging and Spectroscopy 350
        16.2.6. Pulsar Timings 351
        16.2.7. Indirect Effects 352
16.3.   Definitions of Planets and Brown Dwarfs 353
16.4.   Extra-Solar Planets Detected
        or Strongly Suspected 356
        16.4.1. HD 209458b 358
        16.4.2. The Multi-Planet System of v Andromedae 360
        16.4.3. The Multi-Planet System of 55 Cancri 361
        16.4.4. The Multi-Planet System of HD 37124 361
        16.4.5. The Multi-Planet System of HD 69830 362
        16.4.6. The Multi-Planet System of Gliese J 876 363
        16.4.7. The ϵ Eridani System 363
        16.4.8. The TrES-1 System 364
        16.4.9. The XO-1 System 365
        16.4.10. The OGLE-TR-10 System 366
16.5.   Origins of Brown Dwarfs and Planets 367

Index 387

10. Planetary Atmospheres

10.1 Atmospheric Constituents

The constituents of a planetary atmosphere are determined in a general way by the likelihood that a given constituent will be retained, rather than lost by evaporation into space over long periods of time. This likelihood is determined by three factors: (1) the equilibrium temperature of the planet, because the hotter the atmosphere, the larger the mean kinetic energy of its molecules,

$$E = \tfrac{3}{2} kT \tag{10.1}$$

where $k = 1.380658(12) \times 10^{-23}\,\text{J/K}$ is the Boltzmann constant, T is the equilibrium temperature in kelvin, and the number in parentheses is the uncertainty in the last two digits of the constant; (2) the molecular weight, m, of each atmospheric constituent, because equipartition of energy requires that more massive particles have smaller mean velocities,

$$E = \tfrac{1}{2} m v^2 \tag{10.2}$$

and (3) the planet's escape velocity,

$$v_{\text{esc}} = \sqrt{\frac{2GM}{R}} \tag{10.3}$$

where M and R are the planetary mass and radius, respectively. In a region of atmosphere in thermal equilibrium, the distribution of the number of molecules vs. the speed at which they are moving at any instant is said to be *Maxwellian* (See Schlosser et al. 1991/4, Fig. 18.1).

The *molecular weight* of any species of molecule can be written

$$m = \mu m_{\text{u}} \tag{10.4}$$

where $m_{\text{u}} = (1/12)[m(^{12}\text{C})] = 1.6605402(10) \times 10^{-27}\,\text{kg}$ is the unit atomic mass, and the dimensionless quantity μ is expressed in units of m_{u}. It is not uncommon to refer to μ as the *molecular weight*, the units of m_{u} being understood.

From (10.1), (10.2) and (10.4), the root mean square speed of any constituent can be written in terms of μ as

$$v = \sqrt{\frac{3kT}{m}} = 157.94\sqrt{\frac{T}{\mu}} \text{ m/s} \qquad (10.5)$$

A very massive planet will retain even the lightest gases over a wide range of temperatures. Quite generally, if $v/v_{esc} = 1/3$, the atmosphere can be expected to be lost in weeks; if $v/v_{esc} = 1/4$, 10,000 years; if $v/v_{esc} = 1/5$, 10^8 years; and if $v/v_{esc} = 1/6$, it can be considered retained for eons, or billions of years (Gy), assuming no major departure from the assumed equilibrium planetary temperature over this interval of time.

From (10.3) and (10.5), the lower limit for the molecular weight of a molecule that can be *retained* is

$$\mu \geq \left(157.94\frac{\sqrt{T}}{(v_{esc}/6)}\right)^2 = 8.980 \times 10^5 \frac{T}{v_{esc}^2} \qquad (10.6)$$

From (6.31) of Milone & Wilson (2008), the equilibrium temperature for an assumed black body, rapidly rotating planet is:

$$T = \{\mathcal{L}_\odot(1 - A)/[16\pi\sigma r^2]\}^{\frac{1}{4}} = 1.078 \times 10^8 \left[(1 - A)/r^2\right]^{\frac{1}{4}} \text{ K}$$

where A is the bolometric albedo, $\sigma = 5.671 \times 10^{-8}$ W m^{-2} K^{-4} is the Stefan–Boltzmann constant, r is the distance from the sun in meters, and the solar luminosity is $\mathcal{L}_\odot = 3.845 \times 10^{26}$ W (Cox 2000, p. 340).

For a particular molecule to be retained over a significant fraction of the solar system's lifetime (assuming the equilibrium temperature correctly indicates the effective mean temperature of the planet, and there is no significant change in this temperature over this interval),

$$\mu \geq 8.980 \times 10^5 \times 1.078 \times 10^8 \frac{[(1 - A)/r^2]^{1/4}}{2GM/R} = 7.254 \times 10^{23} \frac{R}{M}\left(\frac{1 - A}{r^2}\right)^{1/4} \qquad (10.7)$$

where r, R, M, and $G = 6.6726 \times 10^{-11}$ N m^2/kg^2 are in SI units. The corresponding numerical constant for a slowly (more precisely, synchronously) rotating planet is greater by a factor $2^{\frac{1}{4}} = 1.189$. Finally, with R and M in units of the Earth's radius and mass (6.378×10^6 m, 5.974×10^{24} kg, respectively) and a in au, (10.7) becomes

$$\mu \geq 2.002\frac{R/R_\oplus}{M/M_\oplus}\left(\frac{1 - A}{a^2}\right)^{1/4} \qquad (10.8)$$

As an example, for the Earth, $A \approx 0.307$, $\mu >\sim 2 \times 0.91 = 1.82$, provided T is correctly given by (10.7); i.e., $< T_\oplus >= T_{eq} = 254$ K. In point of fact, the

Earth is slightly warmer because of its atmosphere, so that $< T_\oplus > = 288\,\mathrm{K}$. Using this value directly in (10.6), one obtains, $\mu = 2.07$.

Thus, the Earth is marginally unable to retain hydrogen at present, but should have retained helium ($\mu \sim 4$). It does not, in fact, have large amounts of helium in its atmosphere, suggesting that the Earth's surface was much hotter in the distant past, perhaps as a result of energy of accretion (or reconstitution, if the Mars-sized impact event suggested for the origin of the Moon did in fact occur).

See Schlosser et al. (1991/4, pp. 94–97) for further discussion of the species of molecules retainable in planetary atmospheres.

One of the more interesting concepts in planetary physics is the probability of escape of gases from the sub-solar region, where the instantaneous temperature is much higher than the global or even hemispherical average. How do we know if the evaporation from this location alone is the determining factor?

The answer probably lies in the *mean free path* of a molecule on the surface of a planet. The term refers to the average separation of molecules before colliding with other molecules. If a molecule receives sufficient energy for it to escape, how long does it take before it collides with another molecule? The problem and the solutions to it are described in, for example, Jeans' (1952) *Kinetic Theory of Gases*. The probability of collision depends strongly on the atmospheric density. For a particular molecule, the distance, d, traveled before a collision with some other, stationary, molecule may be written

$$d = \frac{1}{\pi n R^2} \qquad (10.9)$$

where n is the number density [the number of particles (in this case, molecules) per unit volume], and R is a characteristic distance about equal to the mean radius of a molecule. The time it takes to cover this distance depends on the speed of the molecule. The average mean free path of a molecule, the speeds of which are expected to correspond to a Maxwellian distribution, is

$$\frac{1}{\sqrt{2}\pi n R^2} \qquad (10.10)$$

This number is about $6 \times 10^{-8}\,\mathrm{m}$ for Earth's atmosphere, or about $320R$ (Jeans 1952 pp. 44–49).

Thus, a great many collisions may occur prior to escape, the angle of alteration of direction in each collision being a further variable, and, at each collision, energy may be lost as well as gained; obviously, it will be easier to escape in a less dense environment.

Next we discuss the structure of atmospheres, and the behavior of pressure with height.

10.2 Atmospheric Structure

10.2.1 Pressure Variation with Height

For an atmosphere to be a reasonably permanent feature of a planet, the atmosphere must be in hydrostatic equilibrium, i.e., it should neither accelerate outward nor collapse inward. Here, we will assume that the atmosphere can be approximated by spherical symmetry, e.g., only the radial components of force vectors are non-zero.

Figure 10.1 shows a free-body diagram for a small parcel of gas in such an atmosphere, in the form of a short cylinder of thickness dr, cross-section A, volume $dV = A dr$, density ρ, and mass $dm = \rho dV$, located at a distance r from the center of the planet. The unit vector, $\hat{r}$, signifies the positive radial direction; subscripts r or $r + dr$ specify radial location.

For hydrostatic equilibrium, the forces acting on this pillbox-shaped parcel must add to zero (N.B: in this vector equation each term carries its own sign):

$$\vec{F}_r + \vec{F}_{r+dr} + d\vec{F}_g = 0 \tag{10.11}$$

where from Figure 10.1, and with $M(r)$ being the total mass interior to radius r,

$$d\vec{F}_g = -g\,dm\,\hat{r} = -g\rho\,dV\,\hat{r} = -g\rho A\,dr\,\hat{r} = -\frac{GM(r)}{r^2}\rho A\,dr\,\hat{r} \tag{10.12}$$

$$\vec{F}_r = P_r A\hat{r} \tag{10.13}$$

$$\vec{F}_{r+dr} = -P_{r+dr}A\hat{r} = -(P_r + dP)A\hat{r} \tag{10.14}$$

Substituting (10.12)–(10.14) into (10.11) gives

$$P_r A - (P_r + dP)A - g\rho A dr = 0 \tag{10.15}$$

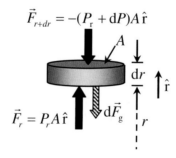

Fig. 10.1. Mechanical equilibrium of a parcel of air in a stationary atmosphere

or, solving for dP,

$$dP = -\frac{dF_g}{A} = -g\rho\,dr \tag{10.16}$$

The quantities g and ρ are positive, so dP is negative when dr is positive, i.e., pressure decreases outward in the atmosphere. The *pressure gradient* is the rate of change of pressure with distance, so

$$\frac{dP}{dr} = -g\rho \tag{10.17}$$

also known as the *equation of hydrostatic equilibrium*.

The pressure can also be expressed in terms of the perfect gas law, which provides an *equation of state*:

$$P = nkT \tag{10.18}$$

Here, n is again the number density, related to the density and the mean molecular weight of the atmosphere through the relation

$$n = \rho/m \tag{10.19}$$

Then

$$P = \rho kT/m = \rho kT/\mu m_u. \tag{10.20}$$

From (10.17) and (10.20), the pressure gradient is then

$$\frac{dP}{dr} = -\frac{\mu m_u g P}{kT} \tag{10.21}$$

from which we obtain the differential equation

$$\frac{dP}{P} = -\frac{\mu m_u g}{kT}\,dr \tag{10.22}$$

The integrated solution to (10.22) is

$$\ln\frac{P}{P_0} = -\frac{\mu m_u g}{kT}(r - r_0) \tag{10.23}$$

or

$$\frac{P}{P_0} = \exp\left\{-\frac{\mu m_u g}{kT}(r - r_0)\right\} \tag{10.24}$$

where r_0 is a reference height (e.g., the base of the atmosphere) and P_0 is the pressure at $r = r_0$.

The pressure *scale height* is defined as

$$H = \frac{kT}{\mu m_u g} \qquad (10.25)$$

k is related to the universal gas constant, R = 8.314472(15) J mol^{-1} K^{-1}, by R = kN$_A$, where N$_A$ = 6.02214179(30) × 10^{23} mol^{-1} is Avogadro's number. Then

$$H = \frac{RT}{\mu m_u N_A g} = (10^3 \text{mol kg}^{-1})\frac{RT}{\mu g} \qquad (10.25a)$$

The factor in parentheses in (10.25a) may be combined with R to obtain a new constant (commonly and somewhat confusingly also written R), R = 8314.472(15) J kg^{-1} K^{-1}; then

$$H = \frac{RT}{\mu g} \qquad (10.26)$$

Then, letting $h = r - r_0$, we get the *pressure scale height equation*,

$$P = P_0 e^{-h/H} \qquad (10.27)$$

Equation (10.27) has been used to compute the mean molecular weight for outer planet atmospheres. In particular, the occultation of a star by Jupiter or another planet provides such an opportunity. The geometry, technique, and many results are reviewed by Elliott and Olkin (1996). As the planet progressively covers the star, astronomers at various sites on the Earth measure the optical depth along the line traversed by the starlight through the planet's atmosphere as a function of time. The *optical depth* is a measure of the absorptive/scattering properties of an atmosphere; an optical depth of one corresponds to the distance required for the transmitted light to decrease in intensity by a factor e. The curve of optical depth vs. time can be converted to a curve of refractive index vs. depth vertically into the atmosphere. Refractive index in turn depends on the number density of absorbers, n, as does the atmospheric pressure by (10.18), $P = nkT$, so that the index of refraction demonstrates similar behavior to pressure as the occultation proceeds. The refractivity scale height allows a determination of the pressure scale height, and from that the mean molecular weight of the atmosphere. In practice, the temperature and therefore pressure scale height vary with height through the atmosphere, and this variation is not generally known, so the profiles are compared to those predicted from atmospheric models with a number of parameters that are adjusted to achieve best fits. An example of the usefulness of an occultation for determining an atmosphere's mean molecular weight is given in Ch. 12.1.1.

10.2.2 Temperature Variation with Height

The temperature structure of a planetary atmosphere, as for the interior, is reached through a consideration of the heat flow.

We consider an atmosphere in which heat flow is dominated by *adiabatic convection*, i.e., parcels of air convect without exchanging heat with their surroundings.

The adiabatic relation between pressure and density in an ideal gas is

$$P = \text{const } \rho^\gamma \text{ or } PV^\gamma = \text{const} \tag{10.28}$$

where $\gamma = c_P/c_V$ is a quantity known as the *ratio of specific heats* (discussed below). Writing n = N/V, where N is the number of particles in volume V, we may solve the perfect gas law (10.18) for V to obtain

$$V = \frac{NkT}{P} \tag{10.29}$$

Then from (10.28) and (10.29),

$$P\left(\frac{NkT}{P}\right)^\gamma = \text{const} \tag{10.30}$$

or

$$(Nk)^\gamma T^\gamma P^{1-\gamma} = \text{const} \tag{10.31}$$

Finally,

$$T = \text{const}' \, P^{(\gamma-1)/\gamma} \tag{10.32}$$

We have shown that P depends on the altitude above the ground; we can now expect T to have such a dependence also.

The *heat capacity*, C, of a system is the heat input per unit temperature increase, i.e., the heat required to raise the temperature of the system by 1°:

$$C \equiv \frac{dQ}{dT} \tag{10.33}$$

Two types of heat capacity are particularly useful:

1. Heat capacity at constant pressure, P,

$$C_P \equiv \left(\frac{dQ}{dT}\right)_{P=\text{const}}$$

2. Heat capacity at constant volume, V,

$$C_V \equiv \left(\frac{dQ}{dT}\right)_{V=\text{const}}$$

If heat is added to a gas, the change shows up as an increase in internal energy and as work done,

$$dQ = dU + P\,dV \tag{10.34}$$

In (10.34), known as the differential form of the first law of thermodynamics, dQ is the heat entering the system, dU is the change in the internal energy of the gas, and $P\,dV$ is the work done by the gas on its surroundings.

We now apply (10.34) to three different processes in a planetary atmosphere.

First, in an *adiabatic* process, such as occurs during adiabatic convection, no heat enters or leaves a parcel of gas as it convects, i.e., $dQ = 0$. Equation (10.34) then shows that, in an adiabatic process, any work done by expansion of the gas is carried out at the expense of internal energy:

$$P\,dV = -dU \tag{10.35}$$

Second, if a process occurs at constant volume (referred to as an *isochoric* process), then no work is done and (10.34) gives

$$dQ = dU \tag{10.36}$$

that is, the heat goes into raising the internal energy because no expansion is permitted. The definition of C_V then gives

$$C_V \equiv \left(\frac{dQ}{dT}\right)_{V=\text{const}} = \frac{dU}{dT} \tag{10.37}$$

for a process at constant volume.

Third, in an *isobaric* process, when the pressure is constant and the volume is allowed to change, (10.34) gives

$$C_P \equiv \left(\frac{dQ}{dT}\right)_{P=\text{const}} = \frac{dU}{dT} + P\frac{dV}{dT} \tag{10.38}$$

Now noting from (10.37) that dU/dT equals C_V, we arrive at

$$C_P = C_V + P\frac{dV}{dT} \tag{10.39}$$

We now write the equation of state, (10.18), $P = nkT$, as

$$PV = NRT \qquad (10.40)$$

where now N is the number of mols[1] and R is the *molar gas constant*,[2] which has the value

$$8.314472(15)\,\text{J mol}^{-1}\,\text{K}^{-1} \cong 1.987\,\text{kcal kmol}^{-1}\,\text{K}^{-1}$$

$$\cong 0.08208\,\text{L atm mol}^{-1}\,\text{K}^{-1}$$

We can differentiate (10.40) to get

$$P\frac{\mathrm{d}V}{\mathrm{d}T} + V\frac{\mathrm{d}P}{\mathrm{d}T} = NR \qquad (10.41)$$

Then for a process at constant pressure,

$$P\frac{\mathrm{d}V}{\mathrm{d}T} = NR \qquad (10.42)$$

and (10.39) becomes

$$C_P = C_V + NR \qquad (10.43)$$

We now define the *specific heat capacity*, also called the *specific heat*, c, by either

$$c \equiv \frac{C}{N} = \frac{1}{N}\frac{\mathrm{d}Q}{\mathrm{d}T} \qquad \text{molar specific heat}$$

or

$$c \equiv \frac{C}{m} = \frac{1}{m}\frac{\mathrm{d}Q}{\mathrm{d}T} \qquad \text{specific heat per unit mass}$$

depending on context. Equation (10.43) then shows that the molar specific heats are related by

$$c_P = c_V + R \qquad (10.44)$$

The specific heat of an ideal gas depends on the number of degrees of freedom, g, of its particles. In general, the molar specific heats are given by

$$c_V = (g/2)R, \qquad c_P = [(g/2) + 1]R$$

---

[1] Or *moles*, gram-molecular weights or the mass equivalent of Avogadro's number $(6.02214179(30) \times 10^{23})$ of molecules of this species.
[2] Or *universal gas constant*. See also Section 10.2.1.

so the ratio of specific heats is

$$\gamma \equiv \frac{c_P}{c_V} = \frac{\left(\frac{g}{2}+1\right)R}{\left(\frac{g}{2}\right)R} = \frac{g+2}{g} \tag{10.45}$$

In an ideal, monatomic gas the particles have only the three translational degrees of freedom in the x, y, and z directions, so $g = 3$ and

$$c_V = (3/2)R, \quad c_P = (5/2)R$$

For the light diatomic gases H_2, N_2, CO and O_2, the molecules have at least five degrees of freedom: three translational and two rotational (about the two axes perpendicular to the long axis of the molecule). If temperatures are high enough, vibration adds a sixth. Thus at the lower temperatures,

$$c_V = (5/2)R, \quad c_P = (7/2)R$$

Consequently,

$$\gamma = \frac{5}{3} \quad \text{for a monatomic gas}$$

and $\quad \gamma = \frac{7}{5} = 1.4 \quad \text{for a diatomic gas}$

For polyatomic gases and chemically active gases (e.g., CO_2, NH_3, CH_4, Cl_2, and Br_2), C_p and C_V, and thus c_p and c_V, vary with temperature in a different way for each gas.

From (10.41) we have

$$P\,dV = NR\,dT - V\,dP \tag{10.46}$$

and from (10.37),

$$dU = C_V\,dT \tag{10.47}$$

Then (10.34) becomes

$$dQ = (C_V + NR)dT - V\,dP \tag{10.48}$$

In the adiabatic case, $dQ = 0$, and, using the relation $C_P - C_V = NR$ from (10.43), we find

$$C_P\,dT = V\,dP \tag{10.49}$$

When a parcel of gas rises through a displacement dr, the pressure change within the parcel is given by (10.16). Substituting from (10.16) and replacing r by the distance, h, above a reference level in the atmosphere then gives

$$C_P\,dT = -V\,g\,\rho\,dh \tag{10.50}$$

Dividing by $V\rho = m$ and using the specific heat per unit mass, $c_P = C_P/m$, we get

$$dT/dh = -g/c_P = -\Gamma \tag{10.51}$$

where Γ is known as the *adiabatic lapse rate*. If the air is dry, the symbol Γ_d is used. Equation (10.51) shows the explicit dependence of the temperature on altitude for an adiabatically convecting atmosphere that we mentioned earlier in this section.

In the Earth's atmosphere, $g = 9.81\,\mathrm{m/s^2}$, $c_P = 1005\,\mathrm{J/kg\,K}$, and

$$\Gamma_d = 0.00976\,\mathrm{K/m} = 9.76\,\mathrm{K/km} \tag{10.51a}$$

Later we will compare the temperature structures of the atmospheres of the Earth, Venus, and Mars, and the *mixing ratio* or relative abundance by volume of specific constituents of the atmosphere.

For a wet atmosphere,

$$c_P = c_{P(\text{water vapor})}\, w + c_{P(\text{dry air})}\, (1 - w)$$

where w is the ratio of the masses of water vapor to dry air for a given volume of air.

Because $c_{P(\text{moist air})} > c_{P(\text{dry air})}$, the variation of T with height is smaller for moist air than for dry air. A lucid and topical expansion of this topic can be found in Seinfeld and Pandis (1998, esp., Chapter 14).

10.3 Circulation in the Atmosphere

10.3.1 Centrifugal and Coriolis Forces

In a rotating frame, such as a planet, there are "apparent" or "virtual" forces and motions that arise solely because of the motion of the frame. These are often referred to as *fictitious forces*.

(In this section, **boldface type** in equations is used to denote vectors.) Consider a displacement vector, $\mathbf{r}$, fixed in a frame of reference, Σ', which is rotating with an angular velocity, $\mathbf{\Omega}$, with respect to an inertial (non-rotating) frame, Σ, as illustrated in Figure 10.2. By virtue of the rotation of Σ', the tip of the vector $\mathbf{r}$ will appear to be moving with a linear velocity

$$\mathbf{v}_\Sigma = \mathbf{\Omega} \times \mathbf{r} \tag{10.52}$$

when viewed from frame Σ.

Fig. 10.2. A vector, **r**, fixed in a rotating frame of reference, Σ', and viewed from an inertial frame, Σ

The quantity $\boldsymbol{\Omega} \times \mathbf{r}$ is a vector cross-product, and is itself a vector, the magnitude of which equals the product of the magnitudes of the two vectors times the sine of the angle between them:

$$|\boldsymbol{\Omega} \times \mathbf{r}| = \Omega r \sin \theta \tag{10.53}$$

and the direction of which is perpendicular to both vectors, as given by the "right-hand rule" (see Figure 10.3):

Point the fingers of your right hand along the first vector in the cross-product, then orient your hand so you can curl your fingers from the first vector to the second vector. Your thumb now points in the direction of the cross-product.

If in Figure 10.2 the vector **r** changes with time as viewed from frame Σ', the effect as viewed from frame Σ will be

$$(D\mathbf{r})_\Sigma = (D\mathbf{r})_{\Sigma'} + \boldsymbol{\Omega} \times \mathbf{r} \tag{10.54}$$

where D is the operator $D = \mathrm{d}/\mathrm{d}t$.

Equation (10.54) gives the relationship between the velocity vectors,

$$\mathbf{v}_\Sigma = \mathbf{v}_{\Sigma'} + \boldsymbol{\Omega} \times \mathbf{r} \tag{10.55}$$

In the example above, $\mathbf{v}_{\Sigma'} = 0$.

In fact, the rate of change of any vector, **A**, as seen in Σ is the rate of change in Σ' plus the cross-product with the angular velocity vector, $\boldsymbol{\Omega}$:

$$(D\mathbf{A})_\Sigma = (D\mathbf{A})_{\Sigma'} + \boldsymbol{\Omega} \times \mathbf{A} \tag{10.56}$$

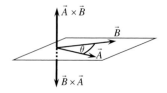

Fig. 10.3. The vector cross-product

Thus the variation of the velocity vector is, from either (10.55) or (10.56),

$$(D\,\mathbf{v}_\Sigma)_\Sigma = (D\,\mathbf{v}_\Sigma)_{\Sigma'} + \mathbf{\Omega} \times \mathbf{v}_\Sigma \qquad (10.57)$$

The quantity on the LHS is the acceleration observed by an observer in the inertial frame, Σ.

Substituting (10.55) into (10.57) gives

$$(D\mathbf{v}_\Sigma)_\Sigma = (D\mathbf{v}_{\Sigma'})_{\Sigma'} + 2(\mathbf{\Omega} \times \mathbf{v}_{\Sigma'}) + \mathbf{\Omega} \times (\mathbf{\Omega} \times \mathbf{r}) \qquad (10.58)$$

Then defining $\mathbf{a} = (D\mathbf{v}_\Sigma)_\Sigma$, $\mathbf{a}' = (D\mathbf{v}_{\Sigma'})_{\Sigma'}$, and $\mathbf{v}' = \mathbf{v}_{\Sigma'}$, we obtain

$$\mathbf{a} = \mathbf{a}' + 2\mathbf{\Omega} \times \mathbf{v}' + \mathbf{\Omega} \times (\mathbf{\Omega} \times \mathbf{r}) \qquad (10.59)$$

where primes denote quantities measured in the rotating frame. If our vectors actually describe the position, velocity, and acceleration of a particle, then multiplying by the particle's mass, we arrive at the force equation:

$$\mathbf{F} = \mathbf{F}' + 2m\mathbf{\Omega} \times \mathbf{v}' + m\mathbf{\Omega} \times (\mathbf{\Omega} \times \mathbf{r}) \qquad (10.60)$$

or, from the standpoint of an observer in the moving frame,

$$\mathbf{F}' = \mathbf{F} - 2m\mathbf{\Omega} \times \mathbf{v}' - m\mathbf{\Omega} \times (\mathbf{\Omega} \times \mathbf{r}) \qquad (10.61)$$

The second term on the RHS,

$$\mathbf{F}'_{Cor} = -2m\mathbf{\Omega} \times \mathbf{v}' \qquad (10.62)$$

is the *Coriolis force*, and the third,

$$\mathbf{F}'_C = -m\mathbf{\Omega} \times (\mathbf{\Omega} \times \mathbf{r}) \qquad (10.63)$$

the *centrifugal force*. The Coriolis and centrifugal forces are so-called *fictitious forces* that are apparent to a non-inertial observer rotating with the planet. Some experimentation with the right-hand rule, and remembering the minus sign in the equation, will show that the centrifugal force is directed perpendicularly outward from the rotation axis of the planet.

The practical consequences of these terms will be examined next.

10.3.2 Physical Effects of the Centrifugal and Coriolis Forces

10.3.2.1 The Centrifugal Force Figure 10.4 shows a parcel of air in the atmosphere of a rotating planet. The view is that of an inertial observer in space, but we analyze the forces as experienced by a non-inertial observer on the planet's surface. The gravitational force on the parcel, $F_g = mg$ (where m is the mass of the parcel), acts directly toward the center of the planet, whereas applying the right-hand rule in Figure 10.4 shows that the centrifugal force, $\mathbf{F'_C}$, acts radially outward from the planet's rotation axis. (Here we retain the primed notation for "fictitious" forces that exist only in the rotating frame.)

From Figure 10.4, the apparent weight of the parcel (i.e., the upward force needed to prevent the parcel from falling) is

$$(F_g)_{\text{eff}} = F_g - (F'_C)_\perp = F_g - F'_C \cos \lambda \tag{10.64}$$

where λ is the latitude of the parcel. Then using (10.52) and (10.63),

$$(F_g)_{\text{eff}} = mg - m\frac{v^2}{r}\cos\lambda = m\left(g - \frac{v^2}{r}\cos\lambda\right) \equiv mg_{\text{eff}} \tag{10.65}$$

where

$$g_{\text{eff}} \equiv g - \frac{v^2}{r}\cos\lambda \tag{10.66}$$

is the effective gravitational acceleration (or *effective gravity*) at the parcel's location. Rotation thus reduces the apparent weight and apparent acceleration due to gravity on a rotating planet.

The component $(F'_C)_{||}$ in Figure 10.4 causes the parcel of air to drift toward the equator, creating an equatorial bulge. The consequent *rotational*

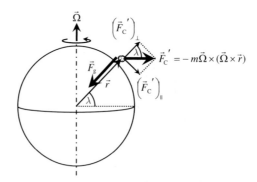

Fig. 10.4. The centrifugal force, $\vec{F}'_C$, on a parcel of air in the atmosphere of a rotating planet, and its components parallel and perpendicular to the planetary surface. λ is the latitude of the parcel

flattening can be seen in images of Jupiter and Saturn as in Figures 12.1 and 12.9, for example. The process is self-limiting, because the bulge in turn creates pressure gradient forces that oppose $(F_C')_{||}$. $(F_C')_{||}$ is larger at a given latitude for faster-rotating planets, so faster-rotating planets are more rotationally flattened.

10.3.2.2 The Coriolis Force From (10.62), the Coriolis force on any object depends on its velocity, $\mathbf{v}'$, relative to the rotating reference frame of the planet. For a planet with a solid surface, e.g., a terrestrial planet, we can take $\mathbf{v}'$ as relative to the horizon plane at a point on the surface.

Some experimentation applying the right-hand rule (Section 10.3.1) to the cross-product in (10.62) combined with appropriate diagrams (see Figure 10.5 for an example) shows that:

1. An object at rest in the rotating frame (e.g., at rest relative to the Earth's surface) experiences no Coriolis force.

2. An object exactly on the equator experiences no Coriolis force when moving due north or south, because $\mathbf{\Omega} \parallel \mathbf{v}'$. If moving in any other compass direction, it experiences no Coriolis force parallel to the surface because $\mathbf{\Omega}$ and $\mathbf{v}'$ are both in the plane of the surface.

3. In both hemispheres, an object moving toward the (nearest) pole experiences a Coriolis force toward the east, and an object moving toward the equator experiences a Coriolis force toward the west.

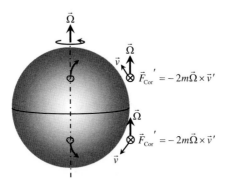

Fig. 10.5. The vectors on the right show the planetary angular velocity, $\vec{\Omega}$, the velocity, $\vec{v}$, and the Coriolis force, $\vec{F}_{Cor}'$ (into the page) on a poleward-moving parcel of air on the limb of a rotating planet, as viewed from space. The *curved arrows* on the central longitude show the resulting motion of the parcel of air *relative to the planet's surface*, in the absence of other constraints. The northern parcel veers right (east) and the southern veers left (also east)

4. In both hemispheres, an object moving due east experiences a Coriolis force perpendicularly outward from the planet's rotation axis; the component parallel to the surface therefore causes the object to veer toward the equator. For an object moving due west, the Coriolis force is perpendicularly inward toward the rotation axis and the object veers toward the pole.

Case (3), above, can be understood intuitively when analyzed in the inertial frame (e.g., in the view from space shown in Figure 10.5). Consider objects at rest relative to the surface of the rotating planet. The linear speed eastward of such objects when measured in the inertial frame is greater for objects closer to the equator because they are further from the rotational axis. In the absence of other constraints, when the object moves poleward it finds itself above points on the ground that are closer to the rotational axis and moving eastward more slowly than itself. Consequently, it drifts toward the east relative to the ground below. An object moving toward the equator finds itself moving over points on the ground that are moving faster than itself, and it therefore drifts toward the west relative to the ground below.

Points (3) and (4) can be summarized by saying that the Coriolis force causes moving objects to veer toward the right in the northern hemisphere and toward the left in the southern, regardless of their direction of travel.

10.3.3 Pressure Gradient Force

Pressure varies horizontally as well as vertically in planetary atmospheres, and the resulting horizontal forces drive winds and atmospheric circulation. Occasionally these winds can be extreme. In Hurricane Wilma (2005), a Category 5 hurricane in the northwestern Caribbean, the surface atmospheric pressure at peak intensity was 1004 mb outside the hurricane and 882 mb at the center of the eye, the latter being the lowest value ever found for an Atlantic-basin hurricane since record-keeping began in 1851. The resulting peak sustained surface wind speed was 160 knots, or 300 km/h. Fortunately, these peak values occurred over open water, but Wilma was still very intense during landfall (Pasch et al. 2006).

In Figure 10.6, a small parcel of air of volume $dx\,dy\,dz$ is situated in a region of a planetary atmosphere containing a pressure gradient. We place the lower front left corner at (x,y,z), so the right-hand face is at $x + dx$, the back face at $y + dy$, and the top face at $z + dz$. The pressure gradient vector, expressed as $\vec{\nabla}P$, where the del (or grad) operator is defined as

$$\vec{\nabla} \equiv \frac{\partial}{\partial x}\hat{i} + \frac{\partial}{\partial y}\hat{j} + \frac{\partial}{\partial z}\hat{k}$$

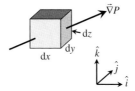

Fig. 10.6. A small volume of air in a region of pressure gradient in a planetary atmosphere

is oriented in an arbitrary direction relative to the x,y,z coordinate axes. The force component, $\mathrm{d}F_x$, in the x-direction on this parcel of air is

$$\mathrm{d}F_x = P_x \, \mathrm{d}A - P_{x+\mathrm{d}x} \, \mathrm{d}A = -(P_{x+\mathrm{d}x} - P_x) \, \mathrm{d}y \, \mathrm{d}z \equiv -\mathrm{d}P_x \, \mathrm{d}y \, \mathrm{d}z$$

where P_x and $P_{x+\mathrm{d}x}$ are the pressures at positions x and $x+\mathrm{d}x$, respectively, and $\mathrm{d}A = \mathrm{d}y \, \mathrm{d}z$ is the cross-sectional area perpendicular to the x-axis. The change in pressure over distance dx is $\mathrm{d}P_x = \partial P/\partial x \, \mathrm{d}x$, so we can write

$$\mathrm{d}F_x = -\frac{\partial P}{\partial x} \, \mathrm{d}x \, \mathrm{d}y \, \mathrm{d}z$$

Applying a similar procedure to the y and z-directions gives

$$\mathrm{d}\vec{F} = -\left(\frac{\partial}{\partial x}\hat{i} + \frac{\partial}{\partial y}\hat{j} + \frac{\partial}{\partial z}\hat{k}\right) P \, \mathrm{d}x \, \mathrm{d}y \, \mathrm{d}z = -\vec{\nabla}P \, \mathrm{d}x \, \mathrm{d}y \, \mathrm{d}z \qquad (10.67)$$

It follows that the pressure gradient force is opposite in direction to the pressure gradient.

10.3.4 Friction

Friction has an important influence on winds and circulation in the lower 1 km of air above the ground.

Within about 1 m of the ground the source of friction is molecular viscosity; above 1 m it is small-scale eddies. As suggested in Figure 10.7, the effect of friction is to limit wind speed progressively closer to the ground.

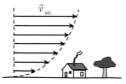

Fig. 10.7. Variation of wind speed with altitude, due to friction. Not to scale

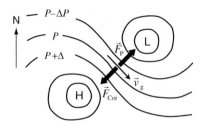

Fig. 10.8. Isobars, forces, and wind near high and low-pressure areas in the northern hemisphere of a planetary atmosphere (see text)

10.3.5 Geostrophic Balance and Geostrophic Winds

Figure 10.8 shows a schematic representation of high- and low-pressure areas in the northern hemisphere of a planetary atmosphere, with illustrative *isobars* (lines of constant pressure). $\vec{F}_P$ is the pressure gradient force, directed from high to low pressure; $\vec{F}_{Cor}$ is the Coriolis force, directed perpendicular to the air motion; and $\vec{v}_g$ is the geostrophic wind, defined and discussed below.

In Figure 10.9, a parcel of air initially begins traveling in the direction of the pressure gradient force. Immediately, it feels a Coriolis force perpendicular to its direction of motion, causing it to veer to the right in the northern hemisphere or to the left in the southern. It continues to veer as long as the total force has a component perpendicular to the wind direction; consequently, the circulation can reach steady state only when the Coriolis and pressure gradient forces exactly balance.

Because of the constraint that the pressure gradient force is always perpendicular to the isobars, the resulting wind direction must be parallel to the isobars (Figures 10.8 and 10.9). This balance of Coriolis and pressure gradient forces is referred to as *geostrophic balance*, and the resulting wind is a *geostrophic wind*.

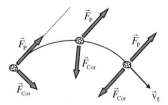

Fig. 10.9. Establishment of geostrophic balance for a parcel of air in the northern hemisphere, initially moving in the direction of the pressure gradient force (see text)

It is also worth noting that the more compressed the isobars, the stronger the pressure gradient force, the stronger the required Coriolis force, and therefore the stronger the geostrophic wind.

10.3.6 Thermal Effects

10.3.6.1 Thermal Circulation A horizontal temperature gradient between one part of a planetary atmosphere and another creates a pressure gradient force that in turn creates a wind. We will look first at distance scales small enough to ignore the Coriolis force (e.g., a few hundred meters), so that the wind direction is in the direction of the pressure gradient.

To see how thermal circulation arises, consider the following thought experiment for a region on the Earth. Consider a vertical cross-section through a volume of calm air above level ground that is at a uniform temperature, T_0. Because the air is calm and the ground temperature is uniform, the pressure scale height, H, is also uniform, as per (10.25) or (10.26), and the isobars are horizontal. The ground is thus both an *isotherm* (line of constant temperature) and an isobar, which we define to be at pressure P_0.

Now allow the air temperature to fall at some location, A, and rise at some other location, B, i.e., $T_A < T_B$. Then by (10.25), $H_A < H_B$. At this point in our experiment the mass of air above each point on the ground remains unchanged, so there is still a horizontal isobar at pressure P_0 at ground level; but because of the larger scale height at B, the isobars are as in Figure 10.10(a). At *any given altitude*, the pressure will be higher above B than above A, and the horizontal pressure gradient will drive a wind from the warmer to the cooler region (B to A), as indicated by the arrow in Figure 10.10(a).

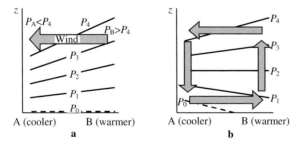

Fig. 10.10. Isobars above a region of uneven temperature ($T_A < T_B$), and the resulting winds. **a** Upper-level wind assuming a horizontal isobar at ground level. **b** Mass transport from B to A by the upper-level wind tilts the lower-level isobars, resulting in a return flow near ground level

The transport of mass from B to A by this wind increases the air mass above A and decreases it above B, causing the air pressure at ground level to rise at A and fall at B. This makes the lower-altitude isobars tilt in the opposite direction from those at higher altitudes. The upper level flow from the warmer to the cooler region is then matched by a return flow from the cooler to the warmer region at low level. These in turn are joined by warm air rising at B and cool air falling at A, creating a complete *thermal circulation* pattern, as in Figure 10.10(b).

An example of thermal circulation is the sea breeze created by solar heating of the land and the water at a seashore, illustrated in Figure 10.11. The water has the higher heat capacity, and by (10.33) the heat input from the sun makes the land hotter than the water. From Figures 10.10(b) and 10.11, on a hot day with otherwise no wind, people on the beach enjoy a cool breeze coming from the water. At night, although there may be few people on the beach to notice, the land cools faster than the water and the circulation reverses: the breeze at night is directed from the land to the water.

10.3.6.2 The Thermal Wind When the scale is larger than in Figure 10.11, then the Coriolis force can no longer be ignored. Horizontal temperature gradients then affect the geostrophic balance and the direction of the geostrophic wind.

Figure 10.12 illustrates isobars at ground level in the vicinity of high and low-pressure regions. For simplicity we assume level ground. The geostrophic wind at ground level, $(\vec{v}_g)_0$, is parallel to the isobars. In the absence of a temperature gradient, the isobars above ground level will be parallel to the ground and the direction of the geostrophic wind will be the same at all altitudes.

Now consider the region between the two isotherms T_1 and T_2 (dashed lines) in Figure 10.12, with $T_1 < T_2$. Figure 10.13(a) plots pressure vs. altitude, z, above a line that follows the P_0 isobar (dotted line) in Figure 10.12, so that ground level is a line of constant pressure (cf. Figure 10.10(a)). At altitudes above the ground the sloping isobars create a pressure gradient force, $(\vec{F}_P)_T$, parallel to the ground-level isobars of Figure 10.12. The subscript, T, indicates

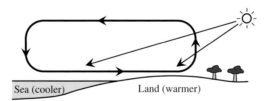

Fig. 10.11. Thermal circulation and the daytime sea breeze. The flow is reversed at night

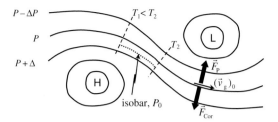

Fig. 10.12. Isobars, forces, and wind near high and low-pressure areas in a planetary atmosphere (see text)

that $(\vec{F}_P)_T$ is produced by the horizontal temperature gradient. $(\vec{F}_P)_T$ is zero at ground level because the isobar is horizontal, and increases progressively as the isobars steepen with altitude.

$(\vec{F}_P)_T$ adds vectorially to the original pressure gradient force, which we now call $(\vec{F}_P)_P$, created by the high and low-pressure areas, as shown in Figure 10.13(b). As described in Fig. 10.9, the Coriolis force, $\vec{F}_{Cor}$, causes the wind to veer until the net force on an air parcel is zero; i.e., until the geostrophic wind (always perpendicular to $\vec{F}_{Cor}$) is perpendicular to $(\vec{F}_P)_{TOT}$ rather than to $(\vec{F}_P)_P$.

Thus, in the northern hemisphere, if the geostrophic wind blows from a colder to a warmer region, it veers to the left at any altitude $z > 0$, compared to its direction at ground level $(z = 0)$, with the amount of veer increasing with increasing altitude. In the southern hemisphere it would veer to the right. If the wind blows from a warmer to a colder region these directions would be reversed: to the right in the northern hemisphere and to the left in the southern.

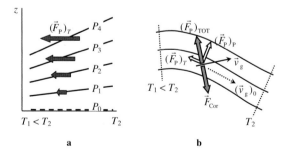

Fig. 10.13. a Isobars above a region with a temperature gradient $(T_1 < T_2)$ and the resulting thermal component of the pressure gradient force, $(\vec{F}_P)_T$ (arrows). **b** Geostrophic wind, $\vec{v}_g$, at altitude $z > 0$ compared to $(\vec{v}_g)_0$ at ground level, in a region of temperature gradient

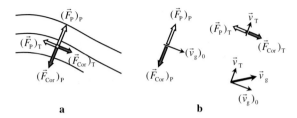

Fig. 10.14. a Components of the pressure gradient force and the Coriolis force on a parcel of air. **b** The corresponding wind components and the resultant geostrophic wind

A more common way of describing the same effect is to picture the wind at altitude as consisting of two components: the geostrophic wind, $(\vec{v}_g)_0$, which is now actually a wind component and remains constant with altitude, and a *thermal wind*, $\vec{v}_T$ (also a wind component), which blows in a direction such the Coriolis force on it, $(\vec{F}_{Cor})_T$, is opposite to $(\vec{F}_P)_T$, as shown in Fig. 10.14(b). The thermal wind increases with increasing altitude, making the resulting geostrophic wind veer increasingly away from $(\vec{v}_g)_0$ as altitude increases.

The thermal wind direction can always be predicted from the isotherms by the fact that $\vec{v}_T$ is directed so that the colder region is to the left of the wind vector in the northern hemisphere and to the right in the southern. $(\vec{v}_g)_0$ can also be predicted from the isobars, because it is directed so that the lower pressure is to the left of the vector in the northern hemisphere, and to the right in the southern.

10.3.7 Global Circulation

10.3.7.1 The Observed Surface Pattern Figure 10.15 shows the major features of the global circulation pattern on the surface of the Earth. The pattern shown is highly smoothed and averaged, and the latitudes are very approximate. The actual pattern on any given day is more complicated. The features also shift north and south somewhat with the seasons, as the direction of illumination by the Sun changes through the year.

The pattern is characterized by alternating high and low-pressure belts, with winds dominantly in one direction in the region between any two belts. (Note that, by tradition, winds are named for the direction *from* which they blow, not the direction *towards* which they blow, i.e., the *westerlies* blow *from* the west, and the *north-east trade winds* blow *from* the north-east (toward the south-west).)

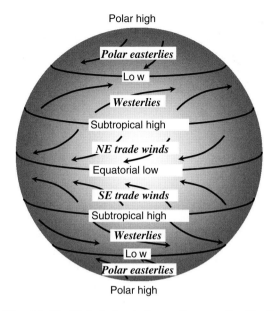

Fig. 10.15. Global circulation pattern on the Earth

The trade winds got their name because they "blow trade," that is, they blow a constant course or direction. This made them very useful to sailing ships carrying trade goods over the world's oceans.

The westerlies are much more variable and unreliable, to the detriment of some sailing ships that did not make it back home.

The area from which the westerlies and the trade winds diverge (the subtropical high) is a region of calm or light winds, and is often called the *horse latitudes*.

In the area near the equator where the NE and SE trade winds converge and the air rises toward the upper troposphere, surface winds not surprisingly tend to be light. This region is often called the *doldrums*; or, more technically, the *inter-tropical convergence zone* (ITCZ). The origin of the word "doldrums" is not known, but it may have been created by contrasting the dullness of, for example, being sick, with the liveliness of having a tantrum. A person getting better from the "flu" would be recovering from the "dull-drums." By extension, any becalmed ship was said to be in the doldrums, and eventually the name became a geographical location, applying to the region of ocean where ships were most often becalmed.

The warm air of the NE trade winds picks up moisture from the ocean and then loses it to condensation as the air rises in the ITCZ. Thus the ITCZ often shows up in satellite photographs of the Earth as a band of clouds near the equator.

In the following sections, we will use the forces and processes described above to try to account for the circulation pattern in Figure 10.15, and discuss a few other topics.

10.3.7.2 The Hadley Cell The basic pattern of westerlies and trade winds was known by the 1600s from data provided by mariners. The cause, however, was unknown. Edmund Halley (of Halley's Comet fame) suggested in 1686 that the trade winds resulted from solar heating at the equator, making the air rise and forcing other air to flow toward the equator from the tropics to replace it. This accounted nicely for the component toward the equator, but not for the easterly component (toward the west).

The English scientist, George Hadley, in 1735, extended Halley's explanation to create a complete circulation cell and suggested that the Earth's rotation deflected the air toward the west. This was 100 years before Gaspar G. Coriolis of France derived the Coriolis force mathematically, in 1835.

The *Hadley cell* is a thermal circulation cell like that causing a sea breeze, but on a global scale. A single Hadley cell on the Earth might (but doesn't!) extend from the equator to the pole in each hemisphere, as shown in Figure 10.16. The air would be heated at the equator, rise, move toward the colder poles for the reasons described in Section 10.3.6, cool there, sink, and return to the equator along the Earth's surface. With a single Hadley cell in each hemisphere, we would have:

Southward-moving surface air which veers toward the west, creating the polar easterlies and the trade winds

A low-pressure area where the air is heated at the equator and a high-pressure area where the air cools at the poles.

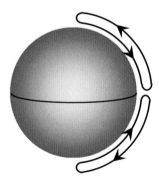

Fig. 10.16. Hypothetical single Hadley cells extending from the equator to the poles

Each cell would also carry warm air from the equator to the pole and cooler air from the pole to the equator, balancing the radiant energy surplus at the equator and the radiant energy deficit at the pole. However, a single Hadley cell in each hemisphere does not account for the subtropical high-pressure belts, the mid-latitude low-pressure belts, or the westerlies.

10.3.7.3 The Ferrel and Polar Cells The Coriolis force makes the single Hadley cell described in Section 10.3.7.2 break up into three cells, as indicated in Figure 10.17. In each cell, the surface winds veer toward the right to create the polar easterlies, westerlies, and NE trade winds. The high-altitude winds also veer toward the right, and are therefore in the opposite sense to the surface winds.

By about 30°N and S latitude, the upper air from the equator has lost much of its northward component and has time to cool and sink. The descending flow splits, as shown in Figure 10.17. Some air flows south, completing the Hadley cell, while some flows north to create a new cell, the *Ferrel cell*.

The warmer and cooler temperatures where the air rises and sinks, respectively, create the equatorial low and the subtropical highs (respectively) in Figure 10.17, as discussed in Section 10.3.7.1.

At about 60°N and S latitude, the surface air from the poles has lost much of its southward component and has time to warm and rise. The warmer temperature creates the mid-latitude low-pressure belt. The rising flow splits, as shown in Figure 10.17. Some air flows north, completing the polar cell, while some flows south, joining the Ferrel cell.

The Hadley and polar cells are thermally driven. The Ferrel cell is thermally indirect (sometimes referred to as *parasitic*) and is driven by the other two cells.

Figure 10.18 shows the observed mean meridional (north–south) circulation pattern in the northern hemisphere during winter and summer. (The word *meridional* refers to the north–south direction, i.e., along a meridian of longitude, whereas *zonal* refers to the east–west direction, along a line of latitude.)

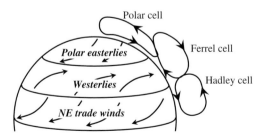

Fig. 10.17. The Coriolis force causes each Hadley cell in Figure 10.16 to break up into three cells. Not to scale

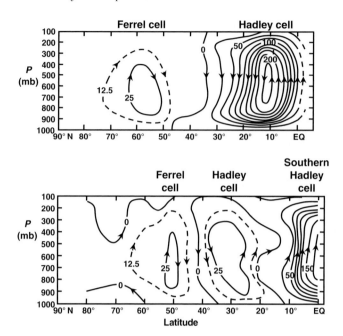

Fig. 10.18. Zonally averaged meridional circulation as a function of latitude and altitude during northern winter (*upper chart*) and northern summer (*lower*). Numbers on flow lines represent mass flux in units of 25×10^6 tons/sec, i.e., the mass transported in the channel between any two adjacent, solid streamlines is 25×10^6 tons/sec. See Figure 10.22 for an approximate conversion chart between pressure and altitude. (After Iribarne and Cho 1980, Fig. VII-22, p. 179.)

Some features evident in Figure 10.18 are:

The Hadley cell is stronger in winter than in summer. This is also true in the southern hemisphere, so the northern cell is stronger than the southern in January and the southern is stronger than the northern in July.

Both Hadley cells move north in the northern summer and south in the southern summer (northern winter). In July, in fact, the ascending branch of the southern cell lies north of the equator.

The polar cell is very weak, and does not show up in Figure 10.18.

10.3.7.4 Eddie Motions in the Westerlies A large-scale view such as that in Figure 10.15 ignores small-scale features that can in fact be important in the poleward transport of heat by the atmosphere. In the westerlies in particular, uneven heating and cooling of the atmosphere, caused, e.g., by the differing heat capacities of continents and oceans, produces *baroclinic instabilities*.

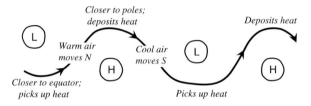

Fig. 10.19. Poleward transport of heat by eddy and wave motion in a planetary atmosphere, illustrated for the northern hemisphere.

Specifically, baroclinic instabilities occur when the isobaric surfaces are inclined to the isothermal surfaces (Section 10.3.6). These instabilities perturb the generally westerly mid-latitude flow, creating large-scale waves. Eddies associated with these waves show up as high and low-pressure areas, the former usually lying equatorward of the flow and the latter lying poleward. As indicated schematically in Figure 10.19 for the northern hemisphere, the alternating northward and southward flow results in a net transport of heat poleward.

10.3.7.5 Air Masses and Fronts An *air mass*[3] is a large volume of air having reasonably unified properties (e.g., temperature, moisture content). A *front* is a boundary between two air masses.

On the Earth, there are three basic types of air mass in terms of temperature: *polar*, *mid-latitude*, and *tropical*, as illustrated schematically in Figure 10.20. Each of these in turn can be divided into two categories in terms of moisture content: *maritime* (moist) and *continental* (dry) air masses.

10.3.7.6 Jet Streams A *jet stream* is a narrow, high-velocity stream of air in the upper troposphere or the stratosphere. Here, we describe in some detail the formation of the *polar jet stream*, formed by the strong horizontal temperature gradient across the polar front. The weaker *subtropical jet stream* forms by a similar process at the more diffuse subtropical front.

Figure 10.21 shows a meridional cross-section of the Earth's atmosphere. The view is facing east, with the North Pole on the left and the equator on the right. The polar and subtropical jet streams, marked by J_p and J_s, respectively, are located at the discontinuities between the tropical and middle and the middle and polar tropopauses.

The strong horizontal temperature gradient across the polar front is evident in Figure 10.21 where, e.g., the temperature on a horizontal line through the polar front at 4 km altitude rises toward the south from -25 to $-15°$C.

---

[3] Not to be confused with *air mass* in astronomical extinction, which is the thickness of a column of air normalized to the zenith value.

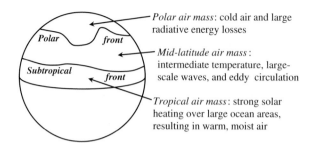

Fig. 10.20. Air masses and fronts in the Earth's atmosphere. A corresponding division occurs in the southern hemisphere

In the northern hemisphere, a horizontal temperature gradient produces a thermal wind directed with the colder air on the left and the warmer air on the right (Section 10.3.6.2). The reverse would be true in the southern hemisphere. Thus the wind becomes stronger westerly (toward the east) as altitude increases through the troposphere in both hemispheres. The effect is strongest where the horizontal temperature gradient is strongest, near the polar front. A similar effect occurs at the subtropical front, but because the front is more diffuse, the subtropical jet stream is weaker.

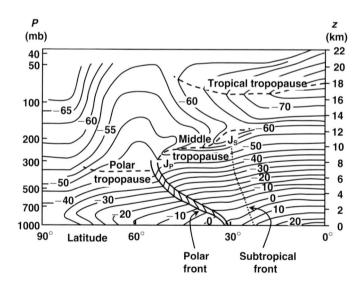

Fig. 10.21. Meridional cross-section of the Earth's atmosphere, showing zonal mean temperatures (°C) in the northern hemisphere on January 1, 1956. The positions of the polar (J_P) and subtropical (J_S) jet streams and the polar and subtropical fronts are marked. The subtropical front is much more diffuse than the polar front. (After Iribarne and Cho 1980, Fig. VII-26, p. 185.)

Figure 10.22 shows a zonally averaged cross-section of the Earth's atmosphere from the North Pole to the South Pole during the northern summer (left) and northern winter (right). The mean zonal winds are marked. Along the surface of the Earth, the wind directions correspond to the trade winds (easterly winds, i.e., *from* the east), the westerlies, and the polar easterlies. However, the winds swing more and more westerly as altitude increases, with the strongest westerly winds occurring near the 200 mb level (about 12 km altitude) in the upper troposphere. These are quite localized and indicate the polar jet streams.

Comparing the two sides of the diagram, it is evident that each jet is strongest during winter in its hemisphere. The entire circulation pattern also shifts

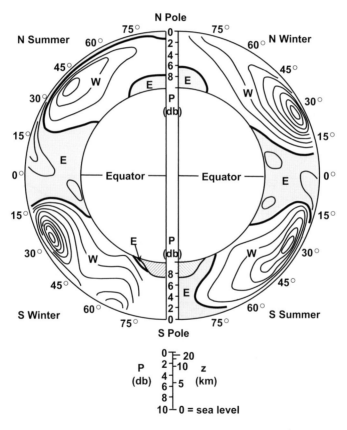

Fig. 10.22. Zonally averaged wind distribution from sea level (*inner circle*) to the top of the atmosphere (*outer circle*), during northern summer (*left*) and northern winter (*right*). Heavy contours = 0 m/s; contour interval = 5 m/s. Regions of easterly (E) and westerly (W) winds are marked. The hatched area near the bottom is Antarctica. An approximate conversion chart is shown from pressure to altitude (1 db = 0.1 bar). (After Iribarne and Cho 1980, Fig. VII-23, p. 180.)

northward during northern summer and southward during southern summer, responding to the changing position of the sub-solar point over the year.

We next discuss the interaction between planetary atmospheres and surfaces.

10.4 Atmospheric Effects on the Heat Budget

Figure 10.23 shows the variation of temperature with altitude in the atmospheres of the Earth, Venus, and Mars. The four main divisions in the Earth's atmosphere are labeled.

10.4.1 The Earth

By far the dominant source of heat for the Earth's atmosphere is the Sun, which approximates very closely a black body of 5800 K. The energy flux incident on

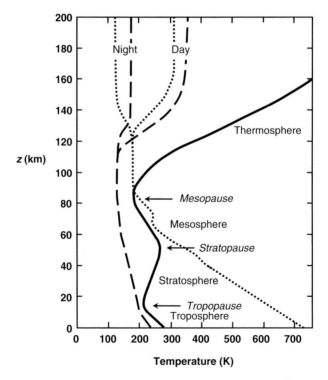

Fig. 10.23. Thermal structure of the atmospheres of the Earth (*solid line*), Venus (*dotted*), and Mars (*dashed*). The atmospheric layers are labeled for the Earth

the Earth is thus almost entirely short wavelength (UV, visible and near-IR, $\lambda \leqslant 2\,\mu\mathrm{m}$); see, e.g., Milone & Wilson 2007, Ch. 4.4.4 for further discussions of blackbody flux, and Ch. 6.3 for the effect of solar radiation on Earth.

The *solar constant* for the Earth (the radiant power from the Sun per square meter perpendicular to the Sun's rays at the Earth's mean distance of 1 au, measured outside the Earth's atmosphere) is $\sim$1368 (Cox 2000, p. 340 gives 1365–1370), W/m^2. The globally and temporally averaged energy flux intercepted by the Earth then equals solar constant $\times \pi r^2 / 4\pi r^2 = 342\,\mathrm{W/m}^2$, where r is the radius of the Earth and $4\pi r^2$ is the surface area of a sphere. For long-term thermal stability, the Earth must reflect or radiate this same globally and temporally averaged energy flux back into space.

On average, of every 100 units of energy incident, approximately 31 are reflected or scattered back to space and 69 are absorbed: three in the stratosphere (primarily by ozone), 17 by cloud droplets, water vapor, dust, and carbon dioxide in the troposphere, and 49 by the ground.

The absorbed energy flux maintains the Earth at a globally and temporally averaged equilibrium temperature near 288 K, for which the emitted black body radiation is almost entirely at long wavelengths ($5\,\mu\mathrm{m} \leq \lambda \leqslant 20\,\mu\mathrm{m}$). Given the same 100 units incident as above, 12 units are radiated directly to space by the ground, 48 are radiated by water vapor and carbon dioxide in the troposphere, and nine are radiated by the clouds; thus the 69 units absorbed from the sun are balanced by 69 units radiated back to space.

10.4.1.1 Troposphere of the Earth The primary absorbers in the troposphere are water vapor and carbon dioxide, which absorb efficiently at $\lambda \geqslant 5\,\mu\mathrm{m}$. Thus, the troposphere allows most of the incoming solar radiation to pass through to the ground, but absorbs most of the IR radiated upward by the ground. (The 12 units radiated directly to space, mentioned above, are only a small fraction of the flux radiated by the ground.) Absorption in lower layers reduces the flux to higher layers, resulting in a temperature that decreases with height, z, above the ground, i.e., the lapse rate from (10.51) in Section 10.2.2,

$$\Gamma = -\frac{\mathrm{d}T}{\mathrm{d}z} \tag{10.68}$$

is positive in the troposphere. The observed average lapse rate in the troposphere is about 6.5°C/km; however, the value at specific locations can deviate significantly from this, even becoming negative (temperature increasing with altitude) in a region of temperature inversion.

If the lapse rate becomes greater than the adiabatic lapse rate then convection begins and the lapse rate decreases to the adiabatic value. As a convecting parcel of air rises, it expands due to decreasing air pressure and cools; falling parcels are compressed and heat up. Consequently, the adiabatic lapse rate is always positive ($\mathrm{d}T/\mathrm{d}z < 0$).

The adiabatic lapse rate for dry air (*dry adiabatic lapse rate*, DALR or $\Gamma_{\rm d}$, cf. Section 10.2.2) is constant at 9.76°C/km from (10.51a). If, on the other hand, the air is saturated, then moisture condenses as the air rises and cools, and evaporates as the air descends and heats. The latent heat of condensation offsets the temperature change, resulting in a *saturated adiabatic lapse rate* (SALR) of typically about 4.9°C/km. Condensation does not occur if the air is not saturated, so air can be treated as dry until the saturation point (or *dew point*) is reached.

The *environmental lapse rate* refers to the lapse rate that would exist if there were no convection. A gravitationally bound gas, whether a planetary atmosphere or a stellar atmosphere or interior, always chooses the smallest lapse rate (usually referred to as the *temperature gradient* in a stellar interior). Thus if the environmental lapse rate is less than the adiabatic lapse rate for the moisture content of the air, then the air is stable against convection. If the environmental lapse rate rises above the adiabatic lapse rate, then the air becomes convective.

Most of the Earth's troposphere is dry (i.e., unsaturated), so comparing the values above shows that the lapse rate in the Earth's troposphere is normally sub-adiabatic, and convection is transient and localized.

10.4.1.2 Stratosphere and Mesosphere Ozone production occurs between approximately 30 and 60 km altitude, primarily by the photodissociation reactions

$$O_2 + h\nu \rightleftharpoons O + O \qquad \lambda \leq 242\,\text{nm (UV)} \qquad (10.69)$$

$$O_3 + h\nu \rightleftharpoons O_2 + O \qquad \lambda \leq 1100\,\text{nm (IR)} \qquad (10.70)$$

(where ν is frequency, $h = 6.62608 \times 10^{-34}\,\text{J s}$ is Planck's constant, and $h\nu$ represents the energy of the photon) and secondarily by collisions,

$$O_2 + O + M \rightleftharpoons O_3 + M \qquad (10.71)$$

$$O_3 + O \rightleftharpoons 2O_2 \qquad (10.72)$$

Here, M represents a third atom or molecule, the collision of which is required to conserve energy and momentum.

The rate of ozone production is determined by a number of competing factors, two of which are:

1. The probability of three-body collisions increases with decreasing altitude because of increasing atmospheric density.
2. The supply of UV photons decreases with decreasing altitude because an increasing amount has already been absorbed at higher altitudes.

Ozone is formed mainly between about 30 and 60 km altitude, where three-body collisions are sufficiently probable and the UV radiation is sufficiently intense.

Ozone is unstable and is destroyed by photochemical reactions involving O_2, NO_x, and the HO_2 radical. Although maximum ozone production occurs between 30 and 60 km altitude, its maximum concentration occurs between 10 and 25 km, depending on latitude (lower at higher latitudes), because of transport to lower altitudes where its destruction is less likely.

The temperature in the stratosphere and mesosphere is determined by the relative rates of energy deposition and loss. Energy deposition is primarily by absorption of near-UV (170 nm $< \lambda <$ 300 nm) by ozone. Energy loss is primarily radiation by CO_2, an efficient emitter of IR. In the stratosphere the ozone concentration is high enough that radiative losses by CO_2 are relatively small, and temperature increases with increasing altitude as the UV intensity increases; whereas in the mesosphere the ozone concentration becomes very low, so absorption of solar UV is relatively low while radiative cooling through IR emission by CO_2 is relatively more important, and temperature decreases with increasing altitude. Because the lapse rate in the stratosphere is negative and therefore always less than the adiabatic value, the stratosphere is convectively stable.

10.4.1.3 Thermosphere Above about 80 km altitude, absorption of far-UV radiation by NO, O_2, O, N, and N_2 partially ionizes and substantially heats the atmosphere. Heat is also deposited by solar x-rays, solar and galactic cosmic rays, and meteoroids.

The energy deposition rate depends on the number density of absorbers, which increases with decreasing altitude, and the number density of UV photons, which decreases with decreasing altitude due to absorption at higher altitudes. The primary heat sink is conduction to the lower atmosphere with subsequent IR emission. As a result of these competing factors, the temperature increases upward from the mesopause to approximately 150 km altitude in the thermosphere. Above this level the rate of energy deposition decreases with increasing altitude, but the atmospheric density is too low for efficient heat loss by conduction, and the thermospheric temperature continues to increase to an altitude of about 200–300 km, above which the atmosphere is approximately isothermal.

The temperature in the isothermal region is highly dependent on solar energy, exhibiting a strong diurnal variation and a dependence on solar activity. Daytime temperatures are in the approximate range of 1000–2000 K, depending on solar activity, and 500–1500 K at night.

The ionization rate also depends on atmospheric density and UV photon flux, and peaks at around 300 km altitude. The ionosphere is thus located in the upper thermosphere.

Because the lapse rate is negative, the thermosphere is convectively stable.

10.4.1.4 Exosphere The *exosphere* is the outer part of the thermosphere, where the density is low enough that the mean free path of a molecule (the mean distance between collisions), L_{mfp}, is greater than the pressure scale height, H. Under these conditions collisions are rare, and particles follow ballistic trajectories.

It is from the exosphere that escape of atmospheric constituents to space takes place. If the particle speed is less than the escape speed, $v < v_{esc}$, the particle remains gravitationally bound to the Earth, whereas if $v > v_{esc}$ then the particle can escape. Root mean square particle speeds are related to temperature by

$$v = \sqrt{\frac{3kT}{m}} = 157.94\sqrt{\frac{T}{\mu}}\,\text{m/s} \tag{10.5}$$

so lighter particles have higher mean speeds and escape at a higher rate, as shown in Section 10.1.

The base of the exosphere (the *exobase*) is defined as the altitude at which $L_{mfp} = H$, and is located at about 500–600 km altitude.

Satellites in low Earth orbit, such as the International Space Station and the Hubble Space Telescope, orbit near the exobase. Although a very good vacuum in normal terrestrial terms, the particle density is high enough that the orbit decays slowly and the satellite has to be boosted to a higher orbit occasionally. For comparison, the first manned space capsules, Yuri Gagarin's Vostok 1, Gherman Titov's Vostok 2, and John Glenn's Friendship 7, orbited at altitudes of about 200–300 km.

The properties of the lower atmospheres of the Earth, Venus, and Mars are compared in Table 10.1.

10.4.2 Mars

The thermal structure of the Martian atmosphere is illustrated in Figure 10.23 (dashed lines). Because of the low thermal inertia of the atmosphere and surface (Section 10.5.4.2) and the planet's axial tilt and orbital eccentricity, the Martian atmosphere is subject to large diurnal, annual, and equator–pole temperature variations. The curve in Figure 10.23 represents a highly smoothed profile above the equator.

Because the Martian atmosphere lacks significant free oxygen, ozone is unable to form in any appreciable quantity and the temperature maximum associated with the stratopause on Earth does not occur. The Martian atmosphere is then divided into three primary levels, the troposphere, a combined stratosphere and mesosphere (often called the *stratomesosphere*), and the thermosphere.

Table 10.1. Properties of the lower atmospheres of Earth, Venus, and Mars

Property	Earth	Venus	Mars[a]
Major constituents (mole fraction)	N_2: 0.781; O_2: 0.209; ^{40}Ar: 0.00934; H_2O: 0–4 × 10^{-2}; CO_2: 2–4 × 10^{-4}	CO_2: 0.965; N_2: 0.035; SO_2: 150 ppm; ^{40}Ar: 70 ppm; CO: 30 ppm; H_2O: 20 ppm	CO_2: 0.953; N_2: 0.027; ^{40}Ar: 0.016; O_2: 0.0013; CO: 0.0027; H_2O: < 3 × 10^{-4}
Mean molecular weight, μ	28.96 (dry air)	43.44	43.45
Surface pressure	1.013 bar	95 bar	6–10 mbar
Mean surface temperature	288 K	730 K	220 K
Surface density	1.225 kg/m^3	65 kg/m^3	0.012 kg/m^3
Pressure scale height at surface	8.43 km	15.75 km	11.3 km
Solar constant	1368	2620 W/m^2	590 W/m^2
Fraction of incident solar energy absorbed	0.7	0.23	0.86
Absorbed solar flux, averaged over the planet	242 W/m^2	151 W/m^2	127 W/m^2
Major absorber of solar energy	Ground (but above 100 km for EUV)	Clouds and above (2.5% reaches the ground)	Ground
$\Gamma_{adiabatic}$ (troposphere)	9.76°C/km (dry); 4.9°C/km (wet)	8°C/km	4.4°C/km
$\Gamma_{observed}$ (troposphere)	6.5°C/km	8°C/km	Varies with dust content

[a] The atmospheric properties of Mars vary seasonally at any given location due to sublimation of CO_2 at the polar caps.

10.4.2.1 Troposphere The Martian troposphere forms in a similar way to that on Earth: absorption of solar radiation by the ground followed by absorption of the IR radiation from the ground by (on Mars) the CO_2-rich atmosphere.

The Martian atmosphere, however, differs from that of both the Earth and Venus in that it always contains silicate dust, raised from the surface by winds. Silicate dust is very effective at absorbing solar radiation. The dust in turn heats the atmosphere, producing a generally stable (sub-adiabatic) lapse rate (Table 10.1).

Dust devils (small, dust-laden whirlwinds) are common in the afternoon, increasing the low-altitude dust content. Stronger winds can produce localized dust storms. When Mars is near perihelion (spring and early summer in the southern hemisphere), the winds can become strong enough to create dust storms that envelope the entire planet. Thus, the amount of dust present in the atmosphere varies considerably with time and location. The tropospheric lapse rate varies with the dust content, generally being greatest (convectively least stable) at night and early morning, and least (convectively most stable) in the afternoon. In the afternoon, the lapse rate can be negative, with the minimum temperature at ground level.

The height of the troposphere varies considerably with ground temperature, and thus with latitude and season.

10.4.2.2 Stratomesosphere As noted above, the lack of ozone on Mars means that there is no distinction between the stratosphere and the mesosphere, and the temperature is roughly constant at about 130 K from the top of the troposphere to the base of the thermosphere.

The lack of an ozone or other such absorbing layer means that the Martian surface, during the daytime, is continuously bathed in sterilizing ultraviolet radiation.

10.4.2.3 Thermosphere Solar EUV is absorbed by CO_2 above $\sim$100 km altitude, producing a hot outer atmosphere (the thermosphere) as on Earth. However, CO_2 also emits IR efficiently, and is thus a good refrigerant. As a result, thermospheric temperatures on Mars vary from around 170 K at night to 300–370 K in the daytime, compared to 500–1500 K and 1000–2000 K, respectively, for the Earth.

10.4.3 Venus

The thermal structure of the Venus' atmosphere is illustrated in Figure 10.23 (dotted lines). Because (1) the massive atmosphere provides a high thermal inertia (Section 10.5.4.2), (2) the rotation axis is almost perpendicular to the plane of the orbit, and (3) the orbital eccentricity is almost zero, the temperature profile below the thermosphere changes little with time.

10.4.3.1 Troposphere Only 2.5% of the incoming solar radiation reaches and is absorbed by the surface of Venus, but the thick CO_2-rich atmosphere is so efficient at absorbing the IR emitted by the ground that, in order for the small fraction that finds its way back to space to equal the amount absorbed by the surface, the surface temperature has to rise to 730 K.

The dry adiabatic lapse rate for an ideal gas is, from (10.51),

$$(\Gamma_d)_i = -\left(\frac{dT}{dz}\right)_i = \frac{g}{(c_P)_i} \tag{10.73}$$

where $g = 8.87 \, \text{m/s}^2$ at the surface of Venus, c_P is the specific heat at constant pressure, and the subscript i denotes values obtained assuming an ideal gas. $(c_P)_i$ is temperature dependent, and equals 1134 J/kg K for CO_2 at 730 K. Equation (10.73) then gives $(\Gamma_d)_i = 7.82 \, \text{K/km}$. However, a triatomic atmosphere generally departs noticeably from an ideal gas, and to obtain the actual dry adiabatic lapse rate, Γ_d, we must multiply $(\Gamma_d)_i$ by a conversion factor, $\Phi \equiv \Gamma_d / (\Gamma_d)_i$. Φ is computed numerically; for CO_2 at 90 bars pressure and 730 K, $\Phi = 1.033$. Then $\Gamma_d = 8.08 \, \text{K/km}$.

The observed value of Γ in Table 10.1 and the approximately steady tropospheric profile in Figure 10.23 suggest that Venus' atmosphere is close to adiabatic throughout most of the troposphere. Specifically, measurements during descent by the Russian Venera landers and the four atmospheric probes released by the American Pioneer Venus craft indicate that the lowest 23 km of the troposphere and a narrow region near 55 km have adiabatic temperature gradients. The other parts of the troposphere appear to be sub-adiabatic.

10.4.3.2 Stratomesosphere Venus' atmosphere, like that of Mars, lacks ozone in any appreciable quantity, and the troposphere and thermosphere are separated by a roughly isothermal stratomesosphere at about 200 K.

10.4.3.3 Thermosphere The dayside thermospheric profile of Venus is similar to that of Mars (Figure 10.23): absorption of solar EUV by CO_2 above $\sim$100 km altitude accompanied by CO_2 infrared emission, producing a thermospheric dayside temperature near 300 K.

The thermosphere of Venus differs from that of Mars and Earth in several important ways arising from the much slower rotation period of the upper atmosphere: 4 days at the cloud tops and 6 days in the thermosphere. Because these are much shorter than the 243-day (retrograde) sidereal and 117-day synodic rotation periods of the solid planet, the atmosphere is said to *super-rotate* (Section 10.5.3.2.3); but thermospheric gases on Venus spend six times longer on the nightside than on Mars or the Earth, and the temperature drops to a little over 100 K (Figure 10.23). For this reason the nightside thermosphere on Venus is often referred to as the *cryosphere* ("cold sphere").

Fig. 10.24. Noon-to-midnight flow pattern on Venus and Mars

The day–night thermospheric temperature difference translates into a much larger pressure scale height on the dayside, and therefore a larger atmospheric pressure on the dayside than at the same altitude on the nightside (recall Figure 10.10).

This pressure difference creates a noon-to-midnight pressure gradient force, strongest at the terminator. The dominant circulation pattern in the thermosphere is therefore a flow from the dayside to the nightside, as illustrated in Figure 10.24.

A similar circulation pattern occurs on Mars.

10.5 Planetary Circulation Effects

10.5.1 Circulation and the Coriolis Force

Basically, circulation in planetary atmospheres is driven by solar heating and modified by planetary spin, topography, and surface friction. Table 10.2 compares some characteristics of circulation in the atmospheres of the Earth, Venus, and Mars.

The *Rossby number, Ro*, equal to the ratio of inertial acceleration ($a = F/m$, due to "real" forces acting) to Coriolis acceleration, a_{COR}, provides a measure of the effectiveness of the Coriolis force (off the equator):

Table 10.2. Circulation in the atmospheres of Earth, Venus, and Mars

Earth, Mars	Venus
Atmospheric circulation is driven from below by solar heating of the ground	Atmospheric circulation is driven from above by absorption of solar energy in and above the clouds
Rapid spin: $\sim$24 h	Slow spin: $(-)243^{d}$ sidereal period

$$Ro \equiv \frac{a_{\text{inertial}}}{a_{\text{COR}}} = \frac{\left(\frac{dv}{dt}\right)}{\left(\frac{4\pi v \sin \lambda}{\tau}\right)} \qquad (10.74)$$

where v is the speed of an air mass over the ground, λ the latitude[4], and τ the planetary spin period (seconds). In the case of circular (e.g., cyclonic) motion of radius r, $dv/dt = v^2/r$, and $2\pi/\tau$ is the planetary angular speed of rotation, Ω. Then $Ro = v/(2\,\Omega\,r \sin \lambda)$. If $Ro \ll 1$ then $a_{\text{COR}} \gg a_{\text{inertial}}$ and the Coriolis force is strong; if $Ro \gg 1$ then $a_{\text{COR}} \ll a_{\text{inertial}}$ and the Coriolis force is weak. Observed values in the lower 10 km of the atmospheres of the Earth, Venus, and Mars at $45°$ latitude are approximately

Earth and Mars: $Ro = 0.1$ ("strong" Coriolis force)
Venus: $Ro = 10 - 50$ ("weak" Coriolis force)

10.5.2 Meridional (N–S) Circulation

The primary element of meridional circulation is the Hadley cell (Section 10.3.7.2), driven by solar heating near the equator and radiative cooling closer to the poles. The situation for the Earth is described in Section 10.3.7.

On Mars, the low thermal inertia results in rapid response to changing solar illumination, and strong heating at the sub-solar latitude. This creates a single Hadley cell from the tropics of the summer hemisphere to the subtropics of the winter hemisphere, as illustrated schematically in Figure 10.25(a).

In the case of Venus, most of the absorption of solar energy takes place in and above the clouds. The main Hadley cell is therefore located at cloud level (the "driver cell"), and because of the small Rossby number it extends from the equator to the poles in each hemisphere (Figure 10.25(b)). Measurements suggest a second, thermally indirect cell below the cloud level, driven by the cloud-level cell, and a third, thermally direct cell between the second cell and the ground.

10.5.3 Zonal (E–W) Circulation

Large-scale zonal flow on the Earth arises initially from deflection of the meridional flow by the Coriolis force. Thermal winds arising from temperature gradients, as well as topography, perturb this zonal flow and create large-scale waves, eddies, and jet streams (Sections 10.3.2–10.3.6 and 10.3.7.4–10.3.7.6).

---

[4] In other parts of Solar System Astrophysics, we have used the conventional astronomical notation of (λ, ϕ) for longitude and latitude, respectively, but in this context, we use λ for latitude in accord with planetary science literature.

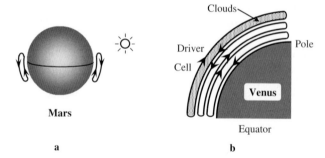

Fig. 10.25. Hadley cells on Mars and Venus. **a** A single Hadley cell from the summer tropics to the winter subtropics on Mars. **b** Solar-driven equator-to-pole Hadley cell at cloud level on Venus, with two possible secondary cells below it

10.5.3.1 Mars On Mars, the low thermal inertia and rapid response to incident solar radiation result in strong meridional temperature gradients only at mid-latitudes in the winter hemisphere. An example is shown by the narrowly spaced, vertical isotherms between approximately 30° and 70°N latitude in the left-hand diagram in Figure 10.26, for late winter in the northern hemisphere of Mars. Pressure in this figure corresponds (non-linearly) to altitude.

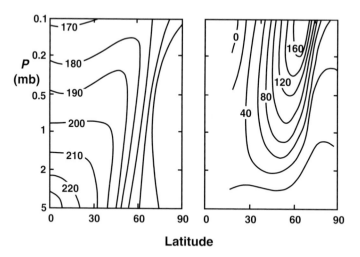

Fig. 10.26. Observed temporally and zonally averaged temperatures (left-hand graph), and the zonally averaged winds deduced from them (right-hand graph) using the thermal wind equation, as functions of latitude and altitude (in pressure units) for a 40-day period in the late northern hemispheric winter on Mars. Temperatures are in K and velocities in m/s, with westerly winds positive. The tropopause lies at about the 0.5-mbar level. (After Conrath 1981, Figs 2 and 3, p. 248.)

In the right-hand diagram in Figure 10.26 the meridional temperature gradient has created a westerly (zonal) flow with speeds near ground level of less than about $20\,\text{m/s}$, and a westerly jet near $60°\text{N}$ latitude extending into the stratomesosphere with speeds up to about $160\,\text{m/s}$. As on Earth, baroclinic waves arise in the westerly flow; thus on Mars these will be found only at mid-latitudes in the winter hemisphere.

10.5.3.2 Venus

10.5.3.2.1 Atmospheric Superrotation The atmosphere of Venus between the first scale height and the ionosphere *superrotates*, i.e., the atmosphere moves in the same direction as the planetary rotation (retrograde, east-to-west), but much faster than the solid planet. The sidereal rotation period for Venus is 243 days, whereas an air mass at cloud level takes only 4 days to circle the rotation axis, or 6 days for an air mass in the thermosphere. Thus, the air at cloud level is moving 60 times faster than the planet under it. Typical wind speeds are of the order of $100\,\text{m/s}$ at the top of the clouds and $65\,\text{m/s}$ in the thermosphere.

10.5.3.2.2 Cyclostrophic Balance On the Earth, with its 24-h rotation period, wind direction and speed on a large scale are determined by geostrophic balance, where the pressure gradient force on a parcel of air is balanced by the Coriolis force (Section 10.3.5).

Venus, however, rotates much more slowly. As a result, the Rossby number for Venus is about 23 (Section 10.5.1), the Coriolis force is negligible, and geostrophic balance does not occur. Instead, the superrotating winds on Venus are in *cyclostrophic balance*, where in a horizontal plane (in a reference frame rotating with the planet) the pressure gradient force on a parcel of air is balanced by the centrifugal force.

Figure 10.27(a) shows the centrifugal force, $\vec{F}'_C = -m\vec{\Omega} \times (\vec{\Omega} \times \vec{r})$, on two parcels of air in the atmosphere of Venus, as Venus rotates (retrograde) about its axis. $\vec{F}'_C$ is directed perpendicularly outward from the rotation axis, and can be separated into components perpendicular to and parallel to the planetary surface, as indicated.

The perpendicular component of the centrifugal force reduces the effective gravity and increases the pressure scale height, i.e., the atmosphere expands vertically compared to the non-rotating case, to maintain pressure equilibrium.

The horizontal component points toward the equator for both air parcels shown, causing the air to move closer to the equator. This in turn increases the atmospheric pressure at the equator and creates a poleward-pointing pressure gradient force. The process continues until the pressure gradient force and the horizontal component of the centrifugal force balance, as indicated in Figure 10.27(b). This is cyclostrophic balance.

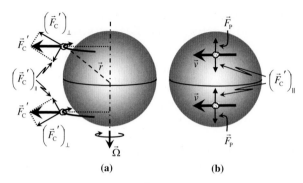

Fig. 10.27. a The centrifugal force, $\vec{F}'_C$, on parcels of air in the atmosphere of Venus, and its components parallel and perpendicular to the planetary surface. **b** Cyclostrophic balance: the pressure gradient force, $\vec{F}_P$, is balanced by the parallel component of the centrifugal force. The resulting cyclostrophic wind, $\vec{v}$, is parallel to the equator

Cyclostrophic balance is familiar on the Earth in intense tropical cyclones, where the pressure gradient force toward the low-pressure area at the center (the eye) is so large that the Coriolis force is negligible by comparison ($Ro \gg 1$). The rotating cyclone contracts, spinning up to conserve angular momentum, until the pressure gradient force is balanced by the centrifugal force, as illustrated in Figure 10.28.

Note: On Venus, the winds are relatively light in the lowest scale height above the ground (containing more than half the mass of the atmosphere). Wind velocities measured at ground level by the Soviet Venera landers were 0.3–1 m/s. Because of the slow wind speeds and the slow planetary spin, the circulation in the lowest scale height is not expected to be either geostrophic or cyclostrophic.

10.5.3.2.3 Atmospheric Angular Momentum A superrotating atmosphere has a greater angular momentum, L, than if it were simply corotating with the planet. The Earth's atmosphere superrotates at mid-latitudes (westerly jets), but this is partly offset by sub-rotation (retrograde compared to the planetary surface) near the equator due to easterly winds (the surface trade winds and

Fig. 10.28. Cyclostrophic balance in an intense tropical cyclone on the Earth

stratospheric easterly summer jet). Also, the mass of the atmosphere is much less than on Venus, and the jets are localized in latitude.

One can gain some appreciation for the amount of superrotation in the atmosphere of Venus by looking at the excess atmospheric angular momentum over that required for corotation: $3.4 \times 10^{28} \, \mathrm{kg \, m^2/s}$ for Venus, or about 200 times that of the Earth's atmosphere $(1.6 \times 10^{26} \, \mathrm{kg \, m^2/s})$. Because of the more massive atmosphere of Venus, the excess atmospheric angular momentum per unit atmospheric mass is $7.1 \times 10^7 \, \mathrm{m^2/s}$, or about twice that of the Earth $(3.0 \times 10^7 \, \mathrm{m^2/s})$. To appreciate superrotation, however, it may be more instructive to express the excess atmospheric angular momentum as a fraction of the planetary angular momentum: 1.6×10^{-3} for Venus, or about 60,000 times that for the Earth (2.7×10^{-8}).

The angular momentum of the Earth's atmosphere changes by several tens of percent over short timescales, due to motions of atmospheric masses. The resulting angular momentum transfer between the atmosphere and the solid planet causes fluctuations (increases and decreases) of the order of milliseconds in the length of the day.

Measurements of zonal wind speeds on Venus made at different times indicate that similar angular momentum fluctuations may occur in Venus' atmosphere. Fluctuations of 20% or more would result in fluctuations on the order of hours in the length of the day on Venus.

Maintaining superrotation in Venus' atmosphere is a concern, because the faster, high-level regions and the slower, lower-level regions exert drag forces on each other. This tends to slow the air at the higher levels and speed it up at the lower levels, producing a transfer of angular momentum downward. Superrotation can be maintained only if there is a competing process transferring angular momentum upward to the cloud level. Although the specific process has not been definitively identified, recent models suggest that meridional circulation in the Hadley driver cell (Figure 10.25) coupled with planetary-scale waves and eddies transport angular momentum equatorward in the lower and middle atmosphere (Yamamoto and Takahashi 2004). Rising air near the equator (Figure 10.25(b)) then transports the angular momentum upward.

10.5.3.2.4 Superrotation vs. Other Circulation Patterns At mid-latitudes on the Earth, the zonal (geostrophic) wind dominates the meridional (Ferrel) flow, so the westerlies are from the west rather than the southwest. In the tropics the zonal and meridional (Hadley) flows are more equal, and the trade winds are from the NE or the SE.

On Venus, the zonal flow (superrotation, speed about 100 m/s at cloud level) by far dominates the meridional (Hadley) flow (about 5 m/s) in all but the lowest scale height, so the flow is primarily parallel to the equator. The Hadley flow still exists, however, and is important in the maintenance of the super-rotation (Section 10.5.3.2.3). The Hadley flow is visible as the polar vortex, a region of converging flow looking a bit like a cyclone at each pole.

In Venus' thermosphere the noon–midnight flow (Section 10.4.3.3) is the dominant circulation pattern, and the 65 m/s superrotation simply rotates the axis of the noon–midnight flow in Figure 10.24 somewhat from midnight toward morning (or noon toward evening).

10.5.4 Other Considerations

10.5.4.1 Latent Heat *Latent heat* is the heat absorbed or released during a phase change. This is an important component of the atmospheric thermody-namics of the Earth, e.g., as saturated air rises and cools, the heat released by the condensation of water vapor to form cloud droplets reduces the adiabatic lapse rate to 4.9°C/km, compared to 9.76°C/km for dry air (Section 10.4.1.1). Latent heat is also important to the formation and maintenance of hurricanes and other weather systems.

Latent heat is not important for atmospheric circulation on Mars and Venus. Mars lacks liquid water on its surface; the very small quantity of water vapor in its atmosphere is maintained primarily by exchange with the water ice in the residual northern polar cap. Precipitable water vapor in the Martian atmosphere is equivalent to a depth of only micrometers of water on the Martian surface. Venus lacks surface water in any form, and Venera and Pioneer Venus measurements suggest a mole fraction of water vapor in the atmosphere 10–20 times less than that on the Earth.

10.5.4.2 Thermal Inertia Massive absorbers exhibit *thermal inertia*, i.e., a large heat transfer produces only a small temperature change. They thus act as a heat reservoir, and reduce the temperature variations in a planetary atmosphere.

The primary heat reservoir on the Earth is the ocean. On Venus, a similar function is provided by the massive lower atmosphere, which is about 50 times denser than on Earth and more than five times deeper (Figure 10.23). As a result there is very little diurnal temperature variation in the troposphere (<1 K at $30°$ latitude) or the stratomesosphere, and only a small temperature decrease from the equator to the poles (5–15 K, compared to 45 K for the Earth and 90 K for Mars).

Mars, with no oceans and only a thin atmosphere, lacks an efficient atmospheric heat reservoir. The low thermal inertia results in strong diurnal and equator–pole temperature variations.

10.5.4.3 Brunt-Väisälä Frequency: When a parcel of air is displaced vertically in a convectively stable region of a planetary atmosphere it returns to its starting point, but overshoots and then oscillates up and down like a mass on a spring. The frequency of oscillation is the Brunt-Väisälä frequency, which also shows up as the buoyant frequency of solar oscillations.

If the period of oscillation is short compared to the damping time (e.g., damping by radiative cooling), then the oscillation is approximately undamped, and combined with horizontal flow produces atmospheric waves as illustrated schematically in Figure 10.29. These waves are often visible as wave-like cloud patterns in the atmospheres of the Earth and Mars.

10.5.4.4 Diffusion and Mixing in Planetary Atmospheres For a planetary atmosphere of any given overall composition, the actual composition at any given height is determined by the interaction of two competing factors:

Mixing: "stirs" the atmosphere and tends to produce a uniform composition, i.e., a composition which is constant with height.

diffusion: lighter gases diffuse to greater heights than heavier gases because of their larger scale height; if diffusion acted alone then the composition would change with height, with lighter gases becoming increasingly dominant with increasing height in the atmosphere.

Table 10.3 provides a comparison of the mole fractions of the most abundant components of the lower atmospheres of the Earth, Venus, and Mars, with the major sources and sinks.

10.5.4.4.1 Diffusion An ideal gas, to which a planetary atmosphere may be approximated for most purposes, is a gas of non-interacting particles, and therefore also non-interacting chemical species. The pressure at any point in the atmosphere is

$$P = nkT \qquad (10.18)$$

where n is the number density. The *partial pressure*, P_i, of any species i (e.g., oxygen, O_2) in the gas equals the total pressure which that species would exert if all other molecules were removed, i.e.,

$$P_i = n_i kT \qquad (10.75)$$

Fig. 10.29. Air flow over a mountain, illustrating oscillations at the Brunt-Väisälä frequency. Clouds often form when the air rises and cools, and evaporate when it sinks and heats, thus outlining the tops of the waves

Table 10.3. Comparative composition of the lower atmospheres of the Earth, Venus, and Mars[a]

Planet	Gas	Mole fraction	Major source	Major sink
Earth	N_2	0.781	Biology	Biology
	O_2	0.209	Biology	Biology
	^{40}Ar	0.0093	Outgassing (^{40}K)	–
	H_2O	<0.04	Evaporation	Condensation
	CO_2	0.00034	Biology, combustion	Biology
Venus	CO_2	0.965	Outgassing	$CaCO_3$ formation?
	N_2	0.035	Outgassing	–
	CO	0.00002	Photochemistry (from CO_2)	Photooxidation
	SO_2	0.00015	Photochemistry	$CaSO_4$ formation
	^{40}Ar	0.000033	Outgassing (^{40}K)	–
	O_2	Undetected in the lower atmosphere	–	–
Mars	CO_2	0.953	Evaporation, outgassing	Condensation
	N_2	0.027	Outgassing	Escape as N
	^{40}Ar	0.016	Outgassing (^{40}K)	–
	O_2	0.0013	Photochemistry (from CO_2)	Photoreduction
	CO	0.0007	Photochemistry (from CO_2)	Photooxidation

[a]From Prinn and Fegley (1987, Tables 1, 2, and 3).

The total number density of the gas equals the sum of the individual number densities, so

$$P = nkT = \left(\sum_i n_i\right)kT = \sum_i (n_i kT) = \sum_i P_i \qquad (10.76)$$

Equation (10.76) shows that, for a mixture of ideal gases, each molecular species contributes a partial pressure proportional to its abundance, and the total pressure is equal to the sum of the partial pressures.

If mixing were to cease in a fully mixed atmosphere, each chemical species would proceed to diffuse upward or downward to establish its own scale height,

$$H_i = \frac{kT}{\mu_i m_u g} \qquad (10.77)$$

where μ_i is the molecular weight of species i. Once this process is complete, the atmosphere is said to be in *diffusive equilibrium*.

The partial pressure of species i as a function of altitude, h, above any given reference level is, from (10.27),

$$P_i = P_0^i e^{-\frac{h}{H_i}} \tag{10.78}$$

where P_0^i is the pressure of species i at the reference level. If the atmosphere is in thermal equilibrium, then at any given altitude the components all have the same temperature. The scale height is then inversely proportional to the molecular weight, i.e., lighter gases have larger scale heights: the partial pressure of heavier gases decreases with increasing altitude faster than that of lighter gases, and for an atmosphere in diffusive equilibrium the composition shifts progressively toward lighter species as altitude increases.

10.5.4.4.2 Diffusion vs. Mixing As seen above, diffusion tends to produce a composition gradient in the atmosphere, whereas convection and other atmospheric circulation tend to mix species, producing a uniform composition (zero composition gradient). Which process dominates at any given altitude depends on the relative timescales for diffusion and mixing. We define the *diffusion time*, τ_D, as the time needed for diffusion to produce a significant change in composition, and the *mixing time*, τ_m, as the time needed for mixing to produce a significant change in composition. The diffusion time is given approximately by

$$\tau_D = 10^{-13}\, n \text{ seconds} = 3 \times 10^{-21}\, n \text{ years} \tag{10.79}$$

where n is the total number density (m^{-3}) at any given altitude.

In the troposphere, $n \sim 10^{25}\,\text{m}^{-3}$ and by (10.79) $\tau_D \sim 30{,}000$ years. The mixing time due to convection and circulation is $\tau_m \sim 1$ month; therefore, mixing by far dominates diffusion and the troposphere is well-mixed (uniform composition). In the stratosphere, $n \sim 10^{23}\,\text{m}^{-3}$ and $\tau_D = 300$ years. The lapse rate is negative (temperature increases with altitude) so there is no convection, but strong winds give $\tau_m \sim 10$ years. Therefore, mixing again dominates diffusion, and the stratosphere is well-mixed.

In the mesosphere, n decreases from $\sim 10^{21}\,\text{m}^{-3}$ at the stratopause to $\sim 10^{20}\,\text{m}^{-3}$ at the mesopause, giving $\tau_D \sim 3$ and 0.3 years (3.5 months), respectively. The mixing time due to convection is $\tau_m \sim 1$ month, so the mixing again dominates diffusion over most of the mesosphere, and the mesosphere is well-mixed except at the top, where diffusion begins to become important. In the thermosphere, $n \sim 10^{13}\,\text{m}^{-3}$ at 100 km altitude, and $\sim 10^{10}\,\text{m}^{-3}$ at 200 km, giving $\tau_D \sim 10$ days at 100 km and 1 h at 200 km. The thermospheric lapse rate is negative so there is no convection, and the mixing times are long. Thus diffusion dominates mixing in the thermosphere.

10.5.4.4.3 The Homopause The *homopause*, or *turbopause*, is defined as the altitude at which the diffusion time equals the mixing time, $\tau_D = \tau_m$. This occurs at about 100 km altitude. Below the homopause is the *homosphere*, where mixing produces a homogeneous composition, and above it is the *heterosphere*, where diffusion produces a composition gradient (heterogeneous composition).

Even in the homosphere, however, local production and loss can be rapid enough to alter the abundance of selected species, e.g., ozone.

10.5.5 Chemical Cycles

A planetary atmosphere is part of an interacting system that includes the solid planet and, where they exist, oceans, other surface liquids, and life. Here we look briefly at several important cycles in this system.

10.5.5.1 Carbon Cycle (Earth) Figure 10.30 shows the terrestrial atmospheric carbon cycle. Juvenile carbon is carbon from the interior of the Earth, being released for the first time in the Earth's history. Almost all of this is in the form of CO_2 from volcanism along mid-ocean ridges.

The largest reservoir of carbon is lithospheric rock, such as limestone. However, the greatest rates of transfer are between the atmosphere and the biosphere by respiration, plant photosynthesis, and decay of organic material. The retention time for carbon (in the form of CO_2) in the atmosphere may be found from

$$\tau = \frac{\text{supply}}{\text{rate of loss}} = \frac{720 \times 10^{12} \text{ kg of C}}{150 \times 10^{12} \text{ kg of C/y}} \cong 5\,\text{y} \qquad (10.80)$$

Thus, atmospheric carbon is recycled every 5 years.

10.5.5.2 Oxygen Cycle (Earth) By far the dominant source of free oxygen in the Earth's atmosphere is photosynthesis; consequently, the oxygen cycle is governed by the biosphere–atmosphere cycle shown by thick arrows in Figure 10.30. The overall reaction for photosynthesis is $CO_2 \rightarrow C + O_2$, so the removal of 150×10^{12} kg of carbon from the atmosphere releases

$$150 \times 10^{12} \, \frac{\text{kg C}}{\text{y}} \times \frac{32}{12} = 400 \times 10^{12} \, \frac{\text{kg O}_2}{\text{y}}$$

into the atmosphere. The amount of oxygen in the Earth's atmosphere is 1.2×10^{18} kg, and from Figure 10.30 the rates of removal and replenishment

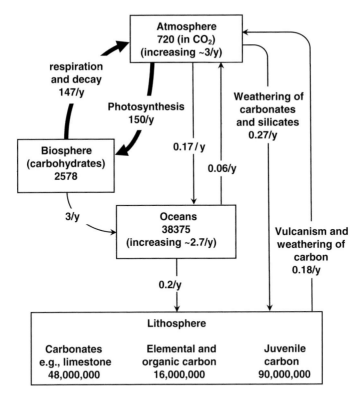

Fig. 10.30. The terrestrial carbon cycle. The *arrows* show the rates of transfer of carbon between the various carbon reservoirs (*boxes*). *Thick arrows* show the dominant paths. The unit for all numbers is 10^{12} kg of carbon. The atmospheric carbon content, in the form of CO_2, is increasing because of the burning of fossil fuels (not shown). (After Wayne 2000, Fig. 1.4, p. 19.)

are approximately equal, so from (10.80) the retention time for oxygen in the Earth's atmosphere is approximately

$$\frac{1.2 \times 10^{18} \text{ kg of } O_2}{400 \times 10^{12} \text{ kg of } O_2/\text{y}} \approx 3000 \text{ y}$$

This compares to only a few years for carbon. The rates of removal of C and O_2 from the atmosphere are of the same order of magnitude, differing only by a factor of $32/12$, so the much larger retention time for oxygen is due almost entirely to the much larger amount of free oxygen in the atmosphere (Table 10.3).

10.5.5.3 Nitrogen Cycle (Earth) Nitrogen (N_2) is the most abundant component of the Earth's atmosphere, taking up a mole fraction of 0.78

(Table 10.3). It is chemically quite inert and does not participate in the cycles of other nitrogen compounds, such as ammonia (NH_3) or nitrous oxide (N_2O).

The fact that Venus and Mars, which lack life, have atmospheres composed almost entirely of CO_2, with N_2 taking up <4% (Table 10.3), indicates that the N_2 in the Earth's atmosphere must be almost entirely biogenic in origin. Figure 10.31 shows the basic cycles involved. The major source of N_2 is denitrifying bacteria in soils and the oceans that convert ammonium and nitrate compounds into N_2; the major sinks are (1) nitrogen-fixing bacteria (e.g., in the root nodules of legumes) that convert atmospheric N_2 into NH_4^+, NO_3^-, and organic nitrogen and (2) lightning and combustion, in which N_2 reacts with free O_2 to produce the oxides NO, NO_2, etc.

If life were to cease on the Earth, then lightning and combustion would provide the major sinks for atmospheric N_2. However, combustion would cease when the supply of organic material was depleted, and both lightning and combustion would become ineffective with the loss of free O_2 from photo-

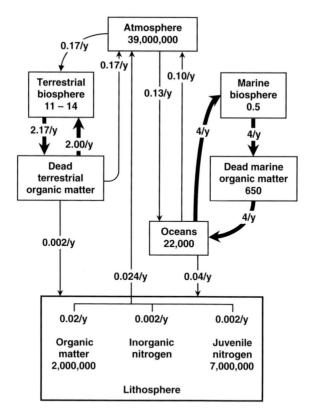

Fig. 10.31. The terrestrial nitrogen cycle; see Figure 10.30 for explication. The unit for all numbers is 10^{12} kg of nitrogen. (After Wayne 2000, Fig. 1.6, p. 26.)

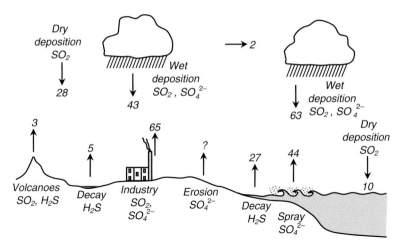

Fig. 10.32. Major sources and sinks of sulfur in the terrestrial atmosphere. The unit for all numbers is 10^9 kg of sulfur per year. The main compounds transferred are indicated in each case. Sources and sinks over land are shown on the left, and over oceans on the right. (After Iribarne and Cho 1980, Fig. II-5, p. 33.)

synthesis. Thus abiotic removal rates would be much less than today, and the retention time for N_2 in the atmosphere is expected to be of the order of Gy.

10.5.5.4 Sulfur Cycle (Earth) Figure 10.32 shows the major sources and sinks in the terrestrial sulfur cycle. The quantities involved are significantly less certain than for the carbon and oxygen cycles, e.g., more recent estimates for some quantities are, in the same units, volcanoes 7; decay 58 over land and 48 over the ocean; and industry 100 (Goody, 1995; Wayne, 2000).

The mixing ratio for SO_2 in the terrestrial atmosphere is 167 p.p.t. (parts per trillion or parts per 10^{12}), and that of H_2S is variable, but between 0 and 100 p.p.t. The corresponding amounts of these sulfur compounds present in the atmosphere are then much smaller than the amounts transferred per year, indicating very short retention times of days or even hours.

Two more sulfur-containing gases are important in the chemistry of sulfur in the Earth's atmosphere: carbonyl sulfide (COS) is thought to be the most abundant sulfur gas in the troposphere at 500 p.p.t., and carbon disulfide (CS_2) is present in variable concentrations of 2–120 p.p.t. Volcanic eruptions contribute some portion of these gases, and oxidation of CS_2 also appears to produce COS, but in general their sources are not known at the present time, or even whether the dominant sources are natural or anthropogenic.

COS survives long enough to reach the stratosphere where it is oxidized to produce sulfate ions (SO_4^{2-}) and sulfuric acid (H_2SO_4), the latter principally in droplet form. This sulfuric acid aerosol layer tends to concentrate in a layer near 20 km altitude, called the *Junge layer*.

10.5.5.5 Sulfur Cycle (Venus) Figure 10.33 shows the major reactions in the sulfur cycle on Venus. The cycle depends in part on the presence of free atomic oxygen in the upper atmosphere, and this is readily produced by photolysis of CO_2:

$$CO_2 + h\nu (\lambda \leqslant 204\,\text{nm}) \rightarrow CO + O \tag{10.81}$$

The recombination reaction

$$CO + O + M \rightarrow CO_2 + M \tag{10.82}$$

is spin-forbidden and slow, so although a considerable amount is lost to recombination to O_2, enough O remains to maintain the cycle.

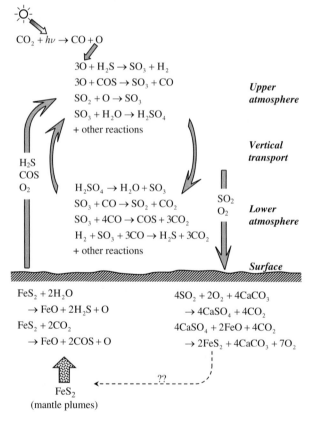

Fig. 10.33. Sulfur cycle on Venus. Major chemical reactions are shown, and the vertical transport that links them

An important component of the cycle is the production of sulfuric acid vapor in the upper atmosphere:

$$SO_3 + H_2O \rightarrow H_2SO_4 \qquad (10.83)$$

The temperature in the region 50–60 km altitude is low enough that the H_2SO_4 vapor readily condenses to form very concentrated droplets of sulfuric acid. The resulting sulfuric acid clouds are much more reflective than water-droplet clouds, and blanket the entire planet, giving Venus a very uniform optical albedo of 0.84.

10.5.5.6 Thermospheric Chemistry of Neutrals (Earth)

Two important thermospheric reactions are photodissociation (AB+$h\nu \rightarrow$ A+B) and photoionization (AB+$h\nu \rightarrow$ AB$^+$ + e$^-$), where $h\nu$ represents the energy of the absorbed photon, and e$^-$ is a free electron.

These reactions do not proceed directly. Instead, the absorbed photon energy creates an excited molecule, AB*, which can decay by any allowed route; dissociation and ionization are just two of these routes. (Collisional de-excitation, or quenching, would be another route, but it is much less important at the low density of the thermosphere.)

Table 10.4 lists several decay routes of interest in a planetary atmosphere.

Table 10.4. Products of molecular photoexcitation in the thermosphere

Reactants	Paths	Products	Description
AB+$h\nu \rightarrow$	AB*		*Photoexcitation*
	$\downarrow$		
	$\rightarrow$	A + B*	*Dissociation* into two neutrals: one in the ground state; one in an excited state
	$\rightarrow$	AB$^+$ + e$^-$	*Ionization*
	$\rightarrow$	AB + $h\nu$	*Radiative de-excitation*
	$\rightarrow$	AB§	*Intramolecular energy transfer*: *luminescence* if by an allowed transition; *phosphorescence* if by a forbidden transition
	+CD $\rightarrow$	AB + CD*	*Intermolecular energy transfer*: radiationless transition to a different excited state of the same molecule.
	+ M $\rightarrow$	AB + M	*Quenching*: the electronic excitation energy of AB* is degraded to vibrational, rotational, or translational energy by collision with another molecule
	+EF $\rightarrow$	AE + BF	*Chemical reaction* mediated by the electronic excitation of AB*

10.5.6 Excess Radiation

Up to now we have discussed the energy budget of the terrestrial planets. The temperatures of Venus and Earth are currently higher than their equilibrium temperatures, as we noted in earlier chapters, but this is understood in terms of greenhouse effects. One of the continuing mysteries of the solar system is the energy excess of the emission of the outer planets: three of the four giant planets appear to radiate more energy that they receive from the Sun. The sources of this energy may not be the same in each of Jupiter, Saturn, and Neptune; Uranus, which should be more closely related to Neptune than to the other two giants, does not show any excess radiation. The thermal emission is independent of the non-thermal radiation which has its origin in the magnetospheres (see Chapter 11), and which in origin is relatively well understood.

Of the possible sources of heat, primordial heat is a possibility, but bulk contraction (conversion of the potential energy of the planets' material into kinetic energy as it falls toward the center) is not favored at present because the liquid must resist compression.

Another related mechanism, however, involves selective "fall out" of some material—such as the "rain" of helium in the interior. This mechanism is thought to be especially important in Saturn, where the composition of the outer atmosphere seems to be deficient in helium relative to that of the Sun. In the case of Neptune, the excess may be due to a greenhouse effect, but here involving the gas methane in the very outer atmospheric layer.

References

Conrath, B. 1981. "Planetary-scale Wave Structure in the Martian Atmosphere," *Icarus*, **48**, 246–255.

Cox, A. N., Ed. 2000. *Allen's Astrophysical Quantities*, 4th Ed. (New York: Springer-Verlag)

Elliott, J. L., and Olkin, C. B. 1996. "Probing Planetary Atmospheres with Stellar Occultations," *Annual Review of Earth and Planetary Sciences*, **24**, 89–123.

Goody, R. 1995. *Principles of Atmospheric Physics and Chemistry* (New York: Oxford University Press)

Iribarne, J. V. and Cho, H.-R. 1980. *Atmospheric Physics* (Boston, MA D. Reidel).

Jeans, J. 1952. *An Introduction to the Kinetic Theory of Gases.* (Cambridge: The University Press). (Reprinting of 1st Ed., 1940).

Kieffer, H. H., Jakosky, B. M., Snyder, C. W., and Matthews, M. S., ed. 1992. *Mars* (Tucson, AZ: University of Arizona Press).

Milone, E. F. and Wilson, W. J. F. 2007. *Solar System Astrophysics: Background Science and the Inner Solar System* (New York: Springer-Verlag).

Pasch, R. J., Blake, E. S., Cobb, H. D. III, and Roberts, D. P. 12 January 2006. *Tropical Cyclone Report Hurricane Wilma 15–25 October 2005* (National Hurricane Center).

Prinn, R. G., and Fegley, B., Jr. 1987. "The Atmospheres of Venus, Earth, and Mars: A Critical Comparison," *Annual Reviews of Earth and Planetary Sciences*, **15**, 171–212.

Schlosser, W., Schmidt-Kaler, Th., and Milone, E. F. 1991/1994. *Challenges of Astronomy: Hands-On Experiments for the Sky and Laboratory* (New York: Springer-Verlag).

Seinfeld, J.H., and Pandis, S.N. 1998. *Atmospheric Chemistry and Physics: From Air Pollution to Climate Change* (New York: John Wiley and Sons).

Wayne, R. P. 2000. *Chemistry of Atmospheres* (Oxford: Oxford University Press).

Yamamoto, M. and Takahashi, M. 2004. "Dynamics of Venus' Superrotation: The Eddy Momentum Transport Processes Newly Found in a GCM," *Geophysical Research Letters*, **31**, L09701 (doi: 10.1029/2004GL019518).

Challenges

[10.1] The phenomenon of the Chinook in Western Canada, the Föhn in Switzerland, and the Sirocco around the Mediterranean, involves a warm wind at lower elevations on the leeward side of mountains, with a higher temperature than is found at comparable elevations on the windward side of the mountains. Explain the physical basis for the phenomenon. (Hint: Consult the end of Section 10.2)

[10.2] From (10.63), find the magnitude of the centrifugal force on an object of mass m on the surface of a planet as a function of Ω, r, and the object's latitude, λ. Also find the magnitude of the horizontal component of this force. At what latitude is the horizontal component largest?

[10.3] From (10.62), find the magnitude of the Coriolis force on an object of mass m travelling horizontally at speed v on the surface of the Earth, as a function of Ω, r, v, and the object's latitude, λ, if the object is travelling (a) due north, (b) due east, and (c) at an azimuth angle α measured eastward from north. In part (c), α will also be in the equation. [Hint: Decompose the velocity into component vectors northward and eastward.] Illustrate the subsequent trajectory of a rocket fired due north from the equator.

[10.4] Compute the relative magnitude of the Coriolis effect on Earth, Venus, and Mars. That is, derive the results of Section 10.5.1.

[10.5] Derive equations (10.7) and (10.8), including the numerical constants in those equations.

[10.6] Compute the tropospheric pressure scale heights for Venus, Earth, and Mars.

[10.7] Consulting Sections 10.1 and 10.2 and Fig. 10.23, discuss the retention of water vapor on the terrestrial planets.

11. Planetary Ionospheres and Magnetospheres

11.1 Earth: Ionospheric Layers

Sunlight interacts with planetary atmospheres to produce excited atoms and ions, and the ions interact with each other and with neutral atoms to produce unique, altitude-dependent populations. As with other aspects of the terrestrial planets, we know the ionosphere of the Earth best, so we start with it. Figure 11.1 shows the electron density and ionospheric layers in the Earth's ionosphere. During the daytime, the D layer forms and the F layer separates into two parts, F_1 and F_2; at night the D layer disappears, and the F layers merge into a single layer.

11.1.1 The F Layer

The F layer is created primarily by solar EUV radiation in the wavelength range $10 \, \text{nm} < \lambda < 80 \, \text{nm}$.

11.1.1.1 Atoms and Ions in the F Layer The dominant neutral atoms are O and N_2. Atomic oxygen is much more abundant than O_2 in the F region

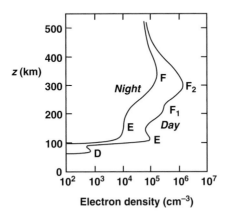

Fig. 11.1. Electron density vs. altitude in the Earth's ionosphere. Typical curves and ionospheric layers are shown for night and day. The curves vary considerably with solar activity, sunspot number, and time of year. From Iribarne and Cho 1980, Figure I-7, p. 10

because most of the O_2 has undergone *photodissociation* (or *photolysis*):

$$O_2 + h\nu \rightarrow O + O \quad \text{photodissociation} \tag{11.1}$$

where, as an aid, we list the mechanism after the reaction. The dominant ions are primarily O^+ with smaller amounts of O, NO^+, N^+.

11.1.1.2 Production Mechanisms The production of the ions is due to several reactions. The dominant (source) reactions are as follows:

A. The primary[1] reactions, the process, and the wavelength range are

$$O + h\nu \rightarrow O^+ + e^- \quad \text{photoionization} \quad \lambda < 91.0\,\text{nm} \tag{11.2a}$$

$$N + h\nu \rightarrow N^+ + e^- \quad \text{photoionization} \quad \lambda < 85.2\,\text{nm} \tag{11.2b}$$

$$N_2 + h\nu \rightarrow N_2^+ + e^- \quad \text{photoionization} \quad \lambda < 79.6\,\text{nm} \tag{11.2c}$$

B. The secondary reactions are

$$N_2^+ + O_2 \rightarrow N_2 + O_2^+ \quad \text{charge transfer} \tag{11.3a}$$

$$N_2^+ + O \rightarrow N + NO^+ \quad \text{ion exchange} \tag{11.3b}$$

Most of the N_2^+ produced in the primary reactions is thus removed by the secondary reactions.

11.1.1.3 Loss Mechanisms The loss mechanisms (sinks) of the ions are as follows:

A. Molecular ions

The most common process is dissociative recombination, where the electron recombines with a molecule, dissociating it into two neutral atoms in the process:

$$O_2^+ + e^- \rightarrow O + O \quad \text{dissociative recombination} \tag{11.4a}$$

$$NO^+ + e^- \rightarrow N + O \quad \text{dissociative recombination} \tag{11.4b}$$

$$N_2^+ + e^- \rightarrow N + N \quad \text{dissociative recombination} \tag{11.4c}$$

---

[1] The words "primary" and "secondary" refer to the sequence of events, not to relative importance, i.e., the primary reactions have to occur first, in order for the secondary reactions to occur.

Dissociative recombination is required instead of simple recombination, e.g.,

$$O_2^+ + e^- \rightarrow O_2 \qquad \text{recombination} \qquad (11.5a)$$

because recombination is an exothermic process: energy is released. $O_2^+ + e^- \rightarrow O_2$ would leave the O_2 with an excess of energy that in fact would dissociate the molecule.

Radiative recombination, e.g.,

$$O_2^+ + e^- \rightarrow O_2 + h\nu \qquad \text{radiative recombination} \qquad (11.5b)$$

has a low probability and can be ignored compared to dissociative recombination.

B. Atomic ions

Simple recombination of electrons with atomic ions does not occur for the same reason described earlier for molecules: the recombination is an exothermic process, and the excess energy given to the atom by the recombination is sufficient to eject the electron again. The principal loss mechanism for atomic ions is charge transfer to a molecule

$$O^+ + O_2 \rightarrow O + O_2^+ \quad \text{charge transfer} \qquad (11.6)$$

followed by molecular dissociative recombination:

$$O_2^+ + e^- \rightarrow O + O \qquad \text{dissociative recombination} \qquad (11.4a)$$

11.1.1.4 Ion Concentration vs. Altitude The concentration of a given species of ion at any given altitude in the ionosphere depends on the following:

A. The production rate, which depends on the concentration of the "source" species (there will be, for instance, little O^+ if there is little atomic O to be ionized) and the amount of ionizing radiation present, which decreases with decreasing altitude because of absorption at higher altitudes.

B. The loss rate, which depends on (among other things) the density of air, because collisions which remove ions through such processes as electron recombination and charge transfer are less frequent at lower densities.

C. The rate of vertical diffusion. Each species of ion tends to redistribute itself according to its own pressure scale height (Section 10.5.4.4), resulting in fractionation because lower mass ions have larger scale heights. We can describe two separate regimes, one below and the other above the peak of the F_2 region at 300 km altitude:

1. $h < 300\,$km altitude. Here, the density is "high," so the mean ionic lifetime is "short," and any ions produced are destroyed before they can diffuse to a different altitude. Therefore, below 300 km altitude, the ion concentrations at any given level are controlled primarily by local production and loss. This condition is called *photochemical equilibrium.*

2. $h > 300\,$km altitude. Here, the density is "low," so the mean ionic lifetime is "long," and ions have time to diffuse to different altitudes before they are destroyed. Thus the ion concentrations above 300 km altitude are controlled primarily by *diffusion*, producing an exponential decrease in the number density of each ion with altitude, according to the scale height.

11.1.1.5 Charge Separation The same processes of diffusion above 300 km altitude and photochemical equilibrium below 300 km control the concentration of free electrons. Because of their small mass, their scale height is almost infinite compared to the positive ions, and their diffusion produces a vertical charge separation in the ionosphere above 300 km. This *ambipolar diffusion* (diffusion resulting in a charge separation) creates an upward force on the positive ions (especially H^+), lifting them to higher altitudes, so that the scale heights for any ion is about twice the scale height of the equivalent neutral atom.

11.1.2 The E Layer

The E layer is distinct from the F layer by virtue of being created by different wavelengths of solar radiation:

F: primarily $10\,$nm $< \lambda < 80\,$nm
E: primarily $80\,$nm $< \lambda < 102.6\,$nm

and is characterized by a different population of neutral atoms and ions.

11.1.2.1 Atoms and Ions in the E Layer The dominant neutrals are N_2, O_2, and O. Because of UV absorption in the F layer, there is less photodissociation of O_2 in the E layer; so O is less abundant than N_2 and O_2 in all but the highest parts of the E region. The dominant ions are NO^+ and O_2^+.

11.1.2.2 Production Mechanisms The production of ions is due to photoionization and other processes: O_2 is more important in the E layer than in the F layer, but otherwise the reactions are similar.

A. The primary reactions (*i.e.*, those that need to occur first) are photoionizations:

$$O_2 + h\nu \rightarrow O_2^+ + e^- \quad \text{photoionization} \quad \lambda < 102.6\,\text{nm} \qquad (11.7a)$$

$$O + h\nu \to O^+ + e^- \quad \text{photoionization} \quad \lambda < 91.0\,\text{nm} \qquad (11.7\text{b})$$

$$N_2 + h\nu \to N_2^+ + e^- \quad \text{photoionization} \quad \lambda < 79.6\,\text{nm} \qquad (11.7\text{c})$$

B. The secondary reactions are as follows:

$$N_2^+ + O_2 \to N_2 + O_2^+ \quad \text{charge transfer} \qquad (11.3\text{a})$$

$$N_2^+ + O \to N + NO^+ \quad \text{ion exchange} \qquad (11.3\text{b})$$

$$N_2 + O_2^+ \to NO + NO^+ \quad \text{ion/atom exchange} \qquad (11.8)$$

As in the F layer, the secondary reactions remove most of the N_2^+ in favor of O_2^+ and NO^+.

11.1.2.3 Loss Mechanisms The loss mechanisms for ions of the E layer are

$$O_2^+ + NO \to O_2 + NO^+ \quad \text{ion exchange} \qquad (11.9)$$

$$NO^+ + e^- \to N + O \quad \text{dissociative recombination} \qquad (11.4\text{b})$$

(11.4b) is the most important charge neutralization step in the E layer. Also, (11.9) is much more important than (11.4b) in the E layer because the concentration of neutral NO is much greater than the concentration of free electrons.

The E layer is in photochemical equilibrium during daytime, with diffusion and other motions also becoming important at night.

11.1.3 The D Layer

The D layer is shielded by the E and F layers above it. Most of the ionizing radiation has been absorbed by photodissociation and photoionization of O_2 in the E and F layers; particularly

$$122\,\text{nm} < \lambda < 180\,\text{nm} : O_2 + h\nu \to O + O \quad \text{photodissociation} \quad (11.1)$$

$$\lambda < 102.6\,\text{nm} : O_2 + h\nu \to O_2^+ + e^- \quad \text{photoionization} \quad (11.7\text{a})$$

11.1.3.1 Dominant Ions The dominant ions in the D layer are NO^+ ($\sim$80% by number) and O_2 ($\sim$20% by number), plus small but important amounts of other positive and negative molecular ions.

Ionization in the D region comes mainly from four sources:

1. "Windows" in the range $102.6 \, nm < \lambda < 122 \, nm$, in which radiation reaches down to as low as 70 km. These windows include the Lyman-alpha (Lyα) emission line at $\lambda = 121.6 \, nm$.

2. Solar X rays in the upper D region, from the quiet (undisturbed) Sun.

3. Cosmic rays in the lower D region. Cosmic rays are particles (electrons, protons, etc.) arriving at the Earth from the Sun (solar cosmic rays) and sources beyond the solar system (galactic cosmic rays).

4. Solar X rays of $\lambda < 0.6 \, nm$ during disturbed conditions, especially from solar flares.

11.1.3.2 Production Mechanisms Production (source) of ions in the D layer is through

A. Primary reactions, which produce mainly NO^+, O_2^+, and O_2^-.

$$NO + h\nu \rightarrow NO^+ + e^- \qquad \text{photoionization} \quad \lambda < 134.1 \, nm \qquad (11.10a)$$

(Ionization of NO by Lyα photons is the main reason for the existence of the D layer.)

Negative ions also form, by electron attachment in a three-body collision:

$$O_2 + e^- + M \rightarrow O_2^- + M \qquad \text{electron attachment} \qquad (11.10b)$$

where M is any molecule or atom. Simple attachment, viz., $O_2 + e^- \rightarrow O_2^-$, cannot occur because the reaction $O_2 + e^-$ is exothermic, and the excess energy ejects the electron again unless a third body carries off the energy excess. The third body also allows both energy and momentum to be conserved. Radiative recombination ($O_2 + e^- \rightarrow O_2^- + h\nu$) is too slow to be important.

Negative ions are important only in the D region, because the probability of three-body collisions depends strongly on the density. In the E and F regions, the density is so low that three-body collisions are extremely rare, and negative ions are essentially absent.

B. Secondary reactions:

NO^+, O_2^+, and O_2^- react with a large variety of neutral molecules. As a result, the D layer is a complex system of many ionic species, such as H_3O^+ and NO_3^-, as well as "cluster ions," such as $H_3O^+ \cdot H_2O$ and $NO_3^- (H_2O)_n$ that form when neutral molecules become attached to an ionized molecule.

11.1.3.3 Loss Mechanisms The loss of ions in the D layer is primarily through dissociative recombinations.

Production and loss in the D layer occur much faster than mixing or diffusion. The concentration of each ion at each altitude in the D region is therefore controlled by photochemical equilibrium.

An important consequence of this photochemical equilibrium is that, with the rapidity of loss processes and the removal of solar ionizing radiation, the D layer disappears at night.

11.1.4 Reflection of Radio Waves

The E and F layers permit long-distance radio communication by reflecting radio waves back to the Earth, particularly in the standard commercial AM broadcast band (approximately 500–1700 kHz) and short wave regions.

Radio waves are emitted by oscillating electric currents in the transmitting antenna. The inverse process takes place in reception: the incoming radio wave creates an oscillating electric current in the receiving antenna.

A similar process happens in the ionosphere. An incoming radio wave encountering a free electron makes the electron oscillate. The oscillating electron in turn behaves as a small transmitting antenna and emits a radio wave of the same frequency as the incoming wave. Each free electron therefore acts as a combination receiver and re-transmitter.

The overall process is complex (see Section 11.4), but the result is that the path of the radio wave is bent back toward the Earth; that is, the wave is "reflected" by the ionosphere, as Figure 11.2 illustrates.

To be reflected by the E and F layers, however, radio waves have to pass twice through the D layer: once on the way up and once on the way back down.

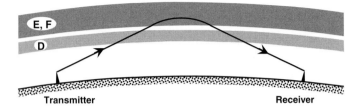

Fig. 11.2. Illustration of the refraction of radio waves by the Earth's ionospheric E and F layers. The diagram is a composite of two scenarios: during the daytime, the waves are absorbed in the D layer and do not reach the receiving antenna, whereas at night the D layer disappears and long-distance reception occurs

The density in the D layer is high enough that the oscillating electrons collide with the surrounding molecules, removing energy from the electrons and increasing the random motion of the molecules. The electrons as a result are unable to retransmit efficiently, reducing the intensity of the radio beam. Thus, energy is taken from the radio waves and converted to heat in the D layer, with the result that the radio waves are absorbed rather than reflected.

During daytime, the D layer limits the commercial AM band to line-of-sight distances. At night, however, the D layer disappears, allowing long-distance reception.

Intense solar flares produce very energetic particles, which can reach the Earth within half an hour after the start of the flare and can penetrate to the D layer. The enhanced ionization in the D layer can then produce radio blackouts by disrupting long-distance radio communication.

11.2 Atmospheric and Ionospheric Chemistry on Mars and Venus

11.2.1 Neutral Atmosphere of Mars

The mixing ratios of H, H_2, O, O_2, CO, and CO_2 in the Martian atmosphere are shown as a function of altitude in Figure 11.3. Some of these are also listed in Table 11.1, for the lower atmosphere. The dominant component at all altitudes below 200 km is CO_2, with a mixing ratio of 0.953 in the troposphere.

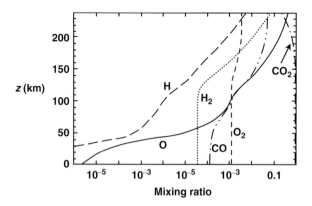

Fig. 11.3. Mixing ratios of H, H_2, O, O_2, CO, and CO_2 in the Martian atmosphere, as a function of altitude, z. Adapted from Nair et al. 1994, Figure 3(a), p. 133, with permission

Because of the thinness of the Martian atmosphere, solar UV penetrates to the ground, and the chemistry of the neutral atmosphere is therefore dominated by photodissociation (photolysis) of CO_2, producing atomic oxygen:

$$CO_2 + h\nu \rightarrow CO + O \quad \text{photodissociation} \quad \lambda < 204\,\text{nm} \tag{11.11}$$

However, as Figure 11.3 shows, molecular O_2 is much more abundant than atomic O below 100 km altitude. The O_2 is created by OH_x radicals derived from water, such as OH (hydroxyl) from

$$H_2O + h\nu \rightarrow OH + H \qquad \text{photodissociation} \tag{11.12a}$$

$$H_2O + O^* \rightarrow OH + OH \tag{11.12b}$$

[where the superscript (*) signifies an excited state] reacting with atomic oxygen:

$$O + OH \rightarrow O_2 + H \tag{11.13}$$

Without an opposing oxidizing process to convert CO back to CO_2, (11.11) and (11.13) would produce the observed CO abundance in less than 3 years, and a mostly CO and O_2 atmosphere in about 2000 years. Figure 11.3 shows that this has not happened. The oxidizing process cannot be direct recombination with O, either by

$$CO + O \rightarrow CO_2^* \rightarrow CO_2 + h\nu \tag{11.14a}$$

or by

$$CO + O + M \rightarrow CO_2 + M \tag{11.14b}$$

because radiative de-excitation of CO_2^* is too slow to compete with (11.11), and (11.14b) is spin-forbidden. However, Mars has significant quantities of water. Various chemical reactions produce OH (hydroxyl) from water, including (11.12a) and (11.12b). The OH then reacts with CO rapidly enough to compete with CO_2 photolysis:

$$CO + OH \rightarrow CO_2 + H \tag{11.15}$$

11.2.2 Neutral Atmosphere of Venus

Venus' clouds shield the regions below them from solar UV radiation, so photochemistry occurs at and above the cloud tops.

The major steps are similar to those on Mars, specifically,

$$CO_2 \text{ photolysis}: \; CO_2 + h\nu \rightarrow CO + O \tag{11.11}$$

$$\text{Formation of } O_2: \;\; O + O + M \rightarrow O_2 + M \tag{11.16}$$

$$\text{Oxidation of CO}: \;\; CO + O \rightarrow CO_2 \;\; \text{(net reaction)} \tag{11.17}$$

Reaction (11.16) replaces (11.13) because Venus lacks significant water.

Reaction (11.17) is required because, in similar fashion to Mars, (11.11) and (11.16) would produce the observed O_2 abundance in only a few years and the observed CO abundance in about 200 years. The result should be a mostly CO and O_2 atmosphere from the cloud tops upward. This is true for the highest levels of Venus' atmosphere, but near the cloud tops the concentrations of CO and O_2 are only 45 and 1 ppm, respectively.

Venus differs from Mars in that the $CO:O_2$ ratio is about 45:1 on Venus, compared to about 1:10 on Mars (Figure 11.3). Thus, an efficient reaction is needed on Venus to break the O–O bond of O_2. The solution is not clear, but several reaction cycles involving chlorine and/or sulfur appear likely.

11.2.3 Ionosphere of Mars

11.2.3.1 Dominant Ions The dominant ion in the Martian ionosphere is O_2^+, with smaller amounts of CO_2^+ and O^+ (Figure 11.4).

It should be noted that, even in the region of greatest ion density ($\sim$130 km altitude), ions are a very small minority of the particles present. A similar situation applies to the ionospheres of the Earth and Venus. Table 11.1 lists some representative values.

11.2.3.2 Production Mechanisms Production is due to the reactions

$$CO_2 + h\nu \rightarrow CO_2^+ + e^- \qquad \text{photoionizaton } \lambda < 93\,\text{nm} \tag{11.18}$$

$$CO_2^+ + O \rightarrow O_2^+ + CO \qquad \text{ion exchange} \tag{11.19a}$$

$$CO_2^+ + O \rightarrow O^+ + CO_2 \qquad \text{charge transfer} \tag{11.19b}$$

$$O^+ + CO_2 \rightarrow O_2^+ + CO \tag{11.20}$$

Not surprisingly, the primary ionospheric reaction is photoionization of CO_2. However, CO_2^+ is *not* the dominant ion (Figure 11.4), because (11.19a) and (11.19b) convert most of the CO_2^+ to O^+ and O_2^+, and (11.20) then converts most of the O^+ to O_2^+.

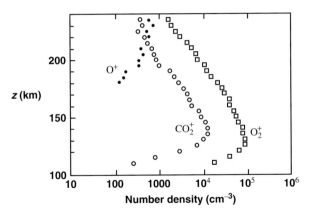

Fig. 11.4. Ion density profiles in the Martian ionosphere. *Open squares* are O_2^+; *open circles* are CO_2^+; and *dots* are O^+. Because each ion contributes one free electron, the electron density is slightly greater than the density of O_2^+. After Chen et al. 1978, Figure 6, p. 3875); copyright 1978, American Geophysical Union. Modified by permission of American Geophysical Union.

11.2.3.3 Loss Mechanisms The most important charge-neutralization reaction is dissociative recombination of the dominant ion:

$$O_2^+ + e^- \rightarrow O + O^* \qquad \text{dissociative recombination} \qquad (11.21)$$

11.2.4 Ionosphere of Venus

11.2.4.1 Dominant Ions The dominant ions in the ionosphere of Venus are different at different altitudes, as on Earth:

Low altitudes (110–200 km): O_2^+ and NO^+
High altitudes (>200 km): O^+
Other ions present include C^+, CO^+, CO_2^+, N^+, N_2^+, H^+, and He^+.

Below 170 km, the ionosphere is in photochemical equilibrium, i.e., abundances are determined by local production and loss. Above 170 km, abundances are determined primarily by diffusion.

Table 11.1. Comparison of the number densities of neutrals and ions in the Martian ionosphere ($\sim$130 km altitude)

Species	Number Density (cm^{-3})	
	Neutral (CO_2, O_2)	Ion (CO_2^+, O_2^+)
CO_2	4×10^{10}	1.5×10^4
O_2	5×10^7	10^5

11.2.4.2 Production and Loss Mechanisms Reactions (11.18)–(11.20) apply on Venus as on Mars. In addition, one of the loss reactions for O_2^+ is the major source for NO^+:

$$O_2^+ + N \rightarrow NO^+ + O \qquad \text{ion exchange} \qquad (11.22)$$

The NO^+ also contributes to the production of neutral O:

$$NO^+ + e^- \rightarrow N + O \qquad \text{dissociative recombination} \qquad (11.4b)$$

Production of O^+ is by

$$O + h\nu \rightarrow O^+ + e^- \qquad \text{photoionization } \lambda < 91.0\,\text{nm} \qquad (11.2a)$$

$$O + e^- \rightarrow O^+ + 2e^- \qquad \text{electron impact} \qquad (11.23)$$

$$CO_2^+ + O \rightarrow O^+ + CO_2 \qquad \text{charge transfer} \qquad (11.19b)$$

Loss of O^+ is by

$$O^+ + NO \rightarrow NO^+ + O \qquad \text{charge transfer} \qquad (11.24a)$$

$$O^+ + CO_2 \rightarrow O_2^+ + CO \quad \text{atomic exchange} \qquad (11.20)$$

$$O^+ + H \rightarrow H^+ + O \quad \text{charge transfer} \qquad (11.24b)$$

Reactions (11.19b) and (11.20) show the importance of CO_2 on Venus.

11.2.5 Atmospheric Escape Mechanisms

11.2.5.1 Jeans Escape The escape of an atmospheric molecule from a planetary exosphere is called *Jeans escape*. It occurs when atoms and molecules have thermal speeds greater than escape speed, and it gives rise to the evaporation phenomenon discussed in connection with long-term retention in Chapter 10.1.

On Mars, $v_{\text{esc}} = 5.5\,\text{km/s}$, and $T = 320\,\text{K}$ in the thermosphere. Then from (10.5) the root mean square thermal speeds of hydrogen ($\mu_H = 1$) and carbon dioxide ($\mu_{CO2} = 44$) are, respectively,

$$(v_{\text{rms}})_H = \sqrt{\frac{3kT}{\mu_H m_u}} = 2.8\,\text{m/s} \cong \tfrac{1}{2}\,v_{\text{esc}}$$

$$(v_{\text{rms}})_{CO2} = \sqrt{\frac{3kT}{\mu_{CO2} m_u}} = 0.43\,\text{m/s}^{-1} \cong \tfrac{1}{13}\,v_{\text{esc}}$$

Thus, from the discussion after (10.5), significant Jeans escape occurs for atomic hydrogen, but very little for CO_2 and therefore for the atmosphere as a whole.

11.2.5.2 Suprathermal Atoms and Ions The term *suprathermal* refers to atoms or ions with characteristic speeds that are much higher than the characteristic local thermal (Maxwellian) speed.

Suprathermal atoms and ions can arise from exothermic chemical reactions where the products have higher KE than the reactants. For example,

$$O_2^+ + e^- \rightarrow O + O^* \qquad \text{dissociative recombination} \qquad (11.21)$$

This reaction releases 2.5 eV of kinetic energy to each oxygen atom ($1\,\text{eV} = 1.602 \times 10^{-19}\,\text{J}$); then, by (10.2), each oxygen atom has a speed $v = 5.5\,\text{km/s}$, equal to the escape speed from Mars.

The exobase of Mars is at $\sim$230 km, where there are significant quantities of both O_2^+ and electrons (Figure 11.4), so the escape of suprathermal atomic oxygen is an important loss mechanism from the atmosphere of Mars.

At the present time, Mars loses $\sim 6 \times 10^7$ oxygen atoms $\text{cm}^{-2}\text{s}^{-1}$ to space. The rate of loss of H is set by the rate of supply of H_2 to the exosphere from below, which in turn is set by chemical reactions involving the dissociation of water. The loss mechanisms for hydrogen and oxygen are in fact mutually self-limiting: two H escape for each O. In effect, water molecules are being lost to space.

With a loss rate of 6×10^7 molecules $\text{cm}^{-2}\text{s}^{-1}$ over the life of Mars ($\sim 4.6 \times 10^9$), a surface layer of water $\sim$2.5 m deep would have been lost. Greater solar UV flux in the early solar system might increase this to $\sim$20 m. In fact, geologic evidence may indicate a layer of water $\sim$500 m deep when Mars was young, so other loss mechanisms may also have been important in the past.

11.3 Solar Wind

The solar wind is composed of charged particles (electrons and positive ions) streaming out from the Sun. More than 99% of the ions are H^+ and $^4\text{He}^{2+}$, and the rest are $^3\text{He}^{2+}$, C, O, and other ions. Of the H^+ and $^4\text{He}^{2+}$, 95–96% is H and 3.5–4.5% is He.

The solar wind behaves as a highly conducting medium, so the relative numbers of electrons and ions maintain charge neutrality.

There is a region of acceleration close to the Sun where the particles are accelerated from low speeds to about 400 km/s (Table 11.2). Thereafter, the speed is approximately constant.

Table 11.2. Typical solar wind values

r	$1.03R$	$1.5R$	$10R$	1 AU
n_{ion} (m^{-3})	2×10^{11}	1×10^{11}	2×10^9	5×10^6
T (K)	1.7×10^6	10^6	4×10^5	4×10^4
B (T)	10^{-4}	5×10^{-5}	10^{-6}	4×10^{-9}
v (km/s)	0.6	3	300	400

For uniform expansion at constant speed, the number density of particles will decrease as one over the surface area, i.e., $n \propto r^{-2}$. In particular, if r is the distance from the Sun in AU, then the number density of ions is given by

$$n_{ion}(m^{-3}) = \frac{5 \times 10^6}{r^2(\text{AU})} \tag{11.25}$$

Variations in these values are produced by high-speed jets ($\sim$600 km/s) of steady flow from coronal holes and by transient events including ejections by eruptive prominences, solar flares, and other active phenomena on the Sun.

11.4 Maxwell's Equations and the Plasma Frequency

A *plasma* is an ionized medium consisting of ions and free electrons. If in a plasma the charges of one sign are displaced relative to those of the other sign, creating a charge separation, the restoring force causes the charges to oscillate relative to each other at the *plasma frequency*, f_0. The behavior of electromagnetic waves traveling through a plasma depends on how the frequency, f, of the waves relates to f_0. We will now examine this dependence, making use of Maxwell's powerful formulation of electromagnetism.

11.4.1 Maxwell's Equations

Two of Maxwell's equations for free space may be written as follows (Griffiths 1999, p. 130):

$$\dot{\mathbf{B}} = -\nabla \times \mathbf{E} \tag{11.26}$$

$$\dot{\mathbf{D}} + \mathbf{J} = \nabla \times \mathbf{H} \tag{11.27}$$

where

$$\mathbf{B} = \mu_0 \mathbf{H} \tag{11.28}$$

$$\mathbf{D} = \varepsilon_0 \mathbf{E} \tag{11.29}$$

$$\mathbf{J} = N e \mathbf{v} \tag{11.30}$$

In (11.30), we define N as the number density of free electrons (unit: $\mathrm{m^{-3}}$) in order to reserve n for the refractive index. Other symbols are

$\mathbf{B}$ = magnetic flux density, unit: tesla $(1\,\mathrm{T} = 1\,\mathrm{Wb/m^2} = 1\,\mathrm{kg\,s^{-2}A^{-1}})$;

$\mathbf{E}$ = electric field strength or electric intensity,
        unit: $\mathrm{N/C = V/m = m\,kg\,s^{-3}A^{-1}}$;

$\mathbf{H}$ = magnetic field strength, unit: $\mathrm{A/m}$;

$\mathbf{D}$ = electric displacement, unit: $\mathrm{C/m^2}$;

$\mathbf{J}$ = current density, unit: $\mathrm{A/m^2}$;

e = charge on the electron $= 1.602177 \times 10^{-19}\,\mathrm{C}$;

ε_0 = permittivity of free space $= 8.85 \times 10^{-12}\,\mathrm{F/m}$;

μ_0 = permeability of free space $= 4\pi \times 10^{-7}\,\mathrm{N/A^2}$;

∇ = gradient or del operator, $\nabla \equiv \frac{\partial}{\partial x}\hat{\mathbf{i}} + \frac{\partial}{\partial y}\hat{\mathbf{j}} + \frac{\partial}{\partial z}\hat{\mathbf{k}}$.

Our task will be to obtain from (11.26) to (11.30) an equation involving only the electric field and its derivatives and to use this equation to study the interaction of the electric vector of a wave with an ionized medium.

Step 1: Take the curl of (11.26) and the time derivative of (11.28), and substitute the latter into the former to obtain

$$\mu_0 \nabla \times \dot{\mathbf{H}} = -\nabla \times \nabla \times \mathbf{E} \qquad (11.31)$$

Step 2: Substitute (11.30) into (11.27), and take the time derivative of (11.29) to eliminate $\dot{\mathbf{D}}$ to obtain

$$\varepsilon_0 \dot{\mathbf{E}} + N\,e\,\mathbf{v} = \nabla \times \mathbf{H} \qquad (11.32)$$

Step 3: Differentiate (11.32) and multiply through by μ_0 to get

$$\mu_0 \varepsilon_0 \ddot{\mathbf{E}} + \mu_0 N\,e\,\dot{\mathbf{v}} = \mu_0 \nabla \times \dot{\mathbf{H}} \qquad (11.33)$$

Step 4: Substitute (11.31) into the RHS of (11.33) to obtain

$$\mu_0 \varepsilon_0 \ddot{\mathbf{E}} + \mu_0 N\,e\,\dot{\mathbf{v}} = -\nabla \times \nabla \times \mathbf{E} \qquad (11.34)$$

Step 5: Use the identity

$$-\nabla \times \nabla \times \mathbf{E} = \nabla^2\,\mathbf{E} + \nabla(\nabla \cdot \mathbf{E}) \qquad (11.35)$$

to obtain

$$\mu_0\varepsilon_0\ddot{\mathbf{E}} + \mu_0 N\, e\, \dot{\mathbf{v}} = \nabla^2\,\mathbf{E} + \nabla(\nabla \cdot \mathbf{E}) \tag{11.36}$$

Generally, the *divergence* of $\mathbf{E}$ is $\nabla \cdot \mathbf{E} = (Nq/\varepsilon_0)$ (Gauss's law). For no net charge, $\nabla \cdot \mathbf{E} = 0$.

The electron current is produced by the $\mathbf{E}$ field initially, therefore the force is electrostatic, and Newton's second law $(\mathbf{F} = m\mathbf{a})$ gives

$$e\mathbf{E} = m\dot{\mathbf{v}} \tag{11.37}$$

where $m = 9.10838 \times 10^{-31}$ kg is the mass of the electron. Whence,
Step 6: Substitute (11.37) into (11.36) to obtain

$$\mu_0\varepsilon_0\ddot{\mathbf{E}} + \mu_0 \frac{Ne^2}{m}\mathbf{E} = \nabla^2\,\mathbf{E} \tag{11.38}$$

Now we are in a position to apply Maxwell's equations to a radio wave passing through the ionosphere.

11.4.2 Application to a Polarized Wave

Assume now that a polarized wave ascends vertically (in the z or $\hat{\mathbf{k}}$ unit vector direction), with the $\mathbf{E}$-field polarized in the x or $\hat{\mathbf{i}}$ direction, as illustrated in Figure 11.5.

Adopt a trial solution to (11.38) of the form

$$\mathbf{E} = \mathrm{E}_x\hat{\mathbf{i}} = \mathrm{E}_0 \exp\left[j\omega\left(t - \frac{nz}{c}\right)\right]\hat{\mathbf{i}} \tag{11.39}$$

where $j = \sqrt{-1}$ and $n = c/v$ is the index of refraction. The last term in (11.39) is equal to

$$\left(\frac{n}{c}\right)\omega z = 2\pi f\left(\frac{n}{c}\right)z = 2\pi f\frac{z}{v} = 2\pi\frac{z}{\lambda} \tag{11.40}$$

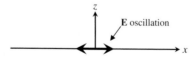

Fig. 11.5. A polarized electromagnetic wave traveling in the z-direction, with its electric field oscillation in the x-direction

Insert (11.39) into (11.38) and define $\alpha \equiv j\omega(t - nz/c)$ to obtain

$$-\mu_0\varepsilon_0\omega^2 E_0 \exp[\alpha]\hat{\mathbf{i}} + \frac{\mu_0 N e^2}{m} E_0 \exp[\alpha]\hat{\mathbf{i}} = -\frac{\omega^2 n^2}{c^2} E_0 \exp[\alpha]\hat{\mathbf{i}}$$

or

$$-\mu_0\varepsilon_0\omega^2 + \frac{\mu_0 N e^2}{m} = -\frac{\omega^2 n^2}{c^2} \tag{11.41}$$

Solving for n and using $c^2 = (1/\mu_0\varepsilon_0)$, we obtain the dispersion relation for an ionizing medium:

$$n^2 = 1 - \frac{N e^2}{m\varepsilon_0\omega^2} \tag{11.42}$$

or, recalling that $\omega = 2\pi f$,

$$n^2 = 1 - \frac{f_0^2}{f^2} \tag{11.43}$$

where

$$f_0^2 \equiv \left(\frac{1}{2\pi}\right)^2 \left[\frac{N e^2}{m\varepsilon_0}\right] \tag{11.44}$$

defines the *plasma frequency*, f_0. From (11.43) and (11.44), we have

1. For a vacuum, $f_0 = 0$ and $n = 1$; or, for $N > 0$,
2. n is real and $0 < n < 1$ for $f > f_0$;
3. $n = 0$ for $f = f_0$; and
4. n is imaginary for $f < f_0$.

We now look at some consequences of (11.43) and (11.44) for an electromagnetic wave of frequency f propagating from an unionized region into a region of increasing ionization, e.g., upward into a planetary ionosphere from below or downward from space (cf. Figures 11.1, 11.2, and 11.4).

In Figure 11.6, a hypothetical ionosphere is divided into layers of constant electron density, N, with N increasing in successive layers upward from $N_0 = 0$ in the region below the ionosphere. An electromagnetic wave of frequency f in the region below the ionosphere (refractive index n_0) encounters the ionosphere at an incident angle α_0, refracts into the first layer (refractive index n_1) at an angle of refraction α_1, the second layer (refractive index n_2) at angle α_2, and the ith layer (refractive index n) at angle α.

Because in (11.44) $f_0 \propto \sqrt{N}$, it follows from (11.44) that $n < 1$ and decreases with increasing N. Thus, n decreases upward in Figure 11.6 and, from Snell's law [(11.45), below], the wave refracts away from the normal, as shown.

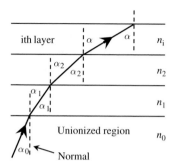

Fig. 11.6. Refraction of an electromagnetic wave in a layered ionosphere of decreasing refractive index, n_i.

From Snell's law,

$$n_1 \sin \alpha_1 = n_0 \sin \alpha_0 = \text{const} \tag{11.45}$$

$$n_2 \sin \alpha_2 = n_1 \sin \alpha_1 \tag{11.46}$$

Therefore,

$$n_2 \sin \alpha_2 = n_0 \sin \alpha_0 = \text{const} \tag{11.47}$$

Continuing this argument to the ith layer and dropping the subscript for the ith layer, we get

$$n \sin \alpha = n_0 \sin \alpha_0 = \text{const} \tag{11.48}$$

Thus, as long as the wave is able to reach the ith layer, the angle of refraction there depends only on the refractive indices n_0 and n and the incident angle α_0 and is independent of the characteristics of the ionosphere in the intervening layers. The result is also independent of the thickness of the layers, so we can take the limit as the thickness approaches zero, i.e., (11.48) applies also to continuous media (real ionospheres).

We now note that for a wave entering the ionosphere from space (vacuum), $n_0 = 1$; and for a wave entering from below, the atmospheric density in the thermosphere is low enough that n_0 can again be taken as (very close to) 1.

It is apparent from Figure 11.6 that if α reaches a value $\pi/2$ then the wave can proceed no further and is reflected (internal reflection). Setting $\alpha_{\max} = \pi/2$ and $n_0 = 1$ in (11.48) gives

$$n = \sin \alpha_0 \tag{11.49}$$

i.e., for a given incident angle, the wave will be reflected if the minimum refractive index in the ionosphere is less than $n = \sin \alpha_0$. Define f_0^{peak} to be the maximum plasma frequency in the ionosphere, in the layer of maximum electron density. Then, from (11.43), a wave will be transmitted through the ionosphere only if its frequency exceeds a critical frequency, f_c, given by

$$n^2 = 1 - \frac{\left(f_0^{\text{peak}}\right)^2}{f_c^2} = \sin^2 \alpha_0 = 1 - \cos^2 \alpha_0 \qquad (11.50)$$

or

$$f_c = \frac{f_0^{\text{peak}}}{\cos \alpha_0} = f_0^{\text{peak}} \sec \alpha_0 \qquad (11.51)$$

For normal incidence ($\alpha_0 = 0$), $\cos \alpha_0 = \sec \alpha_0 = 1$, so from (11.51)

$$f_c = f_0^{\text{peak}} \qquad \text{(normal incidence)} \qquad (11.52)$$

Thus, from (11.43), the wave reflects at the first point (if any) where the refractive index is

$$n = 1 - \left(\frac{f_0^{\text{peak}}}{f_0^{\text{peak}}}\right)^2 = 0$$

It follows that electromagnetic waves cannot propagate in a region of imaginary refractive index.

11.5 The Earth's Magnetosphere

A planet's magnetosphere is the region of space around the planet occupied by its magnetic field. The magnetosphere constrains the motions of charged particles (electrons and ions) within it, sometimes producing very strong electric fields, electric potential differences, and electric currents and is itself shaped by the charged particles within it and by the solar wind.

We begin with a discussion of the Earth's magnetosphere, as illustrative of processes that take place in planetary magnetospheres in general.

The intrinsic magnetic field of the Earth, created by a self-exciting dynamo in the molten metallic core, is approximately dipolar, as illustrated schematically in Figure 11.7. The magnetic axis is tilted relative to the rotation axis, resulting in an offset between the geomagnetic poles and the geographic poles.

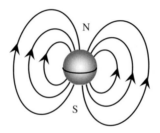

Fig. 11.7. The Earth's intrinsic dipolar magnetic field. (Not to scale)

As of the present writing, the north geomagnetic pole is in the Canadian arctic, but wanders with time, and is currently showing signs of migrating to Siberia.

By convention, magnetic field lines point away from a magnetic north (N) pole and toward a magnetic south (S) pole; but also by convention, the geomagnetic poles are labeled according to the geographic hemisphere (N or S) in which they occur. As Figure 11.7 illustrates, with the magnetic field lines entering the Earth at the North Pole and leaving at the South Pole, these conventions result in the geomagnetic north pole actually being a magnetic south pole and vice versa.

(Historically, the "north" pole of a magnet or compass needle was called the "north-seeking" pole, i.e., the pole that seeks the geographic, or actually geomagnetic, north pole.)

If left to itself, the Earth's dipole field would extend to infinity in all directions. However, the flow of solar wind plasma and interplanetary magnetic field (IMF) past the Earth modifies the dipole field as shown in Figure 11.8.

The *magnetopause* in Figure 11.8 is the boundary between the Earth's magnetic field and the IMF, i.e., the magnetosphere is the region inside the magnetopause. The solar wind is supersonic, but the solar wind particles are slowed to subsonic speeds by their interaction with the Earth. The *bow shock* marks this transition from supersonic to subsonic flow. Between the bow shock and the magnetopause is a region of subsonic solar wind particles and IMF called the *magnetosheath*. Since disturbances cannot propagate upstream in supersonic flow, the bow shock marks the furthest extent of influence of the Earth's magnetic field.

As is evident in Figure 11.8, the solar wind substantially compresses the dipole field on the upstream side of the Earth, while on the downstream side, viscous interaction between the solar wind and the magnetospheric

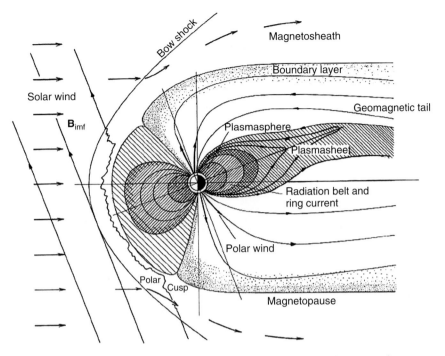

Fig. 11.8. Schematic diagram of the Earth's magnetosphere in the noon–midnight plane. The basic particle and magnetic field features are representative of other internally generated planetary magnetospheres, although the details may be different (from Parks 1991, Figure 1.3, p. 8, reproduced with the author's permission.)

plasma stretches the Earth's magnetosphere into a long *magnetotail*, through processes described below. Because of this stretching, magnetic field lines in the magnetotail are approximately parallel to a warped central plane called the *neutral sheet*. Continuity requires that the magnetic field lines above and below join through the neutral sheet; thus, on the neutral sheet the component of magnetic field parallel to the sheet is zero (this characteristic defines the neutral sheet) and the perpendicular component is extremely weak. Therefore, the neutral sheet can be regarded as a plane of essentially zero magnetic field.

The interaction of charged particles in both the magnetospheric plasma and the solar wind with the Earth's magnetic field produces large-scale currents, as illustrated in Figure 11.9.

We now develop some of the basic physics of the interaction of charged particles with magnetic fields, as it applies to the Earth's magnetosphere in Figure 11.8 and the currents in Figure 11.9.

Solar wind-induced electric currents flowing in the magetosphere

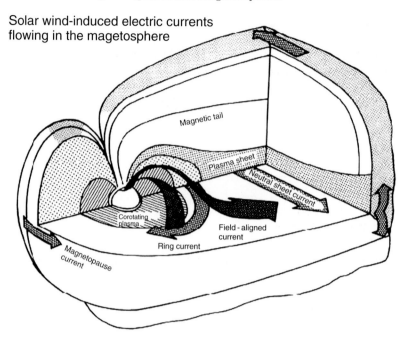

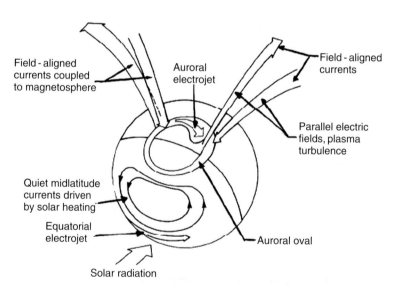

Fig. 11.9. Schematic diagrams showing where currents flow in the Earth's magnetosphere and ionosphere (from Parks 1991, Figure 7.1, p. 244, reproduced with the author's permission.)

11.5.1 Forces Acting on Charged Particles

11.5.1.1 The Lorentz Force The Lorentz force on a charged particle is the vector sum of the electric and magnetic forces:

$$\vec{F} = q\vec{E} + q\vec{v} \times \vec{B} = q\left(\vec{E} + \vec{v} \times \vec{B}\right) \tag{11.53}$$

where

q = electric charge on the particle
$\vec{E}$ = electric field vector
$\vec{v}$ = velocity of the particle
$\vec{B}$ = magnetic field vector

The magnetic force points in the direction of the cross product as given by the right-hand rule (Section 10.3.1) if q is positive, and in the opposite direction if q is negative.

Equation (11.53) shows that

A. electric fields affect all charges, moving or at rest.

B. magnetic fields affect only moving charges.

C. the electric force is parallel or antiparallel to $\vec{E}$.

D. the magnetic force is perpendicular to both $\vec{v}$ and $\vec{B}$.

11.5.1.2 The Gravitational Force

$$\vec{F} = m\vec{g} \tag{11.54}$$

The gravitational force on a charged particle is usually insignificant when electric and/or magnetic fields are present, but there are times when it must be included.

We now apply (11.53) to several situations relevant to the Earth's magnetosphere.

11.5.2 $\vec{E}$ Uniform and Time-Independent; $\vec{B} = 0$

With $\vec{B} = 0$, (11.1) becomes

$$\vec{F} = q\vec{E}. \tag{11.55}$$

If other forces are negligible, then the acceleration of a particle of charge q and mass m is, from Newton's second law,

$$\vec{a} = \frac{q\vec{E}}{m}. \tag{11.56}$$

Thus the electric force acting by itself accelerates particles in a direction parallel or antiparallel to the electric field, sometimes producing very large speeds.

With $\vec{E}$ constant and uniform, finding the motion of a charged particle is a straightforward constant-acceleration kinematics problem (projectile motion) as in Figure 11.10 and is solved the same way as for a rock thrown off a cliff without air resistance. The solution in the direction parallel to $\vec{E}$ is

$$v_{||} = (v_0)_{||} + \left(\frac{qE}{m}\right)t \tag{11.57}$$

$$x_{||} = (x_0)_{||} + (v_0)_{||}t + \frac{1}{2}\left(\frac{qE}{m}\right)t^2 \tag{11.58}$$

11.5.3 $\vec{B}$ Uniform and Time-Independent; $\vec{E} = 0$

With $\vec{E} = 0$, (11.53) becomes

$$\vec{F} = q\vec{v} \times \vec{B}. \tag{11.59}$$

The cross-product ensures that $\vec{F} \perp \vec{v}$ at all times; consequently, the speed, $v = |\vec{v}|$, cannot change. The perpendicular force, however, produces a continually changing direction of travel; thus the particle's velocity changes with no change in speed.

Case 1. If $\vec{v} \perp \vec{B}$, then the magnetic force acts centripetally to produce uniform circular motion (*gyromotion* or *cyclotron motion*) in a plane perpendicular to $\vec{B}$. The radius of gyromotion, r_c, is variously called the *Larmor radius, radius of gyration, gyroradius,* or *cyclotron radius.*

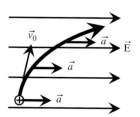

Fig. 11.10. Acceleration of a charged particle by a uniform electric field

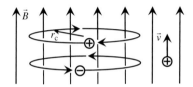

Fig. 11.11. Electric charges moving perpendicular to uniform magnetic field lines execute gyromotion; charges moving parallel to the field lines maintain constant velocity

Because in (11.59) the direction of $\vec{F}$ depends on the sign of q, charges of opposite sign gyrate in opposite directions. Figure 11.11 illustrates this motion for particles of equal mass and speed but opposite charge.

Centripetal acceleration is given by $a = v^2/r$, so if the magnetic force is the only force acting, and with $\vec{v} \perp \vec{B}$, Newton's second law ($\mathbf{F} = m\mathbf{a}$) becomes

$$qvB = m\frac{v^2}{r_c} \tag{11.60}$$

Solving for r_c gives

$$r_c = \frac{mv}{qB} \tag{11.61}$$

Case 2. If $\vec{v} \| \vec{B}$ (Figure 11.11) then $\vec{F} = q\vec{v} \times \vec{B} = 0$. Thus a charged particle moving parallel to a uniform magnetic field feels no $\vec{v} \times \vec{B}$ force and continues to move in a straight line at a constant speed, parallel to $\vec{B}$, unless acted on by some other (non-magnetic) force.

Case 3. If $\vec{v}$ is neither perpendicular nor parallel to $\vec{B}$, as in Figure 11.13, then we can break $\vec{v}$ into components $v_\perp$ perpendicular and $v_\|$ parallel to $\vec{B}$. Using the unit vectors $\hat{i}$, $\hat{j}$, $\hat{k}$ defined in Figure 11.12,

$$\vec{F} = q\vec{v} \times \vec{B} = q\left(v_\|\hat{i} + v_\perp\hat{j}\right) \times B\hat{i} = -qv_\perp B\hat{k} \tag{11.62}$$

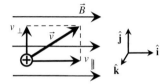

Fig. 11.12. $\vec{v}$ not $\|\vec{B}$

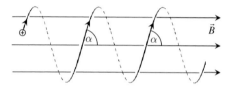

Fig. 11.13. Helical path with constant pitch angle, for a charged particle moving in a uniform magnetic field

because $\hat{\mathbf{i}} \times \hat{\mathbf{i}} = 0$. Thus the force is due entirely to $\vec{v}_\perp$, producing gyromotion with a gyroradius

$$r_c = \frac{mv_\perp}{qB} \tag{11.63}$$

Because the force has no component parallel to $\vec{B}$, $v_\parallel$ remains constant. The particle thus moves at constant speed parallel or antiparallel to the magnetic field lines while executing circular motion perpendicular to the field lines.

The resulting path is a *helix*, as shown in Figure 11.13. The angle, α, between $\vec{v}$ and $\vec{B}$ is called the *pitch angle* of the helix and is constant in a uniform magnetic field.

11.5.4 Guiding Center

The guiding center concept is useful in many situations involving single particles, and in multi-particle situations where the individual particles act independently. It has limitations, however; for example, it is not so useful when the particles act collectively as a fluid.

The guiding center description is based on the idea (illustrated in Figure 11.14) that a particle "spiraling" along magnetic lines of force executes

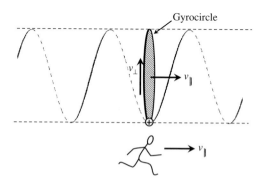

Fig. 11.14. Guiding center concept

purely circular motion as seen by someone moving with a vector velocity equal to $v_{||}$. Thus the particle moves in a gyrocircle, while the center of the gyrocircle is guided in a particular direction at a particular speed by the magnetic field lines.

We now apply the gyrocircle and guiding center concepts to various situations relevant to planetary magnetospheres.

11.5.5 Diamagnetism

A charged particle executing gyromotion is equivalent to a ring of electric current of radius r_c. Conventional current is taken to be in the direction of positive charge flow or opposite to the direction of negative charge flow, as indicated in Figure 11.15. Because positive and negative charges gyrate in opposite directions, the conventional current, I, and the resulting induced magnetic field directions are the same for both positive and negative charges, as shown in Figure 11.15.

Applying the right-hand rule for currents,

point the thumb of the right hand along the direction of conventional current; then the fingers curl around the current in the direction of the induced magnetic field

or current loops,

curl the fingers of the right hand around the current loop in the direction of conventional current; then the thumb points in the direction of the induced magnetic field inside the loop

in Figure 11.15 shows that, *within* each current loop, the induced magnetic field opposes the external magnetic field, whereas *outside* (but

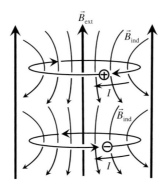

Fig. 11.15. Diamagnetic currents and induced magnetic fields

in the same plane as) the gyromotion, the induced field is in the same direction as the external field. These characteristics identify gyromotion as *diamagnetic.*

11.5.6 $\vec{E} \times \vec{B}$ Drift and Field-Aligned Currents

In Figure 11.16, a particle of charge q and mass m has been released at rest in a uniform, time-independent magnetic field, $\vec{B}$. In addition to the magnetic force, the particle is also subject to a constant force, $\vec{F}$, acting in a direction perpendicular to $\vec{B}$. This force might arise, for example, from a uniform electric or gravitational field in the same region of space.

As illustrated schematically in Figure 11.16, the particle will initially accelerate in the direction of $\vec{F}$. This produces a magnetic force perpendicular to the motion, causing the particle to deviate more and more from its initial direction until it is moving perpendicular to $\vec{F}$.

Up to this point there is a component of the particle's motion in the direction of $\vec{F}$, so $\vec{F}$ does positive work on the particle, the particle's speed increases, and by (11.61) the curve has an ever-increasing radius. After this point the magnetic force deviates the particle's motion back against $\vec{F}$, the work done on the particle by $\vec{F}$ becomes negative, the particle's speed decreases, and the particle moves in a curve of ever-decreasing radius until the motion is again perpendicular to $\vec{F}$. Both forces continue to act, and the subsequent motion is cyclic as illustrated schematically in Figure 11.16. (Note that the curve in Figure 11.16 is flat, and not a helix seen in perspective.)

Thus the motion is somewhat counter-intuitive: the force acting on the charged particle causes the particle to drift in a direction perpendicular to both $\vec{F}$ and $\vec{B}$. Applying the right-hand rule in Figure 11.16 shows that a

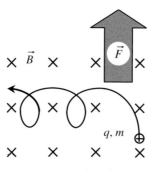

Fig. 11.16. $\vec{F} \times \vec{B}$ drift

positive charge drifts in the direction of $\vec{F} \times \vec{B}$. A mathematical analysis (see the problems at the end of this chapter) gives the drift velocity, $\vec{v}_D$, as

$$\vec{v}_D = \frac{\vec{F} \times \vec{B}}{qB^2} \tag{11.64}$$

If the force is gravitational, then

$$\vec{v}_D = \frac{m\vec{g} \times \vec{B}}{qB^2} = \frac{\vec{g} \times \vec{B}}{(q/m)B^2} \tag{11.65}$$

Thus the drift speed depends on the charge-to-mass ratio of the particle; also, because of the factor q in the equation, charges of opposite sign drift in opposite directions, producing a current.

If the force arises from an electric field, $\vec{E}$, in a region of space having crossed electric and magnetic fields (Figure 11.17), then

$$\vec{v}_D = \frac{q\vec{E} \times \vec{B}}{qB^2} = \frac{\vec{E} \times \vec{B}}{B^2} \tag{11.66}$$

The resulting motion is called $\vec{E} \times \vec{B}$ drift and is perpendicular to both $\vec{E}$ and $\vec{B}$ (in the $\vec{E} \times \vec{B}$ direction). From (11.66), $\vec{v}_D$ is independent of both q and m, so positive and negative charges drift in the same direction and (if there is charge neutrality) $\vec{E} \times \vec{B}$ drift does not produce a current.

The curves in Figures 11.16 and 11.17 are examples of a *trochoid*, the curve traced by a point on the outer edge of a wheel that rolls with or without slippage; e.g., on a side-wheeler steamboat plying a lake, the rim of the paddlewheel turns faster than the boat moves forward, so a person standing on the lakeshore sees the outer edge of each paddle move forward in a long arc at the top, then dip into the water and drop back, producing the loops of the trochoid. If the wheel rolls on a solid surface without slipping, such as the wheel of a bicycle or automobile, the part of the wheel in contact with the

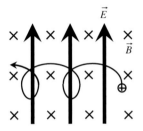

Fig. 11.17. $\vec{E} \times \vec{B}$ drift

ground is momentarily stationary and the loops in Figure 11.17 degenerate to cusps (points). The resulting curve is a *cycloid*.

These analogies suggest that the guiding center concept remains valid for $\vec{E} \times \vec{B}$ drift: like a point on the paddlewheel, the particle moves at constant speed around a gyrocircle while the gyrocircle is guided in the $\vec{E} \times \vec{B}$ direction at the drift velocity, $\vec{v}_D$.

If $\vec{E}$ is not $\perp \vec{B}$ then take components $E_\parallel$ parallel to $\vec{B}$ and $E_\perp$ perpendicular to $\vec{B}$. $E_\perp$ causes $\vec{E} \times \vec{B}$ drift perpendicular to the magnetic field, but (for charge neutrality) no current. $E_\parallel$ accelerates charges of opposite sign in opposite directions along the magnetic field lines, producing a current parallel to the magnetic field, i.e., a *field-aligned current*. $E_\parallel$ creates a potential difference, $V = E_\parallel \ell$, over any distance ℓ along a field line, so magnetic field lines in planetary magnetospheres act like conductors in electric circuits. This idea will be developed further later in this chapter.

11.5.7 $\vec{E} \times \vec{B}$ Drift with Collisions

In a collisionless plasma, $\vec{E} \times \vec{B}$ drift does not produce a current if there is charge neutrality (Section 11.5.6). However, the lower ionosphere is sufficiently dense for collisions to be important, and each collision allows the charge to restart its motion under the influence of the electric field. As shown in Figure 11.18, charges of opposite sign drift in opposite directions parallel or antiparallel to $\vec{E}$ while they continue to $\vec{E} \times \vec{B}$ drift in the same direction perpendicular to $\vec{E}$. The former of these two drifts gives rise to a current, referred to in the ionosphere as the *Pedersen current*, parallel to $\vec{E}$.

In a collision-dominated plasma, $\vec{E} \times \vec{B}$ drift also creates a current, despite charge neutrality, because the smaller gyroradii and higher speeds of the electrons give them greater mobility than the ions. The resulting *Hall current*

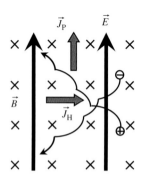

Fig. 11.18. $\vec{E} \times \vec{B}$ drift with collisions. $\vec{J}_P$ is the Pedersen current, and $\vec{J}_H$ is the Hall current

is perpendicular to both $\vec{E}$ and $\vec{B}$, but opposite to the $\vec{E} \times \vec{B}$ direction (Figure 11.18). The Pedersen and Hall currents will be important when discussing ionospheric currents and aurorae (Section 11.6.8).

The Pedersen current is dissipative, because $\vec{J} \cdot \vec{E} > 0$, and the Hall current is dissipationless, because $\vec{J} \cdot \vec{E} = 0$.

11.5.8 Polarization Drift

We now examine the situation where $\vec{B}$ is uniform in space and constant in time and $\vec{E}$ is uniform in space and constant in direction, but changes in magnitude with time, i.e., $dB/dt = 0$ but $dE/dt \neq 0$. Figure 11.19 illustrates the case where E increases with time.

Both particles in Figure 11.19 $\vec{E} \times \vec{B}$ drift toward the left, but because $\vec{E}$ is increasing in strength, each particle's kinetic energy at the midpoint of any cycle is insufficient to allow it to return to its initial level in the diagram at the end of the cycle. Positive charges therefore drift parallel to $\vec{E}$ and negative charges drift antiparallel to $\vec{E}$ while both $\vec{E} \times \vec{B}$ drift perpendicular to $\vec{E}$ and $\vec{B}$. The drift parallel or antiparallel to $\vec{E}$ is called *polarization drift*.

In this case, the drift velocity of the guiding center consists of two components:

1. $\vec{v}_{\mathrm{D}}$, the $\vec{E} \times \vec{B}$ drift velocity perpendicular to $\vec{E}$ and $\vec{B}$, described in Section 11.5.6; but note that v_{D} is increasing with time because E is increasing with time, so from (11.66) the cycles in Figure 11.19 become increasingly stretched.

2. $\vec{v}_{\mathrm{P}}$, the polarization drift velocity parallel or antiparallel to $\vec{E}$, given by

$$\vec{v}_{\mathrm{P}} = -\frac{m}{qB^2}\left(\frac{d\vec{v}_{\mathrm{D}}}{dt}\right) \times \vec{B} \qquad (11.67)$$

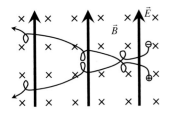

Fig. 11.19. Polarization drift $\|\vec{E}$ with $\vec{E} \times \vec{B}$ drift $\perp \vec{E}$

It follows from (11.66) and (11.67) that

$$\vec{v}_\mathrm{P} = \frac{m}{qB^2}\frac{\mathrm{d}\vec{E}}{\mathrm{d}t} \tag{11.68}$$

Polarization drifts are in opposite directions for opposite charges, so polarization drift produces a current parallel to $\vec{E}$.

11.5.9 Gradient and Curvature Drift

We now investigate inhomogeneous, time-independent magnetic fields, in the absence of electric fields.

It is unusual for a magnetic field to be uniform. Figure 11.20 provides a schematic illustration of a more typical magnetic field (e.g., a magnetic dipole field): the field lines are curved, and the magnetic field strength, B, increases toward the center of curvature.

Thus, real magnetic fields generally show both curvature and gradient.

11.5.9.1 Gradient Drift Figure 11.21 illustrates schematically a region of space containing a magnetic field gradient. Here, we temporarily neglect curvature. From (11.61), the gyroradius is smaller where B is greater and *vice versa*, so a positive particle follows a trochoid, as shown. A negative particle initially traveling downward will curve to the left in Figure 11.21, so positive and negative charges drift in opposite directions and gradient drift produces a current.

If the magnetic field gradient is small ($\Delta B << B$ over a gyroradius) then $\vec{B}$ is approximately uniform over the gyrocircle and the guiding center concept is still valid: the particle moves in a gyrocircle which does not quite close on itself, resulting in a drift of the guiding center perpendicular to both $\vec{B}$ and $\vec{\nabla}B$. Taking the y-axis in the direction of the gradient (upward in Figure 11.21), the drift speed is

$$v_{\nabla B} = \frac{mv_\perp^2}{2qB^2}\frac{\partial B}{\partial y} \tag{11.69}$$

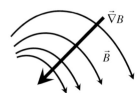

Fig. 11.20. Curvature and gradient in an inhomogeneous magnetic field

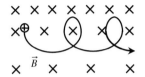

Fig. 11.21. Gradient drift

11.5.9.2 Curvature Drift Figure 11.22 illustrates a hypothetical magnetic field which has curvature, but no gradient. Charged particles follow helical paths along the field lines, as indicated. Because we are neglecting the field gradient, there is no gradient drift in Figure 11.22. However, the particles can follow a curved field line only if there is a centripetal force

$$F_{\text{cent}} = \frac{mv_{\parallel}^2}{R} \tag{11.70}$$

acting on them, directed toward the center of curvature. This force must arise from the motion of the particle and, in fact, is created by a drift of the particle across the field lines with a drift speed v_C (where "C" signifies *curvature drift*). Then from (11.59),

$$\vec{F}_{\text{cent}} = q\vec{v}_C \times \vec{B} \tag{11.71}$$

From the right-hand rule, with q positive in Figure 11.22, $\vec{v}_C$ must be out of the page to give a force toward the center of curvature. From (11.71), negative charges drift into the page, so curvature drift in Figure 11.22 produces a current out of the page. This result is independent of which direction the particle follows along the field line (upward or downward in Figure 11.22), because the direction of the centripetal force must be toward the center of curvature in both cases.

The curvature drift speed is given by

$$v_C = \frac{mv_{\parallel}^2}{qB^2} \frac{\partial B}{\partial y} \tag{11.72}$$

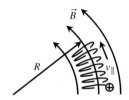

Fig. 11.22. Curvature drift

In fact, in a curved magnetic field the field strength increases toward the center of curvature, so gradient and curvature drift act together. Compare Figure 11.22 to Figure 11.21 (rotated by some appropriate angle) to see that the gradient drift in Figure 11.22 is then also out of the page, and the gradient and curvature drift speeds add. The resulting total drift speed, referred to here as $v_{C\nabla B}$, is

$$v_{C\nabla B} = v_C + v_{\nabla B} = \frac{m}{qB^2}(v_{||}^2 + \tfrac{1}{2}v_\perp^2)\frac{\partial B}{\partial y} \qquad (11.73)$$

We now discuss various current systems that arise from the interactions described in Section 11.5.

11.6 Electric Currents in the Ionosphere and Magnetosphere

11.6.1 The Ionospheric Dynamo

11.6.1.1 The S_q ("Solar Quiet") Current System The S_q current system is caused primarily by solar heating, and to a smaller extent by solar tides. The atmosphere over a given region of the Earth expands as it is heated in the morning and contracts as it cools later in the afternoon. This motion drives ionospheric electrons and ions across magnetic field lines, generating currents in the ionosphere (Figure 11.9). The resulting current pattern remains at rest as seen from the Sun, as the Earth rotates under it.

The term "solar quiet" refers to the absence of solar activity (e.g., at solar minimum) such as solar flares, which can modify the current pattern.

Any current system creates magnetic fields, and these modify the observed magnetic field at a given location on the Earth. Thus, as an observer's location rotates under the ionospheric current system, the observer sees a daily variation in the magnetic field.

11.6.1.2 The L_q ("Lunar Quiet") Current System The L_q current system is caused primarily by lunar tides and remains at rest as seen from the Moon as the Earth rotates under it.

Of the atmospheric tides, the solar thermal tide is stronger than either the lunar or solar gravitational tides, and the lunar gravitational tide is stronger than the solar gravitational tide. The ionospheric dynamo is thus dominated by solar heating, with the next strongest influence being the lunar tide.

11.6.2 Boundary Current

The *boundary current*, a.k.a. *magnetopause current* or *Chapman–Ferraro current*, is a diamagnetic current (Section 11.5.5) formed by the interaction of the solar wind with the Earth's magnetosphere at the magnetopause. Figure 11.23 illustrates schematically the principles involved.

The right-hand rule (Section 11.5.5) shows that positive ions in the solar wind gyrate out of the page when they encounter the Earth's magnetic field, and electrons gyrate into the page. The resulting current direction is out of the page, or from dawn to dusk (see Figure 11.31 for orientations). Individual particles in this current are simply performing an arc of gyration, rather than flowing in the manner of a conduction current.

From the right-hand rule, the induced magnetic field of the boundary current opposes the Earth's magnetic field on the sunward side and strengthens it on the side facing the Earth, in effect compressing the magnetic field on the noon side of the Earth. Above and below, field lines that would have closed on the sunward side are swept back to the magnetotail by the same process, with the solar wind particles creating a dusk to dawn diamagnetic current that provides closure for the dawn-to-dusk current.

A *neutral point*, or a point of zero magnetic field, marks where the field lines diverge above each polar cap. As Figure 11.23 shows, the neutral points are sunward of the Earth. A *polar cusp*, a region of funnel-shaped magnetic field lines, surrounds each neutral point. The polar cusps guide solar wind into the atmosphere, generating dayside aurorae.

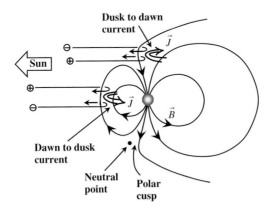

Fig. 11.23. Schematic illustration of the interaction of solar wind particles with the Earth's magnetosphere, and the formation of the Chapman–Ferraro boundary current. In reality, the radius of gyration of positive ions is much greater than that of electrons. The view is from the dusk side of the Earth

An increase in the solar wind flux results in an increase in the boundary current. This in turn causes an increase in the magnetic field strength at the Earth's surface and marks the onset of a magnetic storm (Section 11.6.4).

11.6.3 Ring Current

11.6.3.1 Magnetic Mirrors Figure 11.24 shows a positively charged particle following a helical path into a region of converging magnetic field lines. Using the guiding center concept, we may describe the particle at any given instant as moving at speed $v_\perp$ in a gyrocircle of radius r_c, while the center of the gyrocircle follows the central field line to the right at speed $v_{||}$. We now look at the subsequent motion of this particle.

If the field lines converge at angle θ relative to the central field line, then the Lorentz force, $\vec{F}_B = q\vec{v} \times \vec{B}$, on the particle is also at angle θ to the plane of the gyrocircle, as indicated in Figure 11.24. The components of $\vec{F}_B$ perpendicular and parallel to the central field line are then

$$F_\perp = F_B \cos\theta \tag{11.74}$$

$$F_{||} = F_B \sin\theta \tag{11.75}$$

$F_\perp$ produces gyromotion, while $F_{||}$ acts to change $v_{||}$. The speed of the guiding center thus increases or decreases, depending on the relative directions of $F_{||}$ and $v_{||}$. (In Figure 11.24, $v_{||}$ would decrease.)

$F_{||}$ depends only on q, B, $v_\perp$ (through F_B), and θ and is independent of $v_{||}$. Thus, in Figure 11.24, if the guiding center is brought to rest, it will immediately start to move with increasing speed toward the left, i.e., the converging field lines form a magnetic mirror, reflecting incoming particles back toward the region of weaker field.

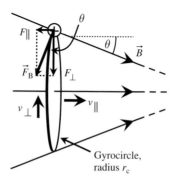

Fig. 11.24. A charged particle in a region of converging magnetic field lines

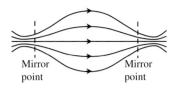

Fig. 11.25. Magnetic bottle

Two magnetic mirrors opposing each other form a magnetic bottle (Figure 11.25), providing one of the methods proposed for containing the million-degree plasma in a fusion reactor.

11.6.3.2 Characteristics of the Motion We now look for constants of the motion for the particle in Figure 11.24. We take the case where:

A. The magnetic field is constant in time as seen by an external observer (the particle, however, encounters a changing magnetic field as it moves through the converging field lines).

B. The field lines are not necessarily straight, but they converge symmetrically about the central field line of the particle's motion, and the magnetic field strength is uniform over any given plane perpendicular to the central field line, although it changes from one plane to the next along the field line.

C. The rate of convergence of the magnetic field lines is "slow"; i.e., the angle θ in Figure 11.24 is small, so that $\cos\theta \approx 1$. Then, over any one gyrocircle, the change in the gyroradius is small compared to the gyroradius itself.

The particle then moves in a gyrocircle, the radius of which changes as the magnetic field strength changes, and the speed and direction of travel of which change because of $F_{\parallel}$.

Gyromotion can be regarded as a current loop in which the current consists of one particle of charge q passing a given point on the gyrocircle f times per second. f is referred to as the *frequency of gyration*. The magnetic moment of a current loop is

$$\mu_B = IA \qquad (11.76)$$

where I is the current and A is the area of the loop. The magnetic moment of the gyromotion can then be shown to be (see Challenges for this chapter)

$$\mu_B = \frac{qr_c v_\perp}{2} \qquad (11.77)$$

The angular momentum of a particle of mass m moving with speed $v_\perp$ in a gyrocircle of gyroradius r_c is

$$L = mv_\perp r_c \qquad (11.78)$$

From (11.77) and (11.78) we obtain

$$\mu_B = \frac{qL}{2m} \qquad (11.79)$$

The torque on a particle equals the rate of change of its angular momentum:

$$\tau = \frac{\mathrm{d}L}{\mathrm{d}t} \qquad (11.80)$$

However, the torque exerted on a charged particle by the magnetic force is zero; hence, from (11.80), the angular momentum of a particle undergoing gyromotion does not change with time and is a constant of the motion. It then follows from (11.79) that the magnetic moment, μ_B, is also a constant of the motion.

Because we are taking the case where B is uniform over the gyrocircle, the magnetic flux through the gyrocircle is

$$\Phi_B = BA \qquad (11.81)$$

It then follows from previous results that

$$\Phi_B = \frac{2\pi m}{q^2}\mu_B \qquad (11.82)$$

Hence, Φ_B is also a constant of the motion.

It can also be shown that the particle's kinetic energy, $K = \frac{1}{2}mv^2$, is a constant of the motion, and hence, from Pythagoras' theorem,

$$\tfrac{1}{2}mv^2 = \tfrac{1}{2}mv_\perp^2 + \tfrac{1}{2}mv_{||}^2 = \text{const} \qquad (11.83)$$

The results above show that, for a particle undergoing gyromotion, Φ, L, and K are constants of the motion. As the magnetic field changes, r_c changes to conserve the magnetic flux through the gyrocircle. As r_c changes, $v_\perp$ changes to conserve angular momentum; and as $v_\perp$ changes, $v_{||}$ changes to conserve kinetic energy.

For the case where the particle is moving toward increasing B, as in Figure 11.24, (11.83) shows that if $v_\perp$ becomes equal to v then $v_{||} = 0$ and further progress is impossible. $F_{||}$, however, is non-zero and pushes the particle back toward weaker magnetic field. For a symmetrical magnetic bottle, as in Figure 11.25, the particle will mirror back and forth inside the bottle without contacting any physical walls, making magnetic bottles very useful for fusion reactors operating at millions of degrees temperature.

11.6.3.3 Trapping and Precipitation The inner magnetic field lines in Figure 11.23 (i.e., those not extending off down the magnetic tail) form a curved analog of Figure 11.25 and cause charged particles to mirror back and forth between the northern and southern hemispheres. In the Earth's magnetosphere, these regions of trapped charged particles are the *Van Allen radiation belts*, consisting of an inner belt of high-energy protons (10–100 MeV) at distances of $\sim$1–3R_E above the Earth's surface and an outer belt of high-energy electrons (0.1–10 MeV) and ions at $\sim$1–10R_E above the Earth's surface, peaking in electron density between 2.3 and 3R_E above the surface. The protons in the inner belt are produced by β-decay of neutrons created by cosmic ray collisions with nuclei in the upper atmosphere.

If one of these particles enters the atmosphere before mirroring then collisions in the atmosphere will prevent it from reflecting. This loss of particles from the trapping region is known as *precipitation*.

The three constants of the motion found in Section 11.6.3.2 allow the magnetic field strength, B_R, to be found at the mirror point for any given pitch angle, α_0, and magnetic field strength, B_0, at the magnetic equator (the location where the magnetic field is weakest). Some relevant parameters are illustrated in Figure 11.26. At the magnetic equator, for a particle of speed v, we can define

$$v_{||}^0 = v \cos \alpha_0 \tag{11.84}$$

$$v_\perp^0 = v \sin \alpha_0 \tag{11.85}$$

At the mirror point,

$$v_{||}^R = 0 \tag{11.86}$$

and therefore, from the constancy of the kinetic energy in (11.83),

$$v_\perp^R = v \tag{11.87}$$

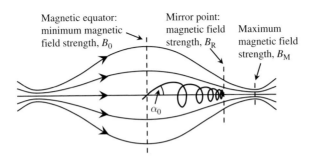

Fig. 11.26. Parameters for a magnetic mirror. α_0 is the pitch angle at the magnetic equator

From (11.78), (11.87), and the constancy of the angular momentum, we obtain

$$v_\perp^0 r_c^0 = v_\perp^R r_c^R = v r_c^R \tag{11.88}$$

Then

$$\frac{r_c^0}{r_c^R} = \frac{v}{v_\perp^0} \tag{11.89}$$

Finally, from the remaining characteristic of the motion, the constancy of the magnetic flux within the gyrocircle, we have

$$B_0 \pi (r_c^0)^2 = B_R \pi (r_c^R)^2 \tag{11.90}$$

from which, with (11.85) and (11.89),

$$B_R = B_0 \left(\frac{r_c^0}{r_c^R} \right)^2 = B_0 \left(\frac{v}{v_\perp^0} \right)^2 = B_0 \left(\frac{v}{v \sin \alpha_0} \right)^2 \tag{11.91}$$

or

$$B_R = \frac{B_0}{\sin^2 \alpha_0} \tag{11.92}$$

Thus, the magnetic field strength at the mirror point depends only on B_0 and α_0 and is independent of the charge, mass, and kinetic energy of the particle.

Equation (11.92) shows that, in a magnetic bottle such as that in Figure 11.26, particles of smaller α_0 mirror at larger B_0, i.e., closer to the point of maximum field strength, B_M. If the field strength, B_R, required at the mirror point [from (11.92)] is greater than B_M then the bottle is unable to constrain these particles, and they escape through the ends of the bottle. The critical pitch angle at the magnetic equator, α_C, such that all particles of $\alpha_0 < \alpha_C$ escape is obtained by setting $B_R = B_M$ in (11.92), yielding

$$\alpha_C = \sin^{-1} \sqrt{\frac{B_0}{B_M}} \equiv \sin^{-1} \frac{1}{\sqrt{R}} \tag{11.93}$$

where the *mirror ratio*

$$R \equiv \frac{B_M}{B_0} \tag{11.94}$$

is the ratio of the strongest to the weakest magnetic field strength. α_C defines the loss cone, as illustrated in Figure 11.27. A similar loss cone exists for particles traveling toward the left.

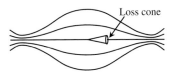

Fig. 11.27. The loss cone. All particles traveling within the loss cone (angular radius α_C) at the magnetic equator escape from the magnetic bottle

In similar fashion, the loss cone in a planetary magnetosphere separates particles that are lost by precipitation into the atmosphere from those that mirror. B_M is then the magnetic field strength at the exobase, where collisions are frequent enough to remove the particles.

If a process injects particles into a planetary magnetosphere with an initially isotropic distribution of pitch angles at the magnetic equator, then the fraction, f, of these particles that are precipitated equals the fraction of half a sphere [2π steradians (sr)] taken up by the loss cone. (There are 4π sr in a complete sphere, but only half of the particles are traveling toward any given end of the magnetic bottle in Figure 11.27.) To find the fraction precipitated we first need to find the solid angle, Ω, taken up by a cone of angular radius α_C, then take the ratio of this result to 2π sr.

In Figure 11.28, the area on the surface of the sphere intersected by the loss cone is

$$
A = \int_{\alpha=0}^{\alpha_C} \mathrm{d}A = \int_{\alpha=0}^{\alpha_C} 2\pi r \, \mathrm{d}s = \int_{\alpha=0}^{\alpha_C} 2\pi (a\sin\alpha)(a\,\mathrm{d}\alpha)
$$
$$
= 2\pi a^2 \int_{\alpha=0}^{\alpha_C} \sin\alpha \, \mathrm{d}\alpha = 2\pi a^2 \left(-\cos\alpha \big|_{\alpha=0}^{\alpha_C} \right)
$$

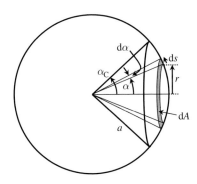

Fig. 11.28. Quantities required to find the solid angle in steradians taken up by a loss cone of angular radius α_C

or

$$A = 2\pi a^2 \left(1 - \cos \alpha_C\right) \tag{11.95}$$

The fraction of particles that are precipitated is then

$$f = \frac{\Omega}{2\pi} = \frac{1}{2\pi} \left(\frac{A}{a^2}\right) = \frac{1}{2\pi} \left(\frac{2\pi a^2 \left(1 - \cos \alpha_C\right)}{a^2}\right) = 1 - \cos \alpha_C$$

$$= 1 - \sqrt{1 - \sin^2 \alpha_C} = 1 - \sqrt{1 - \frac{B_0}{B_M}}$$

from (11.93) or

$$f = 1 - \sqrt{1 - \frac{1}{R}} \tag{11.96}$$

11.6.3.4 The Ring Current Figure 11.29a illustrates schematically the path of a charged particle trapped in a planetary magnetosphere, mirroring back and forth between the northern and southern magnetic poles. Gradient and curvature drift (Section 11.5.9) cause positive charges to drift out of the page and negative charges to drift into the page in this region of the diagram, creating a current, the *ring current*, in the direction shown.

In the Earth's magnetosphere, the ring current is carried by the outer Van Allen radiation belt and is strongest in the region $\sim$2–4R_E above the Earth's surface. As illustrated in Figure 11.29b, the induced magnetic field within the ring current opposes the intrinsic magnetic field in the region between the ring current and the Earth's surface and reduces the measured magnetic field at any point on the Earth's surface.

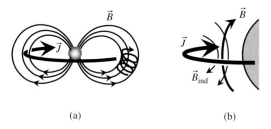

(a)   (b)

Fig. 11.29. The ring current. (**a**) The *helix* shows part of the path of a charged particle mirroring between the northern and southern hemispheres. The *heavy arrow* shows the direction of the gradient and curvature drift of a positively charged particle, and the resulting current direction. (**b**) Within the ring's radius, the induced magnetic field of the ring current opposes the intrinsic magnetic field of the Earth

11.6.4 Magnetic Storms

A *magnetic storm* (Figure 11.30) is a worldwide disturbance in the horizontal component, $B_\parallel$, of the geomagnetic field at the Earth's surface. The initial phase typically begins with a sudden onset in which $B_\parallel$ increases by several to several tens of gammas ($1\gamma = 1\,\text{nT}$) in a few minutes, and lasts from tens of minutes to several hours. The main phase, where $B_\parallel$ decreases below its pre-storm value, typically lasts from several hours to about 2 days, and recovery can take from several days to a month, depending on the strength of the storm. For comparison, the Earth's magnetic field strength at the surface is about 30,000–60,000 nT.

The sudden onset and initial phase occur when an enhancement in the solar wind flux, e.g., from a solar flare, encounters the Earth's magnetosphere, increasing the boundary current and compressing the magnetosphere on the Earth's dayside (Section 11.6.2). After some delay, the main phase of the storm begins as processes in the magnetosphere inject charged particles into the outer van Allen belt, increasing the ring current (Section 11.6.3). Precipitation through the loss cone produces aurorae. The recovery phase occurs as the ring current slowly dies back to its undisturbed state.

Magnetic storms fall into two broad categories, related to the particle source regions on the Sun. Strong magnetic storms occur individually and result from transient high-energy phenomena: solar flares and coronal mass ejections (CME). Weaker magnetic storms often recur with a 27-day period, equal to the synodic rotation period of the Sun, and result from relatively steady streams of particles from coronal holes in the Sun's polar regions sweeping past the Earth as the Sun rotates.

The decay of the ring current during the recovery phase requires particle loss outside the loss cone, which occurs at least in part by charge-

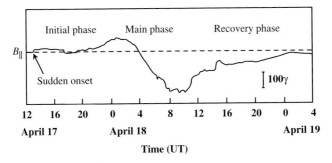

Fig. 11.30. Magnetic storm recorded at Guam. The horizontal component of the Earth's magnetic field is shown as a function of universal time (UT). The tracing has been redrawn and slightly smoothed. After Cahill 1968, Figure 1, p. 264

exchange interactions of ring-current ions with exospheric neutral hydrogen (Section 10.4.1.4). The large pressure scale height of hydrogen results in a density of $\sim 10^9$ H atoms/m^3 at the inner edge of the ring current and $\sim 10^8$ m^{-3} at geosynchronous orbit ($6.6 R_E$). Fast ions capture an electron from a slow-moving H atom, resulting in a fast-moving neutral atom and a slow-moving proton. The spherical volume of neutral H surrounding the Earth is detectable by reflected UV light as the *geocorona*.

The decay processes operate continuously, so the existence of a steady ring current outside magnetic storms shows that, even in quiet times, the outer radiation belt is continuously replenished on timescales of a few days.

11.6.5 Magnetospheric Convection

The solar wind outside the magnetopause exerts a viscous drag on the plasma inside the magnetosphere, arising through waves and particle diffusion across the magnetopause. This interaction creates a boundary layer just inside the magnetopause in which magnetospheric plasma flows antisunward at some fraction of the solar wind speed. Continuity then requires a return flow through the rest of the magnetosphere. These particles pass the Earth and rejoin the boundary particles to complete the circulation.

Currents created by the interaction of this flow with the Earth's magnetic field in turn modify the magnetic field in such a way that the magnetic field lines can be regarded as being carried with the particles as they move, stretching the magnetosphere into a long magnetotail in the antisunward direction. This continuous circulation of particles and magnetic field down and back up the magnetotail is known as *magnetospheric convection*.

Figure 11.31 illustrates magnetospheric convection, looking down on the equatorial plane from above the North Pole of the Earth. The Earth rotates

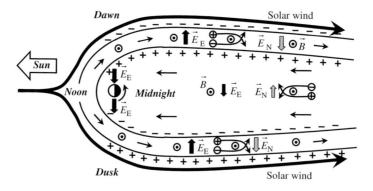

Fig. 11.31. Magnetospheric convection. (Not to scale)

in the direction indicated. Directions in the magnetosphere are denoted by whether they are above the dawn, noon, dusk, or midnight point on the Earth's equator. Illustrative positive and negative charges are shown in each region, and circled dots denote magnetic field out of the page (south to north).

In the boundary layer, convection of charges past magnetic field lines creates a magnetic force in the dawn-to-dusk direction on positive charges and dusk-to-dawn on negative charges. These directions are reversed in the return flow. Magnetic forces can be modeled as arising from *non-electrostatic electric fields*, $\vec{E}_N$ (grey arrows), by treating the magnetic force as if it were a non-electrostatic electric force arising from a non-electrostatic electric field, $\vec{F}_B = \vec{F}_N = q\vec{E}_N$. Then

$$q(\vec{v} \times \vec{B}) = q\vec{E}_N$$

$$\therefore \vec{E}_N = \vec{v} \times \vec{B} \tag{11.97}$$

Thus, $\vec{E}_N$ is in the $\vec{v} \times \vec{B}$ direction, from dawn to dusk in the boundary layer, and dusk to dawn in the return flow (Figure 11.31).

$\vec{E}_N$ produces charge distributions at the interfaces between the boundary layer and the solar wind, and between the boundary layer and the return flow, as shown in Figure 11.31. These charge distributions in turn produce *electrostatic fields*, $\vec{E}_E$ (black arrows), in the dusk-to-dawn direction in the boundary layer and dawn-to-dusk in the return flow, as indicated.

Magnetospheric convection also creates electric fields, and therefore electric currents, in the ionosphere near each pole. This produces the *auroral oval*, the *eastward and westward electrojets* (part of the auroral oval), and other phenomena.

11.6.6 The Magnetotail Current Sheet

Figure 11.32 illustrates schematically (not to scale) a vertical cross-section through the magnetosphere. The dashed line denotes the neutral sheet and the circled dots denote the electrostatic field in the magnetotail, directed from dawn to dusk (toward the viewer).

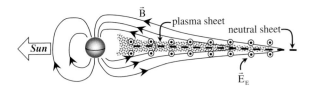

Fig. 11.32. The magnetotail neutral sheet and plasma sheet

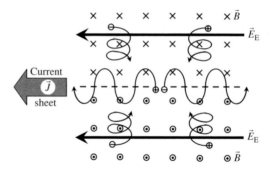

Fig. 11.33. $\vec{E} \times \vec{B}$ drift and the magnetotail current sheet

Figure 11.33 shows a view in the magnetotail, facing toward the Earth (i.e., looking toward the left from the right edge of Figure 11.32). The dashed line again denotes the neutral sheet, with magnetic field away from the observer on the north side (above) and toward the observer on the south side (below).

In regions away from the neutral sheet, positive and negative charges $\vec{E} \times \vec{B}$ drift toward the neutral sheet. Because they drift in the same direction, this motion does not give rise to a current. The resulting collection of electrons and ions near the neutral sheet is the *plasma sheet* (see Figure 11.32) Positive and negative charges in the plasma sheet now drift in opposite directions as they oscillate across the neutral sheet, producing a dawn-to-dusk current density, variously called the *neutral sheet current*, *magnetotail current sheet*, or *cross-tail current*.

The return flow on the outside of the magnetotail in Figure 11.9 is in the same direction as the return flow of the dayside boundary current (Section 11.6.2). These two flows merge near the Earth.

11.6.7 Magnetospheric Substorms

Magnetic substorms appear to be different in character and cause from magnetic storms (Section 11.6.4): (1) they are detected primarily at high geomagnetic latitudes, rather than worldwide as for magnetic storms; (2) the initial increase in $B_\parallel$ is lacking; (3) they tend to be stronger (i.e., deeper) but of much shorter duration than magnetic storms, typically lasting up to about 2 h; and (4) they often occur within a couple of hours after the vertical component of the IMF changes from northward to southward and stays that way. Point (2) shows that, unlike magnetic storms, enhancements of the solar wind are not involved, and (1) shows that the change in the ring current is much smaller than in magnetic storms.

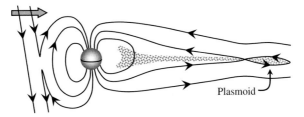

Fig. 11.34. Magnetic reconnection of southward IMF with the Earth's magnetic field at the dayside magnetospheric boundary. Magnetic reconnection also occurs in stretched magnetic field lines in the magnetotail, as described in the text

Magnetic substorms can also occur during magnetic storms, but appear to differ from those described above, e.g., by injecting considerably more charged particles into the inner magnetosphere.

Figure 11.34 illustrates some aspects of the present picture of a magnetic substorm. When the IMF is southward solar magnetic field lines can reconnect with those of the Earth, providing direct entry of solar wind plasma into the Earth's magnetosphere. Reconnection in stretched magnetic field lines in the magnetotail then creates detached regions of plasma and magnetic field called *plasmoids* that convect tailward (away from the Earth). The newly connected field lines convect earthward, injecting fresh plasma into the Van Allen radiation belts, although much less than in a magnetic storm. The newly injected particles within the loss cone precipitate into the atmosphere, causing aurorae and increasing the conductivity in the auroral oval. The latter effect creates increased Birkeland (field-aligned) currents into and out of the auroral oval and an increase in electric current (the *electrojet*) within the auroral oval. The electrojet produces the magnetic signature of the substorm.

11.6.8 Coupling Between the Magnetosphere and the Ionosphere

Aurorae are caused by precipitation of energetic electrons into the ionosphere. The electrons collide with and excite atoms and molecules, notably O, N, and N_2, and visible light is emitted when these species decay back to the ground state.

Time-varying electric currents (the *solar disturbed*, or S_D, current system) associated with the aurora occur at altitudes between 100 and 150 km and cause significant Joule heating in the ionosphere. Their transient magnetic fields can induce disruptive emfs in power grids.

Here, we describe the S_D current system. Figure 11.35 is a perspective view of Figure 11.31, showing the charge separation at the magnetopause boundary

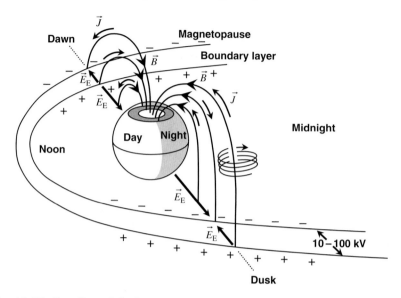

Fig. 11.35. Coupling of the low-latitude magnetosphere and boundary layer to the northern ionosphere. The magnetic field lines act as electrical conduits connecting the boundary layer emf to the polar cap (*white ellipse*) and the magnetospheric emf to the auroral oval (*grey ring*). See also Figure 11.31

layer, the resulting electrostatic fields, and some of the magnetic field lines coupling the magnetosphere to the northern ionosphere. These field lines also connect to the southern ionosphere (not shown in the diagram).

The boundary layer acts as a magnetohydrodynamic (MHD) generator: magnetospheric convection carries plasma antisunward through the Earth's magnetic field, creating a non-electrostatic field, $\vec{E}_N$ that in turn produces a charge separation and a consequent electrostatic field, $\vec{E}_E$ (cf. Section 11.6.5 and Figure 11.31). The emf, ε, of this source may be calculated from

$$\varepsilon = \int \vec{E}_N \cdot d\vec{\ell} = \int (\vec{v} \times \vec{B}) \cdot d\vec{\ell} \tag{11.98}$$

and is typically 10–100 kV. The polarity of the source is as indicated in Figures 11.31 and 11.35. The region just inside the boundary layer also acts as an MHD generator, with the opposite direction of magnetospheric convection and polarity of the emf.

Magnetospheric plasma is essentially collisionless, so the conductivity is very large parallel to $\vec{B}$ and very small perpendicular to $\vec{B}$. The magnetic field lines thus act as electrical conduits connecting the magnetospheric sources of emf to the ionospheric load. *Field-aligned currents* (called *Birkeland currents* in the Earth's magnetosphere, after their discoverer) then flow into and out of the ionosphere in the directions dictated by the polarity of the sources, as shown in Figure 11.35. The current direction inside the magnetopause and magnetosphere in Figure 11.35 is opposite to the electrostatic field, as required in a source of emf.

Figure 11.36 shows the magnetic and electrostatic fields and consequent current systems in the polar cap and auroral oval. Inspection of Figures 11.35 and 11.36 shows that the dusk-to-dawn electric field in the boundary layer maps into the ionosphere as a dawn-to-dusk electric field in the polar cap, and the dawn-to-dusk electric field in the magnetosphere maps into the ionosphere as a dusk-to-dawn electric field in the auroral oval.

The electric and magnetic fields in the ionosphere create the Pedersen current system parallel to the electric field and the Hall current system antiparallel to the $\vec{E} \times \vec{B}$ direction (Section 11.5.7). The Pedersen current system has divergence (currents begin, end, or flow in opposite directions at boundaries) and is closed by the Birkeland currents to and from the magnetosphere, whereas the Hall current system is without divergence.

The Hall currents are more limited in spatial extent and therefore more intense in the auroral oval than in the polar cap and are referred to as electrojets (Figure 11.36). During a moderately sized aurora, the electrojets can carry currents of several million amperes.

The idealized current orientation shown in Figure 11.36 must be modified because electron precipitation results in a greater density of free charge and therefore conductivity in the auroral oval than in the polar cap. Because the Hall current is dominated by electrons drifting opposite to the current direction and the mobility of electrons is higher in the auroral oval (i.e., there is an impedance mismatch between the auroral oval and the polar cap), negative charge occurs where the Hall current enters the polar cap and a

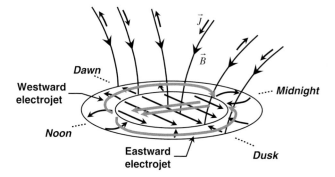

Fig. 11.36. Schematic diagram of ionospheric currents. The *arrows* beside the magnetic field lines are field-aligned (or Birkeland) currents; *black arrows* in the auroral oval and polar cap represent the directions of the electric field and the Pedersen current system; the *gray arrows* represent the Hall current system

corresponding positive charge where the Hall current leaves the polar cap. The electrostatic field of these charges adds vectorially to the one shown in Figure 11.36 to create a resultant electrostatic field directed from a point between midnight and dawn toward a point between noon and dusk. The (final) Pedersen current is parallel to the total electric field and the Hall current is perpendicular to it, so the Pedersen and Hall current systems are rotated clockwise (in the dawn to midnight direction) from the orientation shown in Figure 11.36.

Figure 11.37 shows the S_D current system at 2400 GMT averaged over a 12-month period in 1932–1933. Noon is downward. Note that the S_D current

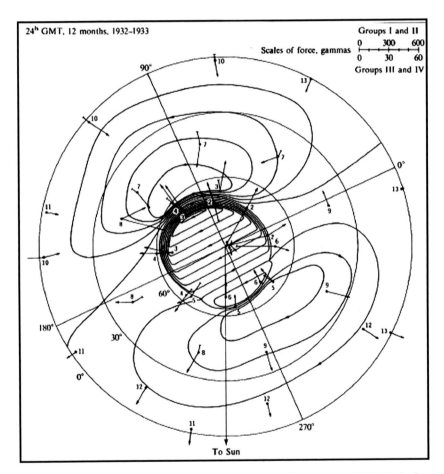

Fig. 11.37. The SD ionospheric current systems. A current of 50,000 A flows between successive contours. Noon is towards the bottom. Current density is maximum shortly after local midnight. (From Silsbee and Vestine (1942), Figure 8; and Parks 1991, Figure 7.11, p. 281, with the latter's permission)

system is fixed relative to the Sun, and the Earth rotates counterclockwise under it. The westward electrojet is much stronger than the eastward and gives a maximum current density in the hours after midnight, above latitudes between 60°N and 70°N.

11.7 Magnetospheres of Mercury, Venus, and Mars

11.7.1 Mercury

Mariner 10 is the only spacecraft to have visited Mercury, flying past the planet three times between March, 1974, and March, 1975. Currently, NASA's Messenger orbiter is *en route* and is expected to achieve orbit insertion on March 18, 2011, after having flown past Earth once, Venus twice, and Mercury three times.

The first and third Mariner 10 flybys passed over the nightside of Mercury at minimum distances of 703 and 327 km from the surface, respectively, and detected an intrinsic magnetic field and magnetosphere. The second pass was at about 48,000 km, too distant to obtain magnetospheric measurements.

Because of the short duration of the observations and the lack of simultaneous observations of the solar wind, the properties of Mercury's magnetosphere are only weakly constrained. The dipole moment is estimated to lie in the range $2\text{--}6 \times 10^{12}\,\mathrm{T\,m^3}$, which, divided by the quantity $\mu_0/(4\pi)$, yields $2\text{--}6 \times 10^{19}$ amp $\cdot$ m^2. This is about 4×10^{-4} that of Earth. The maximum magnetic field strength measured on the third orbit (327 km above the surface) was about 400 nT or about 1% that of the Earth, with the dipole pointing in the same direction as that of the Earth, i.e., the north-seeking pole of a compass on Mercury's surface would point north.

Figure 11.38 shows a model of Mercury's magnetosphere based on the two flybys. The bow shock is about $2.5\mathrm{R_{Me}}$ from the center of the planet, where $1\mathrm{R_{Me}} = 2439\,\mathrm{km} =$ radius of Mercury, and the magnetopause about $1.3 \pm 0.2\mathrm{R_{Me}}$. Because of the weak magnetic field, these boundaries are expected to vary significantly with the solar wind strength. The solar wind may even impinge directly on the planet's surface about 1% of the time.

Because of the weakness of the magnetic field there can be no region of trapped charged particles equivalent to the Earth's Van Allen radiation belts; as illustrated in Figure 11.38, if one were to scale the Earth's magnetosphere to that of Mercury, the trapping region would lie well inside the planet.

Even in the brief time of the flybys, short-term variations were observed that suggested magnetic substorms, including possible magnetic field depolarization, particle injection, field-aligned currents and plasma sheet heating.

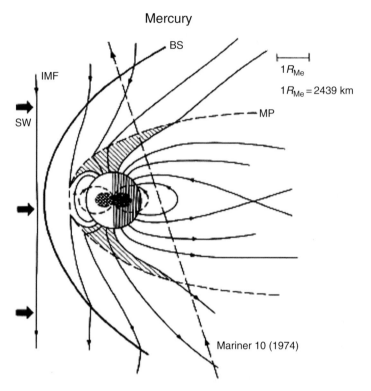

Fig. 11.38. The magnetosphere of Mercury. BS = bow shock; MP = magnetopause; IMF = interplanetary magnetic field. The *hatched region* outside the planet is the polar cusp. The *double hatched region* inside Mercury corresponds to the Earth's plasmasphere scaled to Mercury. The *dashed line* is Mariner 10's trajectory for the third flyby. Magnetic field direction is indicated by *arrows*. (From Bagenal, 1985, Figure 9.6, p. 234, with her permission)

However, it is also possible that these variations are externally driven. Because Mercury's magnetosphere is so small, it can establish a new configuration on the timescale of the solar wind travel-time past the magnetosphere, ~1 min; and the dynamic structure may result from external reorientations of the IMF rather than from internal substorm activity (Luhmann et al. 1998).

The source of the magnetic field is unclear. With only 0.056 the volume of the Earth, Mercury is expected to have cooled sufficiently that, if the core were pure iron–nickel, it should have solidified. However, the possibility that the field is simply induced in such a conducting core by variations in the IMF was ruled out by the observations. Therefore, Mercury's global magnetic field is of internal origin and can arise in one of only two ways: either it is driven by an internal dynamo in a molten outer core (see below) or it is the result of permanent magnetism. The latter would seem to require the former as a precursor, with the dynamo-driven field being frozen in when the core solidified. However, the Curie point for iron, above which permanent magnetism cannot exist, is around $1000\,K$, and the core would be well above this temperature at the time of solidification. (The melting point of iron at STP is $1811\,K$.) Thus, the magnetic field should disappear on solidification, long before the core became cool enough to sustain it.

In the case of the Earth, the global magnetic field results from a *self-sustaining electric dynamo*. As the Earth cools, iron in the molten, iron-rich outer core solidifies onto the solid, inner core, releasing gravitational potential energy and latent heat. If there are impurities present, they are preferentially excluded from the condensate, so the remaining molten material is enriched in the impurity and lighter than its surroundings. (For this reason, the Earth's outer core at the present time is about 5% less dense than the solid, inner core when taken at the same pressure.) This molten, electrically conducting material rises due to thermal and compositional buoyancy, and the Coriolis force from the Earth's rotation creates a helical flow pattern. The motion of conducting material through the existing magnetic field sustains electric currents that in turn sustain the magnetic field. Thus, planetary rotation and both a solid, inner core and molten outer core are necessary ingredients for an internally generated magnetic field in a terrestrial planet.

(To be more precise, the ingredients are planetary rotation and a heat flux greater than can be sustained by conduction alone. Early in the life of a terrestrial planet the outflow of primordial heat from accretion can be sufficient to drive convection in the fully molten core, producing an electric dynamo and planetary magnetic field without the need for a solid, inner core. Such was the case for the Earth, where remanent magnetism from a global field has been measured in rocks that formed long before the core cooled sufficiently to begin to solidify, and it can be expected to have occurred in the other terrestrial planets as well. However, none of the terrestrial planets today have sufficient primordial heat flux to drive convection.)

Mercury's magnetic field therefore appears to require at least a thin, molten, outer core, although the minimum thickness required to produce an electric dynamo is at present uncertain. A molten, outer core in turn suggests that the core contains a significant amount of sulfur or other impurity to lower the melting point. The initial abundance of the impurity could have been relatively small ($\sim$1%), because the remaining liquid is progressively enriched as solidification proceeds; but the uncertainty in minimum thickness carries over to a similar uncertainty in the minimum required initial impurity abundance.

The dynamo model also has to overcome the fact that no sign of internal activity is visible on Mercury's ancient surface from any time in at least the last three billion years. Plate tectonics increases the rate of heat flow through the crust and therefore increases the temperature gradient in the

core, increasing the likelihood of a dynamo (Section 11.7.2). Thus its absence
may argue against a dynamo model for Mercury's global magnetic field.

11.7.2 Venus

No internally generated magnetic field has yet been detected around Venus.
The upper limit on the magnetic dipole moment set by the Magellan orbiter
is 1.5×10^{-5} that of the Earth.

Slow rotation (sidereal rotation period = 243 days, retrograde) has sometimes
been cited as the reason why Venus has no observed intrinsic magnetic field,
but in fact all of the terrestrial planets have more than enough rotation
to create an electric dynamo if the structural and thermal conditions are
met (Stevenson 2003 and references therein). However, because Venus is not
rotationally distorted, its interior structure cannot be investigated by the
evolution of spacecraft orbits. Consequently, the interior structure of Venus
is poorly constrained by observations, and it is not known how much, if any,
of the core is solid.

If the lack of an observed field arises from a lack of convection in the core,
then the two most likely reasons are (1) the core is entirely molten or (2) there
is a solid, inner core, but convection is suppressed. A completely molten core
could occur if the impurity concentration is greater than in the Earth's core
or may simply be a result of the lower central pressure compared to the Earth,
because of the smaller planetary mass and radius (Stevenson et al. 1983).

Even if there is a solid, inner core, the observed absence of plate tectonics can
suppress convection (Nimmo 2002). On the Earth, the heat flux through the
crust is $82 \, \mathrm{mW/m^2}$, 65% of which is carried by plate tectonics. On Venus, the
crustal heat flux is estimated to be in the range $10–40 \, \mathrm{mW/m^2}$, with most
investigations appearing to favor the lower part of this range. The insulating
effect results in a warmer mantle and a smaller temperature gradient in the
molten core.

If the heat flux is $\sim 20 \, \mathrm{mW/m^2}$, then the rate of production of radiogenic heat
from radioactive elements in the mantle may exceed the loss of heat through
the crust by a factor of ~ 1.5, with the result that the mantle temperature may
be increasing at $> 30 \, \mathrm{K/10^9}$ years (Nimmo 2002). This situation, if it exists,
is likely to be temporary. Studies of the history of surface cratering on Venus
suggest a major resurfacing event $\sim 0.6 \times 10^9$ years ago. If this resurfacing
suppressed plate tectonics, then plate tectonics may have occurred prior to
this time, resulting in greater heat flow, convection in the outer core, and
an electric dynamo and global magnetic field. Thus the absence of a global
magnetic field may be a temporary, or possibly cyclic, property of Venus.

Despite the absence of an internally driven magnetosphere, the interaction of
the solar wind and IMF with Venus' ionosphere creates a magnetic obstacle

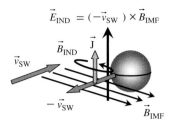

Fig. 11.39. The flow of IMF towards the ionospheric conductor induces a north-wards current in the ionosphere for the IMF direction shown

that deflects the solar wind around the planet. Figure 11.39 illustrates some of the processes involved. The IMF can be regarded as frozen into the solar wind plasma and carried toward Venus at the solar wind velocity, $\vec{v}_{SW}$.

The motion of the IMF toward Venus at velocity $\vec{v}_{SW}$ is equivalent to the motion of the Venus ionospheric conductor through the IMF at velocity $-\vec{v}_{SW}$: an electric field and current are induced in the ionosphere as shown in Figure 11.39. The induced magnetic field, $\vec{B}_{IND}$, adds to $\vec{B}_{IMF}$ on the sunward side of this current, increasing the magnetic field strength above the ionosphere, and opposes it on the side toward Venus. Thus, the magnetic field pressure

$$P_{IMF} = \frac{B_{IMF}^2}{8\pi}$$

increases in the region above the ionosphere, while the ionosphere is compressed by the incoming solar wind plasma and IMF. These processes are evident in passages of Pioneer Venus through the midday ionosphere: with increasing altitude, the ionospheric electron density drops to near zero in the same region where the magnetic field strength increases from near zero to a value somewhat elevated above that far from Venus (see Figure 12, p. 892, Russell and Vaisberg, 1983). Balance is achieved when the inward force due to the magnetic pressure at the ionopause, the outer boundary of the ionosphere, equals the outward force due to the ionospheric plasma pressure. This is illustrated in Figure 11.40, where magnetic field pressure and ionospheric plasma pressure measured by Pioneer Venus are plotted against altitude above the surface of Venus.

If the ionosphere were superconducting, the IMF would be completely excluded, and the magnetic field would be zero below the ionopause. Instead, the spikes in Figure 11.40 suggest that magnetic field enters the Venus atmosphere in localized "flux ropes."

The altitude of the ionopause depends on the solar wind strength. When the solar wind strength is low, the ionopause is at high altitudes where individual ions follow their own trajectories. The result is a diamagnetic current similar in physical cause to the Earth's boundary current, except that instead of

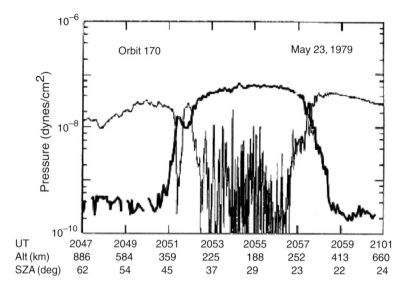

Fig. 11.40. Magnetic field pressure (*light lines*) and ionospheric plasma pressure (*heavy lines*) vs. altitude from one passage of Pioneer Venus through the midday ionosphere (Elphic et al. 1980, Figure 6, p. 7683; copyright 1980, American Geophysical Union. Reproduced by permission of American Geophysical Union.)

solar wind ions moving toward the right into an internally generated magnetic field in Figure 11.23, ionospheric ions in Figure 11.39 move toward the left into the IMF (in the solar wind frame). Because the IMF is horizontal, the diamagnetic current is northward.

When the solar wind strength is high, the ionopause descends to lower altitudes and higher densities, where collisions occur. Whereas diamagnetic currents are dissipationless, collisions produce dissipation, and the current becomes an actual flow of charge in a resistive medium.

With the IMF effectively excluded from the ionosphere, the solar wind is deflected northward and southward past the planet, carrying the IMF with it (Figure 11.41), forming a magnetocavity downwind from the planet where solar wind ions are excluded. The magnetic polarity in this cavity varies with the orientation of the IMF, unlike the magnetotails of planets with intrinsic magnetic fields. As with internally generated magnetospheres, a bow shock forms where the solar wind is slowed to subsonic speeds.

The solar wind–ionosphere interaction on Venus has contributed significantly to atmospheric loss over the life of the planet. The dominant processes are (1) charge-exchange reactions and subsequent ion pick up by the IMF and (2) sputtering. In the former, solar wind ions and electrons ionize atomic H and O in the exosphere of Venus (e.g., $O + e^- \rightarrow O^+ + 2e^-$), and the resulting H^+ and O^+ are picked up by the IMF and carried off in the solar wind flow

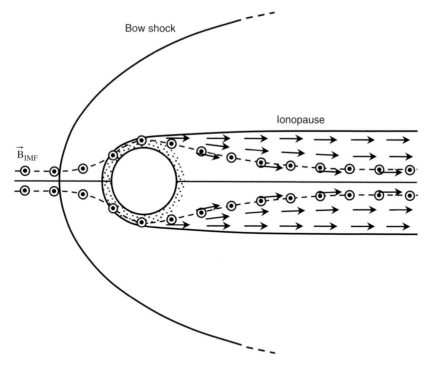

Fig. 11.41. IMF flow around Venus, viewed from the ecliptic plane. The circled dots show magnetic field lines oriented toward the observer, and the arrows show the flow of low-energy ionospheric ions. Based on Russell and Vaisberg 1983, Figure 18, p. 901.

around the planet. In the latter, fast solar wind ions collide with neutrals in the exosphere or near the exobase. In the solar wind impact region, a cascade of collisions results in some atmospheric neutrals moving outward; in the solar wind flow around the planet, motion is tangential and a single collision can eject the atmospheric neutral.

11.7.3 Mars

Of the many spacecraft to have visited Mars, the most accurate magnetic field measurements have been made by Mars Global Surveyor (MGS). MGS used aerobraking to refine its initial, highly elliptical orbit to a final, circular orbit at 400 km altitude, and in this process dipped into the Martian atmosphere, to altitudes as low as 101 km, to slow its orbital speed. In so doing, it became the first spacecraft to carry magnetometers below the Martian ionosphere, where the magnetic fields associated with the Martian crust and interior are stronger and the magnetometers are shielded from the external magnetic fields of the solar wind–ionosphere interaction (Section 11.7.2). Because of a solar panel anomaly, NASA extended the aerobraking period to ~18 months to reduce the force on the solar panels, allowing magnetometer readings to be taken over ~20% of the Martian surface to much greater accuracy and higher resolution than is possible from the final 400 km mapping altitude.

To the limit of observations, Mars has no global magnetic field. MGS established an upper limit of 2×10^{17} A/m^2 for a global dipole moment, corresponding to a magnetic field strength at the Martian equator of 0.5 nT, about 10^{-6} that of the Earth (Acuña et al. 2001); but the observations are in fact equally well represented by a dipole moment of zero. The interaction of the Martian atmosphere and ionosphere with the solar wind and IMF is thus similar to that at Venus (Section 11.7.2).

A surprising and significant discovery by MGS is the existence of intense, localized remanent magnetism in the Martian crust (Connerney et al. 1999). Although some anomalies have been found in the younger, northern lowlands, by far the majority and the strongest occur in the ancient southern highlands. These anomalies appear in the MGS data as linear, approximately parallel, east-west-trending features up to 1000–2000 km in length and ~200 km in width, of alternating magnetic polarity. The 200 km width is imposed by the resolution of the MGS data, so in fact there could be finer lineations. Magnetic field strengths are up to ~1500 nT at periapsis ($\geq$100 km altitude) and ~200 nT at the final mapping altitude of 400 km. These fields dwarf their terrestrial counterparts: the strongest crustal magnetic anomalies on the Earth produce variations of ±10 nT in the dipole field at 400 km altitude. The dipole moment of (any) one 2000 km long anomaly 200 km wide is an order of magnitude greater than that of the Kursk magnetic anomaly in Russia, one of the largest on the Earth.

The intense magnetization is consistent with the fact that the iron content of the Martian crust is much higher than on the Earth (17% in the soil at the Pathfinder landing site and 15–30% in Martian meteorites, compared to 1% on the Earth), but it also requires a relatively intense inducing field, similar to or greater than that of the present-day Earth. The most likely model for this field is a global magnetic dipole created by an electric dynamo in a molten core. The timescale of this dynamo can be estimated from the fact that, within the southern highlands, magnetization tends to be absent in the material covering the large impact basins such as Argyre and Hellas. These are believed to have formed within the first 300 My of the accretion of Mars, suggesting that the dynamo died out during the earliest epoch on Mars. The scarcity and weakness of anomalies in the northern lowlands also suggests that the dynamo died out before the formation of the main north–south dichotomy of Mars; but it then becomes difficult to account for the anomalies that do exist there.

The most straightforward explanation for the linearity and alternating polarity of the anomalies is that they formed during an early period of plate tectonics, in which crustal spreading was accompanied by reversals of the putative Martian magnetic dipole as the spreading crust cooled below the Curie point. Although alternative explanations have not been ruled out, this model has an advantage of self-consistency in that plate tectonics increases the heat flux through the crust and thus the probability of core convection and an electric dynamo (Section 11.7.2). One requirement of the crustal spreading model is a symmetry axis analogous to mid-ocean ridges where new crust is being created on the Earth. None have yet been discovered, but large areas of the highlands have been reworked by impacts, thermal events, and fracturing, and such features may not have survived.

References

Acuña, M. H., Connerney, J. E. P., Wasilewski, P., Lin, R. P., Mitchell, D., Anderson, K. A., Carlson, C. W., McFadden, J., Rème, H., Mazelle, C., Vignes, D., Bauer, S. J., Cloutier, P., and Ness, N. F. 2001. "Magnetic field of Mars: Summary of results from the Aerobraking and Mapping Orbits," *Journal of Geophysical Research* **106**, No. E10, 23,403–23,417.

Bagenal, F. 1985. "Planetary Magnetospheres" in *Solar System Magnetic Fields*, ed. E. R. Priest (Dordrecht: D. Reidel) pp. 224–256.

Cahill, Jr., L. J. 1968. "Inflation of the Inner Magnetosphere" in *Physics of the Magnetosphere*, eds R. L. Carovillano, J. F. McClay, and H. R. Radoski (Dordrecht: D. Reidel) pp. 263–270.

Chen, R. H., Cravens, T. E., and Nagy, A. F. 1978. "The Martian Ionosphere in Light of the Viking Observations," *Journal of Geophysical Research* **83**, 3871–3876

Connerney, J. E. P., Acuña, M. H., Wasilewski, P., Ness, N. F., Rème, H., Mazelle, C., Vignes, D., Lin, R. P., Mitchell, D., and Cloutier, P., 1999. "Magnetic Lineations in the Ancient Crust of Mars," *Science* **284**, (No. 5415), 794–798.

Elphic, R. C., Russell, C. T., Slavin, J. A., and Brace, L. H. 1980. "Observations of the Dayside Ionopause and Ionosphere of Venus," *Journal of Geophysical Research* **85**, 7679–7696.

Griffiths, D. J. 1999. *Introduction to Electrodynamics*, 3rd Ed. (Prentice-Hall, New Jersey).

Iribarne, J. V. and Cho, H.-R. 1980. *Atmospheric Physics* (Dordrecht: D. Reidel).

Luhmann, J. G., Russell, C. T., and Tsyganenko, N. A. 1998. "Disturbances in Mercury's Magnetosphere: Are the Mariner 10 'Substorms' simply driven?" *Journal of Geophysical Research* **103**, No. A5, 9113–9119.

Nair, H., Allen, M., Anbar, A. D., and Yung, Y. L. 1994. A "Photochemical Model of the Martian Atmosphere," *Icarus*, **111**, 124–150.

Nimmo, F. 2002. "Why does Venus Lack a Magnetic Field?" *Geology*, **3**, No. 11, 987–990.

Parks, G. K. 1991. *Physics of Space Plasmas, an Introduction* (Redwood City, CA: Addison-Wesley) 2nd ed., 2003. Harper-Collins, Perseus Books.

Russell, C. T. and Vaisberg, O. 1983. "The Interaction of the Solar Wind with Venus," in *Venus*, eds D. M. Hunten, L. Colin, T. M. Donahue, and V. I. Moroz (Tucson, AZ: University of Arizona Press) pp. 873–879.

Silsbee, H. and Vestine, E. 1942. "Geomagnetic Bays, Their Frequency and Current Systems," *Terrestrial Magnetisn and Atmospheric Electricity*, **47**, 195–208.

Stevenson, D. J. 2003. "Planetary Magnetic Fields," *Earth and Planetary Science Letters*, **208**, 1–11.

Stevenson, D. J., Spohn, T., and Schubert, G. 1983. "Magnetism and Thermal Evolution of the Terrestrial Planets," *Icarus*, **54**, 466–489.

Challenges

[11.1] Derive (11.77) from (11.76). [Hint: Electric current equals the amount of charge passing a given point per second; how is the current associated with a gyrating particle related to the charge of the particle and the period of the gyromotion?]

[11.2] Derive (11.82) from (11.63) and other previous results.

[11.3] Justify, using $\vec{F}_B = q\vec{v} \times \vec{B}$, the statement preceding (11.83) that the kinetic energy of the particle is constant. [A descriptive answer is acceptable, but the description should be physically precise.]

[11.4] Spacecraft orbiting through the trapping region of a planetary magnetosphere, such as the Earth's Van Allen radiation belts, do not necessarily orbit in the plane of the magnetic equator; thus, they sample the magnetosphere at a range of magnetic latitudes. For uniformity of data, measured pitch angles are usually referred to the magnetic equator. If the spacecraft measures a pitch angle, α, for a particle at a location away from the magnetic equator where the magnetic field strength is B, and if *on the same field line* the magnetic field strength at the magnetic equator is B_0, show that the pitch angle, α_0, of the particle as it crosses the magnetic equator is given by

$$\sin^2 \alpha_0 = \frac{B_0}{B} \sin^2 \alpha$$

[Hint: μ_B and Φ_B are constants of the motion. Express the pitch angles in terms of $v_\perp$ at B and $v_\perp^0$ at B_0, and use (11.77) and (11.81).]

[11.5] The loss cone is approximately $3°$ for particles on the geomagnetic equator at $6\,R_E$ above the Earth's surface, i.e., particles of this pitch angle mirror at ionospheric heights:

   a. What is the mirror ratio, i.e., the ratio of the magnetic field strength at the ionosphere to that at the geomagnetic equator, along this field line? [Hint: Refer to problem 11.4. What is the pitch angle at the mirror point?]

b. If particles are injected with an initially isotropic distribution of pitch angles at $6\,R_E$ on the geomagnetic equator, what fraction of these particles are precipitated?

[11.6] Equations (11.64) and (11.66) in Section 11.5.6 give, respectively, the $\vec{F} \times \vec{B}$ drift velocity, $\vec{v}_D$, and its more commonly used form, the $\vec{E} \times \vec{B}$ drift velocity. The purpose of this problem is to derive (11.64) and (11.66).

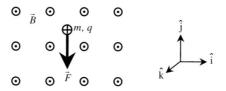

The diagram above shows a uniform magnetic field, $\vec{B}$, out of the page. In addition to any magnetic forces, all charges in this region also feel a downward force, $\vec{F}$, of external origin (e.g., gravitational or electrical) of uniform magnitude regardless of position. ($\vec{F}$ is unrelated to $\vec{B}$.)

a. Suppose you take hold of the charge shown above and run with a constant (vector) velocity, $\vec{v}$, holding tightly onto the charge so it can not execute gyromotion. In what *direction* would you have to run so that the total force (or net force) on the particle is zero? Draw a vector diagram showing and labelling $\vec{B}, \vec{v}$, and all forces. Then write down the vector equation relating $\vec{F}$, $\vec{v}$ and $\vec{B}$. (In this and the other parts, below, keep careful track of minus signs.)

b. Evaluate the quantity $(\vec{v} \times \vec{B}) \times \vec{B}$ for this situation, using the unit vectors shown above. Express your answer in terms of a single unit vector, then rewrite your answer to express the quantity $(\vec{v} \times \vec{B}) \times \vec{B}$ in terms of $\vec{v}$.

c. Obtain (11.64) from your answers to parts a and b.

d. Now assume that the force, $\vec{F}$, is caused by an electric field vertically downward in the reference frame of the diagram (i.e., $\vec{F} = q\vec{E}$). In *your* reference frame, the force on a stationary particle is zero. It follows that, by running, you have placed yourself in a reference frame in which the electric field is zero: if you let go of the particle, it will remain at rest in your reference frame, so evidently there is no electric field.

If a particle is now released at rest in the original reference frame of the diagram, what initial velocity will *you* see for this particle? What subsequent motion do you expect to see for this particle? (Remember, in your reference frame there is only a magnetic field.) By extension, what subsequent motion would an observer see who is at rest in the reference frame of the diagram? What is the $\vec{E} \times \vec{B}$ drift velocity of the guiding center in the reference frame of the diagram? (Also check that the drift velocity *is* in the direction of $\vec{E} \times \vec{B}$.)

12. The Giant Planets

The giant planets contain the bulk of the angular momentum of the solar system and the dominant mass among the planets. The magnetosphere of the Jovian system dwarfs the Sun in size. Jupiter, Saturn, and possibly Neptune radiate more energy than they receive from the Sun. The properties of the giant planets are summarized in Table 12.1 drawn from the Astronomical Almanac, Cox (2000), and other sources, including the NASA-Goddard Space Flight Center website links at http://nssdc.gsfc.nasa.gov/planetary/factsheet/.

It is not difficult to imagine a scenario in which a much more massive Jupiter is a binary star companion to the Sun. In such a case, one would think that planets either would be ejected from the inner system entirely or perhaps find a niche at one of the stable Lagrangian points (Milone & Wilson 2007, Section 3.3). Jovian-type planets thus far detected in extra-solar planetary systems will be considered in Chapter 16; surprisingly, planets are being found in both binary and triple systems as well.

The Voyagers, Galileo, and Cassini missions to the gaseous giants have provided important clues about the structures of their magnetospheres, atmospheres, meteorology, and even their interiors. The remarkable details revealed have been supplemented by changes in atmospheric structures revealed through extended HST observations.

The extensive satellite and ring systems of the giant planets as well as Pluto and its moons will be left to Chapter 13, which picks up from the end of our discussion here of the contributions of Jupiter's Galilean moon Io to Jupiter's magnetospheric composition.

First, we summarize the properties of the gaseous giant planets of our solar system in Table 12.1. Note that we have two types of these gaseous planets: the truly giant planets, Jupiter and Saturn, two orders of magnitude more massive than the Earth and around 10× its diameter, and the two lesser giants, Uranus and Neptune, more massive than Earth by only an order of magnitude and just a few times its diameter.

12.1 Jupiter

Jupiter was the principal god of Rome and among the Greeks the cultural equivalent to Zeus, king of the gods in the pantheon of Mt Olympus. The planet's slow, stately movement in the heavens (12 years to complete a circuit,

Table 12.1. Giant planets data

Property/planet	Jupiter	Saturn	Uranus	Neptune
Orbital data				
Mean distance (au)	5.203	9.577	19.29	30.23
Eccentricity	0.0485	0.0539	0.0428	0.0107
Inclination (to the ecliptic plane)	1°.305	2°.485	0°.773	1°.768
Sidereal period	11ʸ.862	29ʸ.458	84ʸ.678	166ʸ.21
Physical data				
Mass (in units of Earth mass)	317.83	95.159	14.54	17.15
Equatorial radius (in Earth radii)	11.209	9.4491	4.0073	3.8826
Oblateness (flattening)[a]	0.064874	0.097962	0.022927	0.017081
J_2[b]	0.014736	0.016298	0.003343	0.003411
J_3[b]	—	-1.00×10^{-3}	—	—
J_4[b]	-5.84×10^{-4}		—	—
Mean density[c] (kg/m^3)	1327	687	1270	1638
Equatorial surface gravity (m/s^2)	24.79	10.44	8.87	11.0
Equatorial escape velocity (km/s)	59.5	35.6	21.4	23.6
Photometric data				
$V_{\text{opposition}}$ (mean distances)	−2.70	+0.67	+5.52	+7.84
(B−V)	+0.83	+1.04	+0.56	+0.41
(U−B)	+0.48	+0.58	+0.28	+0.21
Radiative energy data				
Visual albedo (geometric)	0.52	0.47	0.51	0.41
Bolometric albedo (geometric)[d]	0.27 ± 0.01	0.24 ± 0.01	0.22 ± 0.05	0.22 ± 0.05
Bolometric albedo (Bond)[d]	0.343 ± 0.032	0.342 ± 0.030	0.300 ± 0.049	0.290 ± 0.067
Absorbed power (10^{16}W)	49.82 ± 2.43	10.18 ± 0.46	0.521 ± 0.037	0.203 ± 0.020
Observed power[d] (10^{16}W)	83.65 ± 0.84	19.77 ± 0.32	0.560 ± 0.011	0.534 ± 0.029
Derived internal power (10^{16}W)	33.8 ± 2.6	9.59 ± 0.47	0.039 ± 0.038	0.331 ± 0.035
T_{eq} (predicted effective temp; K)	109.3 ± 1.3	80.5 ± 0.9	58.1 ± 1.0[e]	46.6 ± 1.1
T_{eff} (from observed power; K)[d]	124.4 ± 0.3	95.0 ± 0.4	59.1 ± 0.3	59.3 ± 0.8

[a] $\varepsilon = (R_{\text{eq}} - R_{\text{pol}})/R_{\text{eq}}$.

[b] See Milone and Wilson (2007, Ch. 5.3)

[c] computed using volumetric radius $[(R_{\text{eq}}^2 R_{\text{pol}})^{1/3}]$.

[d] data from Pearl and Conrath (1991).

[e] "Fast rotator" case; for "slow rotator" case, $T = 69.1$° K.

spending 1 year in each of the zodiacal signs) and its bright appearance convey the impression of majesty and power. It is an appropriate association, since Jupiter is the most massive ($\sim318M_\oplus$) and the largest of all the planets ($\sim11\Re_\oplus$).

12.1.1 Visible Phenomena

The classic source of details about Jupiter's observed features has been the popular book, *The Planet Jupiter* by Peek (1958). Newer detailed sources are Bronshten (1969) and Gehrels (1976). In the 1970s the Pioneer and Voyager spacecraft provided the first high-resolution images of the atmospheric surface structures of Jupiter and Saturn. Voyager 2 also visited Uranus and Neptune. The Galileo spacecraft in the 1990s released a probe which entered the atmosphere of Jupiter in December, 1995. More recently the Cassini space-craft passed Jupiter en route to Saturn and provided useful measurements of the solar wind and Jupiter's magnetosphere that complemented those of Galileo.

An occultation of a star by Jupiter, observed by Baum and Code (1953), provided the first direct evidence of the bulk composition of Jupiter's atmosphere. With a derived effective scale height, $H = 8.3\,\text{km}$, and an assumed T = 86 K, from Kuiper (1952), they derived a mean molecular weight of 3.3. By recalculation, with an observed temperature of the upper clouds (from infrared observations) of $\sim125\,\text{K}$ and a gravitational acceleration of $24.8\,\text{m/s}^2$, the mean molecular weight is found to be

$$\mu = kT/[Hgm_\text{u}] = 1.38 \times 10^{-23} \times 125/[8300 \times 24.8 \times 1.67 \times 10^{-27}] \approx 5$$

revealing hydrogen and helium to be major constituents of the atmosphere. The escape velocity is 59.5 km/s, so from (10.6),

$$\mu_\text{crit} = 8.980 \times 10^5 \times T/v_\infty{}^2 = 0.03$$

and Jupiter is able to retain essentially all of its hydrogen over the age of the solar system.

The mean density of Jupiter ($1327\,\text{kg/m}^3$) is slightly less than that of the Sun, but it (as well as the other giant planets) does not have solar compo-sition. In particular, it is slightly deficient in helium compared to the Sun, as we note below. Jeans escape is not the relevant mechanism to explain the apparent deficiency, but rather a fractionation of the material in Jupiter's deep atmosphere and its strong gravitational field.

The measured rotation periods, P_{rot}, depend on latitude, ϕ, and there is another period, associated with a deeper source, perhaps in the solid body of the planet, determined from radio emissions. Three systems are recognized:

$$P_{\text{rot}}^{\text{I}} = 9^{\text{h}}\,50^{\text{m}}\,30.0^{\text{s}}\ (|\phi| < 10^{\circ})$$

$$P_{\text{rot}}^{\text{II}} = 9^{\text{h}}\,55^{\text{m}}\,40.6^{\text{s}}\ (|\phi| > 10^{\circ})$$

$$P_{\text{rot}}^{\text{III}} = 9^{\text{h}}\,55^{\text{m}}\,29.75^{\text{s}}\ (\text{System III; presumably solid body})$$

The planet's flattening is 0.0649, considerable when compared to the terrestrial planets, but it must be remembered that what is being viewed is the exterior of a largely fluid object that is rapidly rotating, and therefore necessarily spheroidal.

The circulation pattern in Jupiter's extensive gaseous atmosphere demonstrates strong geostrophic effects. The Rossby number (Chapter 10.5.1; Lewis 1997, p. 182), as it applies to giant planet circulation, is

$$Ro = [dv/dt]/[4\pi v \sin\phi/P] = v_{\text{x}}/fL \qquad (12.1)$$

where v_{x} is the eastward wind component, $f = 2\Omega \sin\phi$ is referred to as the *Coriolis parameter*, and L is a "characteristic length" for horizontal motions, the largest of which are up to about the planetary radius. The Coriolis acceleration is generally important in the giant planets, especially for large-scale motions, except near the equator (where $\phi = 0$), so the Rossby number is small.

In steady-state situations, the Coriolis force is balanced by horizontal pressure gradient forces, creating geostrophic winds. Near the equator and at latitudes $\pm 30^{\circ}$, the zonal wind speeds are in excess of 100 m/s, whereas the belt winds tend to show retrograde rotation relative to the planet (i.e., currents moving from E to W; on Earth, these would be called "easterlies"). With $v_{\text{x}} = 100$ m/s and $fL = 12,600$, $Ro \approx 0.01$, indicating Coriolis effects to be highly important.

Jupiter presents a banded appearance (see Figures 12.1 and 12.2, Plates 1 and 2), with dark bands (belts) interspersed among light bands (zones). Note that the strongest winds are in the equatorial zones and are "westerlies." At first glance, this seems opposite to what is expected, because Figure 10.31 indicates that the corresponding equatorial zone on Earth has easterly "trade winds." The difference comes about because in the case of the Earth, we are looking at surface winds; at altitude, at the top of the Hadley cell, these winds are westerly. The winds on Jupiter, and on the other giant planets, are similarly seen at the tops of their circulation cells. The zones represent cooler, higher regions of the atmosphere and the belts warmer and lower regions. The white colors of the zones are due to NH_3 crystals condensing in the troposphere from

Fig. 12.1. A composite image of Jupiter in three passbands, as viewed from the Cassini Orbiter on October 8, 2000, at a distance of 77.6 million km from Jupiter. See Plate 1 for color. Note the bright zones, dark belts, Great Red Spot, white ovals and brown barges. NASA/JPL/CICLOPS/University of Arizona image PIA02821

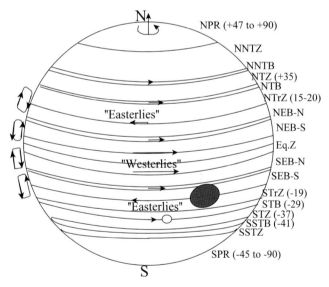

Fig. 12.2. The zones (Z) and belts (B) of Jupiter's atmosphere. Other abbreviations are N (north), S (south), T (temperate), Tr (tropical); NPR and SPR are North and South Polar Regions, respectively. The parenthetical numbers are the approximate latitudes in degrees. The *arrows* indicate roughly the wind direction and relative speed. The Jupiter that we see in visual light is dominated by high-altitude effects: We see the tops of the convective cells. The zones are *lighter bands*, marking ascending (high-pressure) columns of gas; the belts are *darker* and indicate descending (low-pressure) columns. The Great Red Spot and a white oval are also shown

rising gas. The belts are dark, reddish-brown, and deep purple and indicate regions which are free of the overlying ammonia clouds. UV photolysis of H_2S may explain the colors. For detailed discussions of low-resolution Jovian atmospheric phenomena, see Bronshten (1967). For a detailed account of Jupiter's photochemistry, see Lewis (1997, pp. 185–202).

The Great Red Spot (*GRS*), seen in Figures 12.1 and 12.3, has been observed, arguably, since the seventeenth century, with occasional fading and with return to former dark red intensity, and considerable variation in longitude (about its own mean position), but little variation in latitude. It is located in the Southern Tropical Zone at ∼20°S. It is ∼50,000 km across, has a CCW rotation, and its top is significantly cooler and hence higher than all other features of the surrounding atmosphere. Thus it represents a region with strongly upwelling material, the buoyancy of which has carried it to more than 8 km above the highest cloud deck. Long thought to be a consequence of a *Taylor instability*, an atmospheric disturbance produced by heavy winds crossing over a surface obstruction (Hide 1961), it is now viewed as an intense storm, albeit a long-lasting one. Ingersoll et al. (2004, Sections 6.3 and 6.7) summarizes both observations and theories.

In addition to the GRS, there are *brown barges* (probably NH_4SH clouds), *white ovals* (NH_3 clouds), some of which can be seen in Figures 12.1 and 12.4, *plumes* (high-altitude cirrus clouds of NH_3, spread out by high-altitude winds), and *blue-grey regions*. The latter occur most often in the North Equatorial Belt and show the strongest infrared emission at 5μm. They are thought to be the lowest regions of atmosphere that can be viewed from outside the planet.

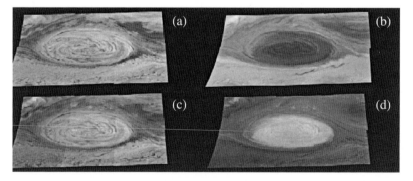

Fig. 12.3. Mosaics of six images of the Great Red Spot in each of four passbands centered on **a** 415 nm, **b** 757 nm, **c** 732 nm, and **d** 886 nm. From the Galileo mission Solid State Imaging System, obtained on June 26, 1996. Credits: NASA/JPL/California Institute of Technology. Image PIA00721

Fig. 12.4. Structure in the equatorial zone and belts of Jupiter, as viewed by the Cassini Orbiter near its closest approach. Note the blue-grey clouds (see Plate 2). The picture is a modified composite of images in three passbands: 728, 752, and 889 nm, in bands of methane and the continuum. The Great Red Spot is seen at the lower right. Bright white points may be high-altitude lightning. NASA/JPL/CICLOPS/Univ. of Arizona image PIA02877

The atmosphere is expected to behave as a perfect gas to a depth of $\sim$400 km down from the cloud tops, with increasing deviation from the perfect gas law below this depth.

Jupiter was visited by Pioneer 10 (1973), Pioneer 11 (1974), Voyager 1 and 2 (1979), Ulysses (1992 and 2003–2004), and Galileo (1995–1999). Galileo carried a probe which entered the atmosphere of Jupiter and studied the properties of the outer atmosphere. The probe gave evidence of decidedly drier atmosphere than expected, but one which confirmed the interior-driven circulation model of the atmosphere, i.e., that the circulation does not depend at least primarily on solar heating: winds as high as 650 km/s extend down to 130 km below the cloud tops.

The presence of ammonia and methane in the upper atmosphere of Jupiter and of methane in the other giant planets has been known since the 1930s, but subsequent studies have revealed the presence of hydrogen sulfide (H_2S), phosphine (PH_3), germane (GeH_4), and CO in Jupiter. Aside from H_2S, which is thought to be in equilibrium, these molecules must be newly convected into the high atmosphere because in equilibrium other gases are expected which are not found (e.g., P_4O_6 by reaction of PH_3 with H_2O). In this way, vigorous convection is seen to be present in Jupiter, Saturn, and Neptune, but less so in Uranus. The presence of likely chromophores, as certain colorful compounds are known, such as sulfur, hydrogen sulfide (H_2S), ammonium hydrosulfide

(NH$_4$SH), and their photochemical byproducts, such as hydrazine (N$_2$H$_4$), and possibly involving hydrocarbons, is needed to explain the redder colors.

As noted above, the white clouds seen in Jupiter's zones are likely ammonia which condenses out at high altitudes in that planet's atmosphere. Because gases condense out when their equilibrium vapor pressures are equal to their partial pressures, the altitude at which this happens varies from planet to planet. Thus, ammonia condenses out below the visible cloud decks in Uranus and Neptune, and the dominant spectral signatures from those planets are of methane, the clouds of which are seen in those planets. Indeed, clouds are main features of Figures 12.1 through 12.10!

12.1.2 Jovian Atmospheric Structure

One model for the structure of Jupiter (from top-down) is as follows:

1. Topmost layer: solid NH$_3$ particles; $P \approx 1$ bar ($\sim 10^5$ Pa); $T \approx 150$ K

2. More massive cloud: solid NH$_4$SH (ammonium hydrosulfide) particles; $P \approx 2$–5 bars; $T \approx 210$ K

3. Dense layer of H$_2$O ice crystals

4. Thin layer of dilute aqueous ammonia solution; $P \approx 7$ bars; $T \approx 280$ K

5. Gaseous hydrogen (H$_2$ + He)

6. Liquid hydrogen (layers 5 + 6: to 0.78$\Re_J$)

7. Metallic hydrogen, to 0.2$\Re_J$

8. Solid core of ice, rock, Fe–S

Some of this structure was studied by the Galileo Probe (see Figure 12.5, Plate 3).

A classic work on the interiors of the giant planets is Wildt (1961). Since then both observations and theoretical models have progressed, as has knowledge of the Earth's interior and the properties of the phases of hydrogen, helium, and of ices generally; but for its elegance, scientific approach, and clarity this work is still worth reading. Guillot et al. (2004) provide a good post-Galileo summary.

The fractional abundance of He relative to everything else, including H, in Jupiter's atmosphere, 0.234 ± 0.005 by mass, is only slightly less than that in the Sun, 0.275 (Lunine et al. 2004). At the pressures that are thought to exist within Jupiter (> 3 Mbar), hydrogen should become metallic. Helium has been thought to be immiscible in this material; if it is, helium drops can "rain" down toward the center of the planet but this has been challenged,

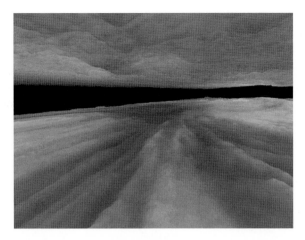

Fig. 12.5. NASA/JPL/CalTech 3D visualization PIA01192 of the structure of the atmosphere of Jupiter near the equator, as revealed by the Galileo Probe. The empty layer is clear dry air; above it is haze and below are thick clouds. The *blue area* marks a region of dry, descending air, while the *whitish region to the right* is ascending, moist air. The *blue–green–red colors* indicate enhanced reflection/emission in the 889, 727, and 756 nm passbands, respectively (see Plate 3)

both as a source for the excess heat flux and on the basis that the diffusion coefficient is too low to permit this. For the other gas giants:

- On Saturn, the temperatures are lower, and He should separate more fully than on Jupiter; also, the smaller size of Saturn means that the separation would occur sooner than on Jupiter. Both arguments suggest that the helium abundance in Saturn's atmosphere should be lower than in Jupiter's. However, the current Cassini result of 0.18–0.25 by mass (Conrath and Gautier 2000) is not yet sufficiently precise to settle this issue. The model of Anderson and Schubert (2007) predicts a transition from H_2 to metallic H at about 0.5R for Saturn.

- The internal pressures on Uranus and Neptune are too low for the metallic hydrogen transition, so there is no separation and the He/H ratio is, as a consequence, close to the solar ratio (0.262 ± 0.048 for Uranus).

Both lightning (see Figure 12.6) and aurorae have been seen in Jupiter's atmosphere by spacecraft.

Jupiter has an extensive magnetosphere, to be discussed separately below. The other giant planets also have extensive magnetospheres, although the sources of the magnetic fields in the lesser giants is not clear. The presumption is that in Jupiter it arises in the metallic hydrogen region.

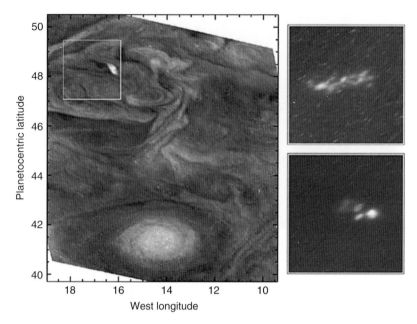

Fig. 12.6. Lightning near the Great Red Spot on Jupiter. The image on the *left* was taken during the Jovian day. The *box* indicates a bright ammonia cloud in an updraft region and to the *immediate left*, a clear downdraft region. The *insets* indicate extensive lightning in nightside images of the boxed area; they were taken 110 min after the dayside view and were taken with exposures of 167 sec for the upper and 39 sec with higher gain for the lower. The nightside images were taken only 218 sec apart! The lightning strikes may appear fuzzy because they appear at the level of water vapor, ~75 km below the cloud decks, and ammonia clouds overlay them, acting as a diffusion screen. Credits: NASA/JPL/California Institute of Technology

12.2 Saturn

Saturn, the farthest and slowest moving planet known in antiquity, was the Roman god of agriculture and the father of the other gods. In Greece, the planet was called *Kronos,* noted for its slow motion among the stars. The Greek word for time, *chronos,* is found in such English words as *chronometer* and is, at least sometimes, identified with it (van derWaerden 1974, pp. 188–197). The planet's sidereal period of $29\overset{\text{y}}{.}46$ implies a mean motion of only $0.033°/\text{d}$ or about a degree a month.

Known for its beautiful ring system, visible even in small telescopes, Saturn presents a yellow appearance in the sky. Its colors indicate Saturn to be the reddest of the giant planets with (B-V) = 1.04 and (U-B) = 0.58.

Saturn is the least dense of all the planets and the most flattened with an oblateness of nearly 10%.

There are three regimes or "systems" of rotation assigned to the planet, although the rotational velocities vary in detail from zone to zone and belt to belt. Prior to the Cassini mission, the velocity regions were

$$P_{rot}^{I} = 10^{h}\,14^{m} \text{ (equator)}$$

$$P_{rot}^{II} = 10^{h}\,38^{m} \text{ (temperate zones)}$$

$$P_{rot}^{III} = 10^{h}\,39^{m}\,22^{s} \text{ (System III, from radio data)}$$

However, Cassini's measures of the radio period have yielded a range from 10^{h} $47^{m}\,06^{s}$ in 2004 to $10^{h}\,32^{m}\,35^{s}$ in 2007, implying that the P_{rot}^{III} as observed is not that of a solid body but of a magnetic field structure that may be modulated by Saturn's plasma disk (Gurnett et al. 2007). A determination by Anderson and Schubert (2007) from occultation, Doppler, and gravity data gave $10^{h}\,32^{m}\,35^{s} \pm 13^{s}$ for Saturn's rotation period and was consistent with a period they found from Voyager 1 and 2 data: $10^{h}\,32^{m}\,55^{s} \pm 30^{s}$.

Saturn has a series of zones and belts but they are 100 km deeper in the atmosphere than in Jupiter, and a high-altitude haze masks the colorful materials below. See Figures 12.7–12.9. Consequently the features on Saturn suffer from poor contrast.

Fig. 12.7. An image of Saturn at 728 nm wavelength, as seen by the Cassini Orbiter wide-angle camera on July 31, 2005, from a distance of 1.3 million km. The South Pole is seen at the bottom. Note the presence of white ovals and turbulent features, similar to those on Jupiter. NASA/JPL/Space Science Institute image PIA07585

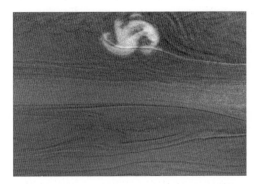

Fig. 12.8. An intense storm visible even through the extensive upper haze layer to viewers on Earth. This image was captured on the nightside of Saturn by the Cassini Orbiter. The atmosphere is illuminated by sunlight reflecting off the rings. NASA/JPL /CICLOPS/ Space Science Institute image PIA07789

Fig. 12.9. Saturnian aurorae seen simultaneously at both poles, captured in UV light with the STIS instrument onboard the Hubble Space Telescope. Note the auroral oval, similar to that seen on Earth. Note also the similarity in alignment of both magnetic and rotational poles in Saturn. During the impacts of the components of the Comet Shoemaker–Levy 9, aurorae were triggered at both magnetic poles of Jupiter. Credits: HST/ESA/NASA/JPL/J.T. Trauger. Courtesy, John Trauger.

The velocities of regions within about 15° of the equator are extremely high, ~400 km/s. Curiously, the speeds of corresponding latitudes in the N and S hemispheres do not match. In particular, winds near 70°N are ~100 km/s, but the same latitude in the south has slight negative (retrograde) speeds.

Saturn has been visited by Pioneer 11, Voyager 1, and Voyager 2. The spacecraft *Cassini*, comparable to Galileo in many respects, arrived at the Saturnian system in 2004. The images of the planet itself have been overshadowed by the magnificent images of the rings and moons of the planets (see Chapter 13), but are spectacular, nonetheless, as evidenced by Figures 12.7 and 12.8.

Saturn has visible aurorae (see Figure 12.9, Plate 4), indicating the presence of a magnetosphere and thus a strong magnetic field, only ~20× weaker than Jupiter's (but still of order 10^3 times that of Earth). The upper atmosphere also exhibits weak, collision-excited UV emission from H and H_2, termed *electroglow*. Proposed sources of the colliding particles include precipitated electrons from the magnetosphere and locally-produced ions and low-energy electrons in the ionosphere.

Spectra of Saturn show a helium depletion compared to Jupiter, a circumstance that may be coupled to Saturn's radiative emission, which exceeds that received from the Sun by more than a factor of 2. Further discussion of Saturn's properties can be found in Gehrels and Matthews (1984), but the Cassini mission has already resulted in much additional detail about Saturn's properties.

12.3 Uranus

Uranus was discovered by William Herschel (1738–1822) in 1781 with a $6\frac{1}{2}$ in. telescope during a sky survey. Uranus appeared as a round, nebulous disc with a motion of $1''.75$ per h, and Herschel assumed he had discovered a new comet. Later, he realized that it had a circular and therefore non-cometary orbit, and its 84^y period indicated a mean distance of 19 au. He called it *Georgium Sidus*, "George's Star" after his royal patron, George III. Lalonde called it "Herschel," but Bode's suggestion, Uranus, after the sky god of the ancient Greeks, became generally accepted.

Bode subsequently discovered that Uranus had been charted by Flamsteed, the first Astronomer Royal, on December 23, 1690, and labeled 34 Tauri; he also found that it had been cataloged as number 964 by Tobias Mayer (1723–1762) of Göttingen in 1756. The earlier observations were sufficient to compute an orbit. In addition, Pierre Charles Lemonnier (1715–1799) found in 1788 that he himself had made observations of Uranus in 1764 and 1769.

Later it was discovered that Flamsteed had observed it on other occasions also, and so had James Bradley (on December 3, 1753). More than a dozen other pre-discovery observations were detected over the next 30 years.

See Grosser (1962) or Turner (1963) for accounts of the discovery of Uranus and its importance for the subsequent discovery of Neptune.

Uranus's magnitude at opposition is $V = 5.52$, and thus detectable, in principle, to the naked eye. There is, however, no evidence of pre-telescopic observations of it. It has a maximum angular size of $3''.9$, so little detail has been discerned on its disk as observed with ground-based telescopes.

Uranus's density is less than that of the Sun, and it is clearly a lesser giant, with a radius only $\sim 4\Re_{\oplus}$ and a mass of only $14.5 M_{\oplus}$.

Telescopically, Uranus is somewhat green in appearance. Methane was identified in its atmosphere spectroscopically by Wildt (1931, 1932). Uranus is a very dark planet with little contrast among its features, and the contrast in the images returned by Voyager 2 (which arrived at Uranus in 1986) needed to be stretched to show banded structure near the pole. E–W motion was seen in a sparse number of spots and streaks; some of these can be seen in Figure 12.10 (Plate 5). Observations of selected features in the atmosphere showed that the rotation rate varied with latitude, as on Jupiter and Saturn.

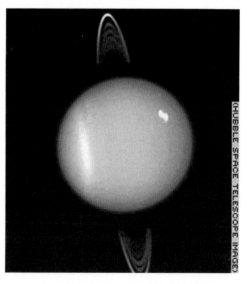

Fig. 12.10. Uranus as seen by Voyager 2, showing backscattered rings, polar clouds, and a relatively high cloud, probably of methane, the red absorption of which gives rise to the greenish blue hue of the planet's atmosphere. Credits: NASA/JPL

Uranus' rotation axis is oriented 97°.86 to its orbital axis so that the planet's north pole[1], when viewed "pole-on," is rotating CW, or retrograde to its orbital motion. Its rotation period is given as $-0.71833\,d = -17^h\,14^m.40$.

The fact that Uranus' rotation axis lies almost in its orbital plane creates bizarre "seasons" for Uranus. Over an 84-year interval, the Sun will illuminate successively lower latitudes beginning at one of the rotation poles until the other pole is facing the Sun, and then increasing latitudes again until the original orientation is reached again. Nevertheless, when the south pole was pointed almost directly towards the Sun, Voyager 2 found remarkably uniform temperatures throughout the atmosphere even at the winter side. This justifies to some extent the isothermal or rapid rotator assumption for the computation of the equilibrium temperature.

Uranus' atmosphere has abundant hydrogen and the helium abundance is essentially solar, $Y = 0.262 \pm 0.048$ by mass. Superheated water is expected to be present in its extensive gaseous atmosphere, perhaps in an ocean 10,000 km deep.

The planet has a magnetic field, and some evidence for lightning in the form of whistlers was detected by radio receivers on Voyager. The Voyager data also displayed a very strong UV emission (called the *electroglow*; see Section 12.2) from a region 1500 km above the cloud tops on the sunlit side. Similar, though fainter, electroglows have been seen on Jupiter, Saturn, and Titan.

12.4 Neptune

Almost immediately, it became clear that something was wrong with the theory of Uranus' motion. By 1829, the theory of mutual perturbations (developed by Simon de Laplace) was applied to the orbit of Uranus but its positions could not be reproduced to better than $\sim 1/2'$. Moreover it was clear that the residuals were not due to observational error. Five hypotheses were put forward in explanation:

1. The (hypothetical) cosmic fluid of Descartes, causing a drag force; but there was no evidence for such a fluid, or for a mechanism by which it could produce the perturbations in Uranus but in nothing else.

2. Perturbations due to an unobserved massive satellite; but the time scale of such perturbations would be expected to be shorter than those seen in Uranus' motion.

3. A comet had struck Uranus close to the time of its discovery, changing its orbit; but perturbations continued; moreover, two elliptical orbits were still insufficient to recover the motion (that cometary masses were insufficient to cause such effects was not known at the time).

---

[1] i.e., the pole on the north side of the planet's orbital plane, as per the terrestrial analog.

4. The law of gravity was either not effective or changed at the (great) distance of Uranus; but all previous challenges to the Newtonian law had been refuted. The hypothesis in this case was impossible to refute except empirically.

5. Finally, the existence of a trans-Uranian planet.

The Royal Academy of Sciences in Göttingen offered a prize for the resolution of the problem, with a deadline of November 1, 1846. The problem was independently taken up by John Couch Adams (1819–1892) and Urbain J. Leverrier (1811–1877). Adams communicated his prediction of a trans-Uranian planet to the Astronomer Royal in October, 1845, and Leverrier communicated his to the French Academy of Sciences on November 10. James Challis, Director of the Cambridge Observatory, searched based on Adams' prediction starting in July, 1846, but was unsuccessful. Galle and d'Arrest compared Leverrier's predicted position to recent charts and found Neptune in September, 1846. Both sets of calculations assumed the Titius–Bode law to find the mean distance of the planet, and the predictions, as Airy, the Astronomer Royal conceded, were similar. Nevertheless Neptune was found within a degree of the prediction by Leverrier, and with about the predicted angular size.

According to Leverrier, the Bureau des Longitudes in Paris proposed the name *Neptune* (it now appears that they did no such thing; it was Leverrier's suggestion). A suggestion by Galle that it be called Janus was rejected because Leverrier said in a letter to Galle that "the name Janus, would imply that this is the last planet of the solar system, which we have no reason at all to believe" (cited in Grosser 1962, p. 123). See Grosser (1962) and Turner (1963) for more historical details.

Subsequently, it was found that it had been observed in 1795 by Lalonde, and even by Galileo in 1612–1613. Its mean motion is only 0.006° per d for a period of 165^y.

Neptune's size ($3.9\Re_\oplus$) and mass ($17M_\oplus$) make it a lesser giant and near twin of Uranus. Its density is the highest of the giant planets, $1760\,\mathrm{kg/m^3}$, but it must still be composed mainly of H and He.

Neptune's maximum angular diameter is $2''.3$, so little detail is discernible from Earth. What we know of the planet's cloud decks was obtained from the sole space probe flyby, that of Voyager 2, which arrived at Neptune in 1989.

Neptune, like Uranus, has abundant methane in its atmosphere. Its most characteristic feature in Voyager 2 imagery is the Great Blue (or Dark) Spot, seen at disk center in Figure 12.11 (Plate 6), comparable in shape and scale to the GRS on Jupiter. It was located at 20°S and rotated CCW, indicating that it was a similar meteorological phenomenon. It had an associated high-altitude haze. In Hubble imagery five years later, however, it had vanished, and a new dark spot appeared in the northern hemisphere within the next year.

The Voyager imagery also shows additional dark spots at higher latitudes (55°S) and a highly variable, irregular feature, the "scooter" is seen at 42°S. Neptune also has higher altitude, white methane "cirrus" clouds amidst a more general methane haze. They are ~100 km or more long and cast shadows down on the cloud decks below a 50 km clear region.

Wind velocities in Neptune's atmosphere vary from about 450 m/s retrograde at the equator to about 300 m/s prograde at 70 S latitude, giving Neptune the greatest range of planetary wind velocities in the solar system. Consequently, rotation periods derived from atmospheric features vary considerably with latitude. Infrared photometry from ground-based observation has revealed a rotational period of ~17–18^h at 30° to 40° S latitude.

Neptune, like Jupiter and Saturn, appears to have a higher temperature than thermal equilibrium predicts. It has been suggested that here the explanation may be a greenhouse effect, caused by absorption of methane, which traps infrared radiation below the CH_4 clouds. However, most sources suggest that an internal source of heat is responsible. See Section 12.6 and, for an alternative explanation, Lunine (1993).

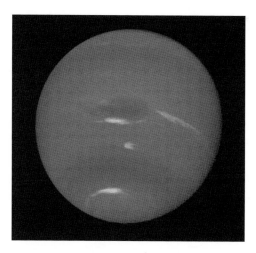

Fig. 12.11. A Voyager 2 image of Neptune showing the main discernible features of its atmosphere: The Great Dark Spot (GDS) at disk center, bright streaks of cirrus, and Spot D2, below the GDS. As on the greater giants, belts and zones are present, if subdued due to overlaying haze. NASA/JPL image PIA02245. See Plate 6

Neptune also has a magnetic field and a magnetosphere. The rotation rate of radio emissions, determined by Voyager 2, is $16^h.11$; it probably represents the rotation of a deep atmospheric layer—an electrolytic sea below the Neptunian cloud decks.

12.5 Internal Pressures

The steady-state pressure is related to the radius of a planet through the equation of hydrostatic equilibrium (10.12):

$$dP/dr = -g\rho$$

and (through an equation of state) to (10.21):

$$dP/P = -(\mu m_u g/kT)dr$$

In order to find the exact pressure at the center of a planet, the density is needed as a function of depth. Unfortunately, unlike the Earth, where the Adams–Williamson equation can be used to integrate over all radii from center to the surface (Milone & Wilson (2007), Chapter 5.4.2), we do not have the acoustical wave speeds at various points in the interiors of the other planets, at present. Therefore the models are less well constrained than for the Earth.

One may, however, make use of the mean density to estimate the central pressure. The simplest form that can be used is

$$P_0 - P_R = - < g\rho > [0 - R] \qquad (12.2)$$

where the subscripts indicate the values of r. Obtaining $< g\rho >$ is perhaps easier said than done. One could take $(g/2)R < \rho >$ as an approximation,

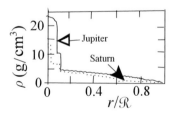

Fig. 12.12. Density profiles from models of Jupiter and Saturn, after Marley (1999)

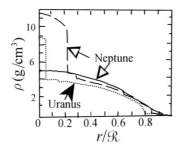

Fig. 12.13. Density profiles from two of several models of Neptune and a model for Uranus, after Marley (1999)

but this leads to a computed central pressure of 1.7 Mbar for the Earth, and 10.9 Mbar for Jupiter, compared to 3.5 and 40 Mbar, respectively, from the models.

Another approximation is to substitute for g and then M in (10.17), to give

$$\mathrm{d}P/\mathrm{d}r = -GMr^{-2}\rho = -(4\pi G/3)r^{3-2}\rho^2 = -(4\pi G/3)r\rho^2 \qquad (12.3)$$

with, perhaps, the approximations, $<\rho^2> = <\rho>^2$, and $P_R = 0$, giving

$$P_0 = -\tfrac{1}{2}(4\pi G/3)<\rho>^2 [0 - R^2] = (2\pi G/3)<\rho>^2 R^2 \qquad (12.4)$$

Calculation of the central pressure, P_0, with this equation gives about the same result for the Earth, but a slight improvement for Jupiter (12.6 Mbar). Models of the interiors of the giant planets have been produced by, among others, Zharkov and Trubitsyn (1978) and Stevenson (1989). The latter computes central pressures of the order of 4×10^{12} Pa ($\sim 4 \times 10^7$ bars or 40 Mbar) for Jupiter, $\sim 10^{12}$ for Saturn, and less than 10^{12} for Uranus.

Recent models for the march of density with radius in the interiors of the giant planets are shown in Figures 12.12 and 12.13, taken from Marley (1999).

12.6 Excess Radiation

Table 12.1 indicates the energy received, the albedos, and the equilibrium and effective temperatures. The effective temperature for Uranus is that for a rapidly rotating planet. During the Voyager flyby, one pole faced the Sun. One would expect that the emission would come from only the one polar region, and the effective temperature should be larger by the factor $2^{1/4} = 1.19$, or $T_s = 69.1$K. In fact, Uranus was seen to be nearly isothermal, and so to have a very efficient circulation system. Thus there is no strong evidence for radiation excess for Uranus, to one sigma or less.

As noted, one of the mysteries about some of the gas giants is an energy excess: three of the four giant planets appear to radiate more energy than they receive from the Sun, as seen in Table 12.1 (but Lunine, 1993, has noted that the situation for Uranus and Neptune is more complex than the mere comparison of IR flux temperatures with expected equilibrium temperatures, indicating that there may be no excess from Neptune, either). In any case, there is no dispute about the excess radiation from Jupiter and Saturn. Of course, the source of the excess may not be the same in each planet in which it is seen. It also needs to be mentioned that the thermal emission in which the excess is seen is independent of the non-thermal radiation, which has its origin in the magnetospheres. The origin of the latter is relatively well understood.

Of the possible sources of a gas giant's internal heat flux, primordial heat is a possibility, but contraction (conversion of the potential energy of the planet's material into kinetic energy as it falls toward the center) is not favored at present because the giants have liquid mantles and liquids must resist compression.

Another suggested possibility involves selective "fall out" of some material—such as a "rain" of helium in the interior (see the review of Hubbard et al. 2002). This mechanism is thought to be important in Saturn, where the outer atmosphere may be more deficient in helium than is Jupiter, as noted in Section 12.1. Because Uranus does not have an observed deficiency of helium, this may explain its lack of excess infrared emission. Questions have arisen about the effectiveness of the diffusion of He in this mixture (one estimate giving a diffusion time scale that greatly exceeds the age of the solar system), and some calculations show that even if all of the He that appears to be missing in the upper atmosphere of Jupiter had settled onto the core, the energy released would still not account for the observed energy excess. We avoid questions about the efficiency of the settling process, both because of these uncertainties and because there are disagreements between the models of the interior and of the sound profiles deduced from measurements of Jupiter's global oscillations (one of the ways to reconcile the data is to have the H and He immiscible; notwithstanding work on equations of state and simulations of such mixtures [Levashov et al. (2006) and references therein]. See Guillot et al. (2004) for further discussion and recent models.

An alternate theory for the excess in Jupiter has been advanced by Ouyed et al. (1998): that high enough temperatures in the early core accretion stages of giant planet formation in the early solar system allowed deuteron–deuteron nuclear reactions to take place; for this to happen efficiently, the deuterons needed to have been stratified in a deep layer within the proto-gas giants. Ouyed et al. (1998) indicate that this could have been achieved in other giant planets (of Jupiter mass or greater) also, but the theory does not seem to have found wide acceptance. At the time of writing, the matter is unsettled.

Whatever the source of radiated energy from their interiors, the origins of the giant planets are also strongly debated. The leading theories, core accretion, and gravitational instability, both have impediments to full acceptance. The problems have to do mainly with the time scales for disk dissipation and formation and the mass of the protostellar disk in which they must have been formed. Because we have yet to discuss the important insights gained from the small bodies of the outer solar system and because there are now many planetary systems known, we will take up the discussion of formation in the last two chapters.

12.7 Ionospheres of the Giant Planets

In Chapter 11 on planetary ionospheres we discussed the cause, structure, and composition of the ionospheres of the inner planets (see especially Section 11.4). Recall that the index of refraction in the ionosphere, n, decreases for greater ion density, N, because, from (11.42), (11.43), and (11.44),

$$n^2 = 1 - Ne^2/[(2\pi f)^2 m\varepsilon_0] = 1 - (f_0/f)^2 \tag{12.5}$$

where f is the frequency of the radio wave and f_0 is the plasma frequency. Thus, as a radio "ray" moves higher into the ionosphere from a point below, so that N increases, the ray is bent *away* from the normal, as in Figure 11.6. From (11.50),

$$n^2 = 1 - \cos^2 \alpha_0 \tag{12.6}$$

so

$$f_0{}^2 = f^2 \cos^2 \alpha_0 \tag{12.7}$$

If f is the lowest frequency that can pass through the ionosphere for a given incident angle, α_0, we can invert (12.7) to define the quantity

$$f_c{}^2 = f_0{}^2 \sec^2 \alpha_0 \tag{12.8}$$

where f_c is called the *critical frequency* for the particular angle of incidence, α_0. This is equivalent to (11.51). The critical frequency is thus a cut-off frequency for rays entering with angle α_0. Because the secant function can be no smaller than 1, the critical frequency must be greater than the plasma frequency. The result is a cone of visibility determined by the ionosphere.

Note that the detection of Jovian decametric radio noise depended on the fact that Jupiter's radiation was able to penetrate the Earth's ionosphere!

In the presence of magnetic fields, the situation is even more complicated, leading to birefringence, causing one of the planes of polarization of the wave to be removed. The index of refraction in this case becomes

$$n^2 = 1 - f_0^2/[f(f \pm f_H)] \qquad (12.9)$$

where

$$f_H = eB/(2\pi m) \qquad (12.10)$$

the *cyclotron frequency*. The "+" case represents the *ordinary ray*, the "−" case, the *extraordinary ray*.

The presence of strong magnetic fields led the discoverers of Jovian decametric radiation (Burke and Franklin 1955) to conclude that the extraordinary ray was internally reflected.

Later it was realized that the motion of the inner Galilean satellite, Io, can trigger Jovian decametric radiation through a dumping of high-energy particles into the Jovian ionosphere; this is described in Section 12.8.4.

12.8 The Jovian Magnetosphere

The most extraordinary of all the bizarre properties of the giant planets may be their extensive magnetospheres. The extent of a giant planets' magnetospheric volume is enormous, dwarfing the Sun in scale. Although the full extent and complexity of Jupiter's magnetosphere were realized only with the Voyager approaches, the main phenomena, especially the decametric radio emissions, have been known from ground-based radio astronomy since the 1950s. Consequently, the spacecraft were equipped with extensive particle detectors.

Most of the Jovian magnetosphere corotates with Jupiter, in contrast to that of the Earth, with which only the innermost magnetosphere corotates.

Jupiter's magnetosphere is divided into three main regions according to the field geometry and the processes operating.

The sizes of all three regions vary considerably according to solar wind conditions.

12.8.1 Inner Magnetosphere of Jupiter

Its characteristics (illustrated in Figure 12.14) are

- Quasi-dipolar field produced by Jupiter's internal dynamo:

    – It is tilted $\sim 10°$ to Jupiter's rotation axis.

    – It has opposite polarity to that of the Earth.

        $\Rightarrow$ The N rotational pole is near the N magnetic pole.

    – It has stronger quadrupole and octupole moments than Earth.

    This implies that the dynamo is closer to the surface than is Earth's; this, in turn, suggests that it is produced (most likely) in the liquid metallic hydrogen mantle (Figure 12.15).

- This region gives rise to *Decimetric* radiation (*DIM*), illustrated in Figure 12.16, and as observed in Figure 12.17 (color Plate 7). It is characterized by

    – Radio wavelength radiation at $\lambda \approx 0.1m$

    – Synchrotron radiation from trapped relativistic electrons with energies ~ 10–$40\,\mathrm{MeV}$ mirroring back and forth near the magnetic equator

    – A source region which is located at ~ 1.6–$3R_\mathrm{J}$ and with thickness $\sim 1R_\mathrm{J}$

12.8.2 Middle Magnetosphere of Jupiter

The field lines in this region are oriented approximately radially outward from Jupiter and roughly parallel to the equatorial plane.

Parallel field lines require a sheet of current: this implies the existence of a disk-like current sheet around Jupiter, the *magnetodisk*, illustrated in Figure 12.18. The precise geometry is uncertain: close to Jupiter it appears

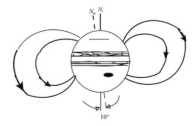

Fig. 12.14. The inner magnetosphere of Jupiter

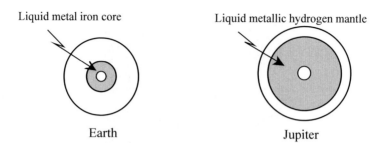

Liquid metal iron core   Liquid metallic hydrogen mantle

Earth   Jupiter

Fig. 12.15. The sources of global magnetic fields on Earth and Jupiter; figures not to scale

to be parallel to the *magnetic* equator; further out, it appears to be parallel to Jupiter's *rotational* equator, so it wobbles as Jupiter rotates, as seen in Figure 12.19.

12.8.3 Outer Magnetosphere of Jupiter

This is the region between the magnetodisk and the magnetopause. It includes a *magnetotail* a few au long, with cross-sectional radius $\sim$150–300R_J, the upper and lower parts of which are separated by a current sheet that is a continuation of the magnetodisk, and is $\sim$5R_J thick. The configuration is sketched in Figure 12.20.

The outer magnetosphere is highly variable, with inner edge $\sim$30–50R_J, outer edge $\sim$45–100R_J (dayside). It is inflated by plasma pressure of energetic ions that have energies of 20–40 keV, similar to energies of particles in the Earth's ring current. The dominant ions are H, He, S, O. The number densities of the sulfur and oxygen ions are approximately that of hydrogen, $n_{S,O} \approx n_H$.

The plasma pressure inflates the outer magnetosphere to about twice the size expected from the planetary dipole alone. This implies that the

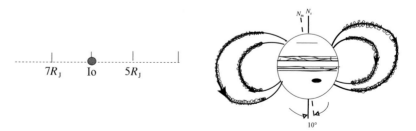

$7R_J$ Io $5R_J$

Fig. 12.16. The DIM source region, the radiation belts, and the locations of the orbit of the Galilean satellite Io

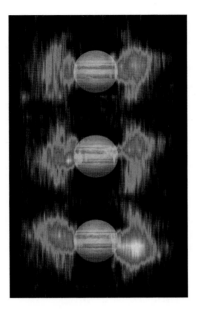

Fig. 12.17. The inner magnetosphere of Jupiter is revealed by the radiation belts at three moments in the 10 h rotation period of Jupiter. The particle density is coded so that *light colors* indicate higher density. They were observed by Cassini during its flyby of Jupiter en route to Saturn. NASA/JPL image PIA03478. See Plate 7

magnetosphere is soft, like a partially filled balloon, and that it is easily compressed by variations in the solar wind. In fact, the distances of the bow shock and the magnetopause from Jupiter can vary by about a factor of 2.

12.8.4 Interaction with Io

It is appropriate to close this chapter with a discussion of Io, because satellites and rings are considered next, in Chapter 13. Io has a major influence on Jupiter's magnetosphere, as illustrated in Figure 12.21.

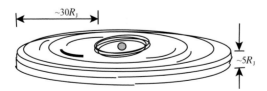

Fig. 12.18. The Jovian magnetodisk

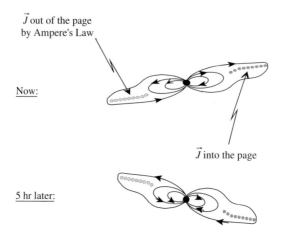

$\vec{J}$ out of the page
by Ampere's Law

Now:

$\vec{J}$ into the page

5 hr later:

Fig. 12.19. The wobbling of the Jovian magnetodisk

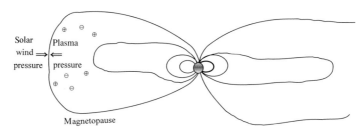

Solar
wind
pressure

Plasma
pressure

Magnetopause

Fig. 12.20. The outer magnetosphere of Jupiter. This is the largest structure in the solar system, dwarfing the Sun in scale

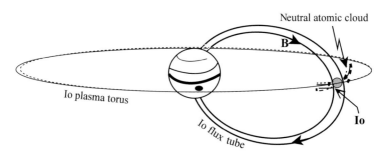

Neutral atomic cloud

B

Io plasma torus

Io flux tube

Io

Fig. 12.21. The interaction between Io and the magnetosphere of Jupiter. Note the flux tube along which ions stream between the Jovian atmosphere and Io

12.8.4.1 DAM This is the source region for *Decametric* radiation (*DAM*), radio radiation of wavelength $\lambda \approx 10\,\text{m}$.

It is highly variable, with time scales of seconds to tens of seconds and structure down to milliseconds, and is produced by electrons in the tens of keV range. Other characteristics are

- High intensity:

  Collective phenomena (not just individual electrons executing gyromotion)
  Strongest near N and S magnetic poles
  $\sim 10^{10}\text{W}$ emitted power

- Some DAM is strongly influenced by the longitude of Io

  The source seems to be in the Io flux tube where it converges toward
  Jupiter's magnetic poles

- Other DAM is tied to Jupiter itself and is fixed with respect to the central
  meridian longitude (CML) of Jupiter

12.8.4.2 Neutral and Ionized Population of the Jovian Magnetosphere The species observed in the magnetosphere are

H; H_2 and H_3 molecules; He; C; O; Na; S; K

The species concentrated in the Io plasma torus and neutral clouds:

$$
\begin{array}{ll}
\text{sodium: Na (neutral);} & \text{sulfur:} \quad S^+, S^{2+}, S^{3+}; \\
\text{potassium: K (neutral);} & \text{oxygen:} \quad O^+, O^{2+}, O^{3+}.
\end{array}
$$

Abundance ratios and sources

$$\frac{C}{He} \sim \text{solar}$$

$\therefore$ The source of C and He is the solar wind;

$$\frac{H}{He} > \text{solar}$$

$\therefore$ The major source of the hydrogen is the Jovian ionosphere;
   (Jupiter has approximate solar abundances of H and He, but H has a
   larger scale height than He; consequently, there is a preferential escape of
   H relative to He from the ionosphere into the magnetosphere).
   Jupiter's ionosphere is also the most likely source for the H_2 and H_3.

$$\frac{S}{He} \gg \text{solar, and increases toward Jupiter and Io}$$

$\therefore$ the source of S is Io.

A. $\dfrac{O}{He} \gg$ solar, in the region close to Jupiter and Io

∴ the source of O is Io.

B. $\dfrac{O}{He}$ is nearer to solar in the outer magnetosphere

∴ the solar wind may also be an important source.

Io is the source for at least half the number of particles and 98% of the mass in the Jovian magnetosphere.

12.8.5 Io as a Source of Particles

12.8.5.1 Rate of Supply Sodium D-line emission (589.0, 589.6 nm) is seen: it is bright near Io (neutral sodium cloud), and faint emission is seen throughout the magnetosphere.

This requires $\sim 10^{27}$ sodium atoms per second from Io.

$$\text{The ratio } \frac{Na}{O,S} \approx 10\% \text{ in the middle magnetosphere;}$$

$$\approx 5\% \text{ in the outer magnetosphere.}$$

If we assume the same ratio for ions leaving Io, then Io loses

$$\sim 10 \times 10^{27} = \sim 10^{28} \text{ atoms (total) per second}$$

O, Na, S: average mass $\approx 24\,u$

$$10^{28}\frac{\text{atoms}}{s} \times 24\frac{u}{\text{atom}} \times 1.66 \times 10^{-27}\frac{kg}{u} = 400\frac{kg}{s} \text{ from Io}$$

12.8.5.2 Loss Mechanisms from Io Io has a surface covered mostly with S and SO_2 frost, and a tenuous SO_2 atmosphere.

The escape velocity from Io is $V_{esc} = 2.6\,km/s$. This enables us to eliminate three loss mechanisms right away:

A. *Jeans escape*

Jeans escape is the slow, thermal escape of atoms and molecules from an atmosphere (Section 10.1). The surface temperature of Io is 90 K – 100 K over most of the nightside and about 120 K over most of the dayside, so over most of Io's surface $\bar{v} < 0.44$ km/s $\sim 1/6\,V_{esc}$ for O (less for Na, S, SO_2). The temperatures of some silicate lava lakes can approach 2000 K, at which temperature $\bar{v} \sim 1.8$ km/s for O, but these regions are very localized. Thus the rate of Jeans escape is expected to be far too small.

B. Hydrodynamic outflow

A high temperature produces rapid Jeans escape of the lighter atmospheric components, and also a large scale height that ensures rapid replenishment of the lost material. The hydrodynamic outflow that results drags heavier atmospheric components with it that would not be lost through classical Jeans escape.

Io's pressure scale height is ∼10 km. This is far too small to produce a hydrodynamic outflow.

C. Volcanic ejection

Observed plume velocities ≈ 1 km/s ($≈ 1/3V_{esc}$)
Therefore the loss cannot be due to direct ejection by volcanic eruption.

The dominant process appears to be

D. Sputtering from Io's surface

High-energy ions from Jupiter's magnetosphere strike the surface of Io and eject atoms (Figure 12.22).

This requires an atmosphere which is tenuous enough that the ions can get in and the atoms can get out relatively unimpeded: $P < 10^{-11}$ bar, $n < 10^{15}$ m^{-3}.

Above this pressure the gas is dense enough to be controlled by molecular collisions, and thus forms a true atmosphere. Below 10^{-11} bar, the exosphere reaches right down to the surface, allowing magnetospheric plasma to eject ions.

Io's atmosphere is very non-uniform in density and pressure. The dominant gas ejected in volcanic plumes is SO_2, so the dominant atmospheric gas should be SO_2. The atmospheric pressure of SO_2 close to or under a plume can be up to about 10^{-7} bar. Away from volcanic plumes, sublimation of SO_2 frost on the dayside can produce pressures perhaps as high as 10^{-8} bar at noon, decreasing to 10^{-11} bar or less at the terminator.

At night, the surface temperature of Io plummets to 90 K, or even as low as 61 K over thick deposits of SO_2 ice. The ice is colder than other

Fig. 12.22. Sputtering as a mechanism for ejecting atoms from Io's surface

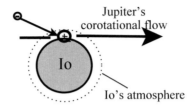

Fig. 12.23. Entrainment of Io's atmospheric atoms by Jupiter's corotational flow

surface regions during the daytime because of its higher reflectivity, and hence it reaches a colder temperature at night also. At these temperatures, essentially all SO_2 is frozen on the surface, and the atmospheric pressure drops to around 10^{-16} bar.

Thus at night the surface appears to be unprotected from the Jovian magnetosphere, and magnetospheric ions can sputter atoms and molecules from Io's surface far into space. They become entrained in the corotational flow and get caught up in the torus (see Figure 12.23).

E. Other possibilities

Other gases may be present, although none have been detected by Voyager or from Earth.

Small amounts of frozen H_2S have been detected in the surface deposits and appear to vary with time. H_2S has a higher vapor pressure than SO_2, so even though it is less abundant it could produce a significant atmosphere at the terminator. The observed upper limit to the pressure of H_2S gas is about 10^{-10} bar.

O_2 could be produced photochemically from SO_2 and could provide an atmosphere of up to 10^{-10} bar even on the nightside.

We now move on to a full discussion of the moons and rings of the solar system.

References

Anderson, J. D. and Schubert, G. 2007. "Saturn's Gravitational Field, Internal Rotation, and Interior Structure," *Science*, **317**, 1384–1387.

Baum, W. A. and Code, A. D. 1953. "Photometric Observations of the Occultation of σ Arietis by Jupiter," *Astronomical Journal*, **58**, 108–112.

Bronshten, V. A., ed. 1967. *The Planet Jupiter* (Moscow: USSR Acad. of Sciences) (TR. by A. Moscana, Israel Program for Scientific Translations, 1969).

Burke, B. F. and Franklin, K. L. 1955. "Observations of a Variable Radio Source Associated with the Planet Jupiter," *Journal of Geophysical Research*, **60**, 213–217.

Gehrels, T., ed. 1976. *Jupiter* (Tucson, AZ: University of Arizona Press).

Gehrels, T. and Matthews, M. S. 1984. *Saturn* (Tucson, AZ: University of Arizona Press).

Grosser, M. 1962. *The Discovery of Neptune* (Cambridge: Harvard University Press).

Guillot, T., Stevenson, D. J., Hubbard, W. B., and Saumon, D. 2004. "The Interior of Jupiter," in *Jupiter: The Planet, Satellites and Magnetosphere*. eds. Bagenal, F., Dowling, T. E., and McKinnon, W. B. (Cambridge: Cambridge University Press) pp. 35–57.

Gurnett, D. A., Persoon, A. M., Kurth, W. S., Groene, J. B., Averkamp, T. F., Dougherty, M. K., and Southwood, D. J. 2007. "The Variable Rotation Period of the Inner region of Saturn's Plasma Disk," *Science*, **316**, 442–445.

Hide, R. 1961. "Origin of Jupiter's Great Red Spot," *Nature*, **190**, 895–896.

Hubbard, W. B., Burrows, A., and Lunine, J. I. 2002. "Theory of Giant Planets," *Annual Review of Astronomy and Astrophysics*, **40**, 103–136.

Ingersoll, A. P., Dowling, T. E., Gierasch, P. J., Orton, G. S., Read, P. L., Sanchez-Lavega, A., Showman, A. P., Simon-Miller, A. A., and Vasavada, A. R. 2004. "Dynamics of Jupiter's Atmosphere," in *Jupiter: The Planet, Satellites and Magnetosphere*, eds. Bagenal, F., Dowling, T. E., and McKinnon, W. B. (Cambridge: The University Press), pp. 105–128.

Kuiper, G. P. 1952. "Planetary Atmospheres and their Origin," in *The Atmospheres of the Earth and Planets*, ed. Kuiper, G.P. (Chicago: University of Chicago Press), 306–405.

Levashov, P. R., Filinov, V. S., Bonitz, M., and Fortov, V. E. 2006. "Path Integral Monte Carlo Calculations of Helium and Hydrogen-Helium Plasma Thermodynamics and the Deuterium Shock Hugoniot," *Journal of Physics A: Mathematical and General*. **39**, 4447–4452.

Lewis, J. S. 1997. *Physics and Chemistry of the Solar System* (San Diego, CA: Academic Press). Revised edition.

Lunine, J. I. 1993. "The Atmospheres of Uranus and Neptune," *Annual Review of Astronomy and Astrophysics*, **31**, 217–263.

Lunine, J. I., Coradini, A., Gautier, D., Owen, T. C., Wuchterl, G. 2004. "The Origin of Jupiter," in *Jupiter: The Planet, Satellites and Magnetosphere*, eds. Bagenal, F., Dowling, T.E., and McKinnon, W.B. (Cambridge: The University Press), pp. 19–34.

Marley, M. S. 1999. "Interiors of the Giant Planets" in *Encyclopedia of the Solar System*, eds P. R. Weissman, L.-A. McFadden, and T. V. Johnson (San Diego, CA: Academic Press).

Ouyed, R., Fundamenski, W. R., Cripps, G. R., and Sutherland, P. G. 1998. "D–D Fusion in the Interior of Jupiter," *Astrophysical Journal*, **501**, 367–374.

Pearl, J. C., Conrath, B. J. 1991. "The Albedo, Effective Temperature, and Energy Balance of Neptune, as Determined from Voyager Data," *Journal of Geophysical Research*, **96**, Supplement, 18921–18930.

Pearl, J. C., Conrath, B. J., Hanel, R. A., Pirraglia, J. A., and Coustenis, A. 1990. "The Albedo, Effective Temperature, and Energy Balance of Uranus, as Determined from Voyage IRIS Data," *Icarus*, **84**, 12–28.

Peek, B. M. 1958. *The Planet Jupiter: The Observer's Handbook*, (Rev by P. Moore, 1981) (London: Faber).

Stevenson, D. J. 1989. In *The Formation and Evolution of Planetary Systems*, eds H. A. Weaver and L. Danly (Cambridge: University Press), p. 82ff.

Turner, H. H. 1963. *Astronomical Discovery* (Berkeley: University of California Press).

Waerden, B. L. van der 1974. *Science Awakening II. The Birth of Astronomy* (Leyden: Noordhoff).

West, R. A. 1999. "Atmospheres of the Giant Planets" in *Encyclopedia of the Solar System*, eds P. R. Weissman, L.-A. McFadden, and T. V. Johnson. (San Diego, CA: Academic Press).

Wildt, R. 1931. "Ultrarote Absorptionsbande in den Spektren der Grossen Planeten," *Naturwissenschaften*, **19**, 109–110.

Wildt, R. 1932. "Methan in den Atmosphären der Grossen Planeten," *Naturwissenschaften*, **20**, No. 47, 851.

Wildt, R. 1961. "Planetary Interiors" in *Planets and Satellite*, eds G. P. Kuiper and B. M. Middlehurst (Chicago, IL: University of Chicago Press), Ch. 5, pp. 159–212.

Zharkov, V. N. and Trubitsyn, V. P. 1978. *Physics of Planetary Interiors* (Tucson, AZ: Pachart Publishing House) (Tr. W. B. Hubbard).

Challenges

[12.1] Derive an expression for the central pressure and density of a planet, and use it to calculate the central pressure of each of the giant planets.

[12.2] Work out an expression for and compute the central temperature of each of the giant planets.

[12.3] The 21 fragments of Comet Shoemaker–Levy 9 impacted Jupiter in 1994 at speeds of ~60 km/s. Compute the explosive energy in each of the impacts of the ~1 km diameter chunks of cometary material. Assume densities of $250 \, \text{kg/m}^3$ for the comets.

[12.4] (a) Verify the values for the absorbed power and their uncertainties in Table 12.1, using the Bond albedos[2] and the solar luminosity (3.845×10^{26} W $\pm 8.000 \times 10^{23}$ W). The uncertainties may be found from the square root of the sum of the squares of the individual relative uncertainties.

(b) Verify the predicted and observed temperatures and their uncertainties in Table 12.1, using the absorbed and observed powers.

(c) Compute the "energy budget" (power emitted / power absorbed) and its uncertainty for each planet in Table 12.1. Discuss the numerical significance of the individual values, given their uncertainties. The energy budget of Saturn is greater than that of Jupiter. Is this difference significant? What about Neptune?

---

[2] The Bond albedo is formally defined by Lewis (VI, p. 235) for a reflecting astronomical body over a passband (characterized by $\lambda = \lambda_{\text{eff}}$), such that: $A_B = [\int F_{\lambda, \text{ reflected}} \, d\lambda / \int F_{\lambda, \text{ incident}} \, d\lambda]$.

13. Satellite and Ring Systems

13.1 Satellites

Satellites have been discussed for decades, for example, in a number of large books on planetary satellites published at the University of Arizona (such as, Burns 1977), and spectacular images from the Voyager missions appear in cached collections on optical disks. The Galileo and Cassini missions have revealed even more detail and have probed the structure, composition, and environs of the moons even more thoroughly.

The radii of the satellites of the giant planets vary tremendously, with the largest moons (Ganymede and Titan) exceeding Mercury in size. The Galilean moons, Earth's moon, Titan, and Triton are the largest satellites and constitute a class of giant moons. Most of these moons (and even some slighter ones) possess atmospheres; Titan's is so extensive that it is optically thick to visible radiation. Study of its surface has been mainly through radar, but the photometric passbands of the instruments aboard the Cassini Orbiter have revealed some of the surface features and the Huygens probe that it released revealed a surprisingly familiar-looking world as it approached the surface.

It is interesting that each satellite has its own particular properties and its own thermal history. The satellites demonstrate a variety of color, as evidenced by their color indices. This information, along with reflectance spectroscopy, can provide significant details about the surface structure and composition of the satellites.

A number of lesser moons, typically hundreds of kilometers in radius, make up the bulk of the Saturnian and Uranian systems. These moons too are diverse: the very dark moon known as Umbriel in the Uranian system contrasts with the bright, icy moons Ariel, Oberon, and Titania.

The moons of the giant planets show a surprising amount of activity; several display bright, whitish smears across reworked landscapes, sharply contrasting with ancient cratered terrains. The smallest moon which shows these properties, Miranda, displays amazing landforms: at least three large, ovoidal or trapezoidal coronae made up of an outer belt of approximately parallel ridges and troughs around an inner, smoother region; tilt-block style normal faults with cliffs up to 3 km high; and fault canyons up to 20 km deep. Such an amazing disparity of surface features could have been produced by the recent reconsolidation of a shattered moon after a major impact, in which the coronae formed as heavier material sank after reconsolidation. Such violent events must have been numerous since the origin of the solar system. Alternatively, the outwardly-facing tilt blocks and other features suggest an extensional origin in which the crust was forced apart by plumes of material rising from the interior (Pappalardo et al. 1997 and references therein). Tidal heating during a short-lived orbital resonance with another Uranian satellite could have provided the internal heat to drive the plumes.

The major moons, and some of the planets and dwarf planets—Pluto–Charon, Mercury, and perhaps Mars, as well—may be the last of a larger population of intermediate-sized objects which were abundant in the early solar system. Most of them were either ejected from the solar system or had a violent end, such as the hypothetical Mars-sized body that has been proposed to have impacted the Earth to form the Moon.

Fragments of larger moons may make up many of the rest of the current satellite population; and many of the outliers seem to be captured asteroids. The two Martian moons, Phobos and Deimos, are likely in this category.

The outer moons of the solar system tend to be icy bodies, but there are some bodies that have densities less than $1 \, \text{gm/cm}^3$! Of course some of the densities are ill-determined, but the trend of lower density with distance outward from the sun is certainly striking.

In the Jovian system, the Galilean satellites show a similar trend of decreasing densities with distance from Jupiter. There is some suggestion that this is because Jupiter was much hotter earlier in its history, but the inner ones could have lost much of their volatiles through tidal heating subsequent to their formation.

The relative resurfacing ages of the satellites can be determined by the cratering densities, if correct assumptions are made about impact frequency with time across the solar system. The various ages of the surfaces on the Galilean moons can be seen in Johnson (1990, p.176, Fig. 7) where the numbers of craters are plotted against crater diameter for the Galilean satellites, Earth, Mars, and the Moon. The absence of cratering implies a youthful, that is, reworked surface. For example, the cratering densities for Io and Europa are artificially low because of the very active surfaces of these bodies. Both have two heat sources:

1. Radioactivity
2. Tidal action

The latter is much stronger than the former in Io, with its abundant volcanism and limited size. On Europa, on the other hand, tidal action is weaker because it depends on r^{-3}, so that radioactivity may play a more important role in maintaining its watery mantle (or lower crust).

Magnetometer data gathered by the *Galileo* orbiter, show evidence for a deep, salty ocean below the thick, icy crust of Callisto. Tidal heating is negligible, so radioactivity must be the dominant source of its heat.

Although Jupiter's main satellites very likely accreted from ices as well as rock, another important source of icy materials has been cometary bombardment, a process that is continuing today, as Comet Levy–Shoemaker 9 illustrated in 1994 when its fragments collided with Jupiter.

If we count as "major moons" those with a radius exceeding 1000 km, there are seven: the Moon; the Galilean satellites Io, Europa, Ganymede, and Callisto; Titan; and Triton. Relaxing the definition to moons that exceed 1000 km in diameter we can add four moons of Saturn (Tethys, Dione, Rhea, and Iapetus), four moons of Uranus (Ariel, Umbriel, Titania, and Oberon), and, significantly, Charon. Finally, if we relax the definition somewhat more to include objects with smaller diameters, say roughly 500 km, we may add Enceladus in the Saturnian system and Miranda (470 km) in the Uranian system. Mimas is just under 400 km diameter; Nereid is only 340 km in diameter and the dark moon Proteus, also in the Neptunian system, is an irregular object with dimensions $436 \times 416 \times 402$ km.

Table 13.1 lists the orbital characteristics of the major moons of the planets. Table 13.2 does the same for the physical characteristics. In both tables, one or more colons after an entry indicate levels of uncertainty. The sources for both are the Astronomical Almanac for 2005 & 2008 and the Natural Satellite Data Center online files; additional data are from links from the University of Hawaii's Institute for Astronomy website, http://www.ifa.hawaii.edu/research/research.htm and from the current literature. The satellite numbers in Column 1 are usually expressed in Roman numerals. The unnamed satellites (and some of the more recently named ones) are identified in terms of the discovery notation: *S/YYYY PN*, where the initial S = "Satellite"; Y denotes discovery year; P is the planet initial; and N is the sequence number of discovery within the year. Relatively newly named satellites have their discovery designations noted in the alternative final column under "Comment." Values of orbital elements are usually mean quantities, and the inclinations are with respect to the planet's equator, unless noted otherwise (note the variation in some cases). The many unfilled columns, especially of masses and photometric characteristics, demonstrate how much work still needs to be done to study these objects. Note the similarities among some of the satellites, strongly suggesting a common origin for groups of them. As of late 2004, there were 63 known satellites of Jupiter, 31 of Saturn (at current writing, 60 have been reported!), 24 of Uranus, and 13 of Neptune; moreover, two of the "dwarf" planets have moons. Eris has one: Dysnomia (Brown et al. 2006), and Pluto has three moons: Charon, Nix, and Hydra (Weaver et al. 2006). More solar system satellite discoveries can be expected.

Table 13.1. Orbital elements of solar system natural satellites

No. Name	P (d)	a (10^3 km)	e	i (°)	$\mathrm{d}\,\Omega/\mathrm{d}t$(°/y) [or $\mathrm{d}\varpi/\mathrm{d}t$]/comment
Earth					
1. Moon	27.321661	384.400	0.0549	18.28	19.34[a]
				−28.58	
Mars					
1. Phobos	0.31891023	9.378	0.015	1.0	158.8
2. Deimos	1.2624407	23.459	0.0005	0.9	6.614
				−2.7	
Jupiter					
1. Io	1.769137786	421.8	0.004	0.036	48.6
2. Europa	3.551181041	671.1	0.009	0.469	12.0
3. Ganymede	7.15455296	1070.4	0.001	0.170	2.63
4. Callisto	16.6890184	1882.7	0.007	0.187	0.643
5. Amalthea	0.49817905	181.4	0.003	0.388	914.6
6. Himalia	250.5662	11480	0.162	27.496	524.4
7. Elara	259.6528	11737	0.217	26.627	506.1
8. Pasiphae	743.63	23624	0.409	151.431	185.6
9. Sinope	758.90	23939	0.259	158.109	181.4
10. Lysithea	259.2	11717	0.112	28.302	506.9
11. Carme	734.17	23404	0.253	164.907	187.1
12. Ananke	629.77	21276	0.244	148.889	215.2
13. Leda	240.92	11165	0.164	27.457	545.4
14. Thebe	0.6745	221.9	0.018	1.070	−
15. Adrastea	0.29826	128.9	0.002	0.027	−
16. Metis	0.294780	128.1	0.001	0.021	−
17. Callirrhoe	758.77	24103	0.283	147.158	− = S/1999 J1
18. Themisto	129.71	7387	0.204	45.67	− = S/2000 J1[b]
19. Megaclite	752.88	23493	0.528	151.7	− = S/2000 J8
20. Taygete	732.41	23280	0.246	163.5	− = S/2000 J9
21. Chaldene	723.70	23100	0.155	165.6	− = S/2000 J10
22. Harpalyke	623.31	20858	0.200	149.3	− = S/2000 J5
23. Kalyke	742.03	23566	0.318	165.8	− = S/2000 J2
24. Iocaste	631.60	21061	0.269	149.9	− = S/2000 J3
25. Erinome	728.51	23196	0.346	160.9	− = S/2000 J4
26. Isonoe	726.25	23155	0.281	165.0	− = S/2000 J6
27. Praxidike	625.38	20907	0.146	146.4	− = S/2000 J7
28. Autonoe	760.95	24046	0.334	152.9	− = S/2001 J1
29. Thyone	627.21	20939	0.229	148.5	− = S/2001 J2
30. Hermippe	633.90	21131	0.210	150.7	− = S/2001 J3
31. Aitne	730.18	23229	0.264	165.1	− = S/2001 J11
32. Eurydome	717.33	22865	0.276	150.3	− = S/2001 J4
33. Euanthe	620.49	20797	0.232	148.6	− = S/2001 J7
34. Euporie	551	19302	0.144	145.6	− = S/2001 J10
35. Orthosie	623	20721	0.281	145.9	− = S/2001 J9
36. Sponde	748.34	23487	0.312	151.0	− = S/2001 J5
37. Kale	729.47	23217	0.260	165.0	− = S/2001 J8
38. Pasithee	719	23004	0.267	165.1	− = S/2001 J6
39. Hegemone	739.60	23947	0.328	155.2	− = S/2003 J8
40. Mneme	625	21069	0.227	148.6	− = S/2003 J21

41. Aoede	761.50	23981	0.432	158.3	— = S2003 J7
42. Thelxinoe	629	21162	0.221	151.4	— = S2003 J22
43. Arche	723.90	22931	0.259	165.0	— = S/2002 J1
44. Kallichore	681.94	22335.4	0.223	163.9	— = S/2003 J11
45. Helike	634.77	21263	0.156	154.773	— = S/2003 J6
46. Carpo	455.07	17056.0	0.430	51.4	— = S/2003 J20
47. Eukelade	746.39	23.661	0.272	165.5	— = S/2003 J1
48. Cyllene	737.80	23544.8	0.412	141.0	— = S/2003 J13
49. Kore	807.20	24974.0	0.222	140.9	— = S/2003 J14

Unnamed satellites of Jupiter (at present writing)

S/2000 J11	288.5	12623.0	0.215	28.55	— = S/1975 J1
S/2003 J2	983	28493.8	0.380	151.8	—
S/2003 J3	505	18290.7	0.241	143.7	—
S/2003 J4	722	23195 6	0.204	144.9	—
S/2003 J5	761	24019.6	0.2095	165.0	—
S/2003 J9	684.0	22381.5	0.269	164.5	—
S/2003 J10	768.4	24184.6	0.214	164.1	—
S/2003 J12	533.0	18951.5	0.376	145.8	—
S/2003 J15	667.17	22011.8	0.113	140.8	—
S/2003 J16	596.76	20434.4	0.269	148.6	—
S/2003 J17	689.98	22510.6	0.187	163.7	—
S/2003 J18	607.68	20683.0	0.138	146.5	—
S/2003 J19	700.83	22745.9	0.334	162.9[a]	—
S/2003 J23	762	24055.5	0.309	149.2[a]	—

Saturn

1. Mimas	0.942421813	185.52	0.0202	1.53	365.0
2. Enceladus	1.370217855	238.02	0.0045	0.000	156.2[c]
3. Tethys	1.887802160	294.66	0.0000	1.86	72.25
4. Dione	2.736914742	377.40	0.00223	0.02	30.85[c]
5. Rhea	4.517500436	527.04	0.00100	0.35	10.16
6. Titan	15.94542068	1221.83	0.02919	0.28	0.5213[c]
7. Hyperion	21.2766088	1500.9	0.0274	0.630	—
8. Iapetus	79.3301825	3561.3	0.02828	7.49	—
9. Phoebe	550.31	12947.78	0.1635	174.751[a]	—
10. Janus	0.6945	151.472	0.007	0.14	—
11. Epimetheus	0.6942	151.422	0.021	0.34	—
12. Helene	2.7369	377.40	0.007	0.21	—
13. Telesto	1.8878	294.66	0.000	1.18	—
14. Calypso	1.8878	294.66	0.001	1.50	—
15. Atlas	0.6019	137.670	0.000	0.3	—
16. Prometheus	0.6130	139.353	0.003	0.0	—
17. Pandora	0.6285	141.700	0.004	0.0	—
18. Pan	0.5750	133.583	—	—	—
19. Ymir	1305.8	23041	0.335	173.1	— = S/2000 S1
20. Paaliaq	681.98	14943	0.464	47.24	— = S/2000 S2
21. Tarvos	842.71	17207	0.619	34.86	— = S/2000 S4
22. Ijiraq	456.22	11430	0.364	49.10	— = S/2000 S6
23. Suttungr	993	19185.7	0.145	174.6	— = S/2000 S12
24. Kiviuq	442.80	11205	0.154	48.74	— = S/2000 S5
25. Mundilfari	911.50	18131	0.284	169.41	— = S/2000 S9

(Continued)

Table 13.1. (Continued)

No. Name	P (d)	a (10^3 km)	e	i (°)	$d\Omega/dt(°/y)$ [or $d\varpi/dt$]/comment
26. Albiorix	785.5	16182	0.478	33.98	− = S/2000 S11
27. Skadi	738.32	15755	0.206	148.51	− = S/2003 S1
28. Erriapo	913.64	18160	0.625	33.50	− = S/2003 S1
29. Siarnaq	895.55	17531	0.296	46.0	− = S/2000 S3
30. Thyrm	1079.46	20295	0.513	174.98	− = S/2003 S1
31. Narvi	978.	19007	0.431	145.8	− = S/2003 S1
32. Methone	0.784	194	0.00	0.0	− = S/2004 S1
33. Pallene	1.144	211	0.00	0.0	− = S/2004 S2
34. Polydeuces	2.373	377.4	0.00	0.0	− = S/2004 S5
35. Daphnis	0.595	136.5	0.00	0.0	− = S/2005 S1
36. Aegir	1005	19350	0.241	167.0	− = S/2004 S10
37. Bebhionn	824	16950	0.336	41.0	− = S/2004 S11
38. Bergelmir	959	18750	0.180	156.9	− = S/2004 S15
39. Bestla	1028	19650	0.795	147.4	− = S/2004 S18
40. Farbauti	1040	19800	0.235	157.6	− = S/2004 S9
41. Fenrir	1235	22200	0.135	163.0	− = S/2004 S16
42. Fornjot	1235	22200	0.213	168.0	− = S/2004 S8
43. Hati	1052	19950	0.292	162.7	− = S/2004 S14
44. Hyrokkin	912	18217	0.360	153.3	− = S/2004 S19
45. Kari	1245	22350	0.341	148.4	− = S/2006 S2
46. Loge	1314	23190	0.139	166.5	− = S/2006 S5
47. Skoll	869	17610	0.418	155.6	− = S/2006 S8
48. Surtur	1237	22290	0.368	166.9	− = S/2006 S7
Unnamed satellites of Saturn (as of present writing)					
S/2004 S7	1040	19800	0.580	165.1	−
S/2004 S12	1028	19650	0.401	164.0	−
S/2004 S13	936	18450	0.273	167.4	−
S/2004 S17	947	18600	0.259	166.6	−
S/2006 S3	1142	21132	0.471	150.8	−
S/2006 S4	905	18105	0.374	172.7	−
S/2006 S6	942	18600	0.192	162.9	−
S/2007 S1	895	17920	0.107	49.86	−
S/2007 S2	800	16560	0.218	176.7	−
S/2007 S3	1100	20578	0.130	177.2	−
S/2007 S4	1.04	197.70	0.001	0.100	−
Uranus					
1. Ariel	2.52037935	191.02	0.001	0.041	6.8
2. Umbriel	4.1441772	266.30	0.004	0.128	3.6
3. Titania	8.7058717	435.91	0.001	0.079	2.0
4. Oberon	13.4632389	583.52	0.001	0.068	1.4
5. Miranda	1.41347925	129.39	0.001	4.338	19.8
6. Cordelia	0.3350338	49.77	0.00026	0.08	550
7. Ophelia	0.376400	53.79	0.0099	0.10	419
8. Bianca	0.43457899	59.17	0.0009	0.19	229
9. Cressida	0.46356960	61.78	0.0004	0.006	257
10. Desdemona	0.47364960	62.68	0.00013	0.113	245
11. Juliet	0.49306549	64.35	0.00066	0.065	223

12. Portia	0.51319592	66.09	0.0000	0.059	203
13. Rosalind	0.55845953	69.94	0.0001	0.279	129
14. Belinda	0.62352747	75.26	0.00007	0.031	167
15. Puck	0.76183287	86.01	0.00012	0.319	81
16. Caliban	579.6	7170.4	0.081	139.8[a]	− = S/1997 U1
17. Sycorax	1289.	12216	0.512	152.7[a]	− = S/1997 U2
18. Prospero	1948.13	16089	0.328	146.3[a]	− = S/1999 U3
19. Setebos	2303.	17988	0.512	148.3[a]	− = S/1999 U1
20. Stephano	675.71	7942.4	0.146	141.5[a]	− = S/1999 U2
21. Trinculo	758	8571.015	0.208	166.3[a]	− = S/2001 U1
22. Francisco	267	4276	0.146	145.2	− = S/2001 U3
23. Margaret	1641	14345	0.661	56.6	− = S/2003 U3
24. Ferdinand	2886	20901	0.368	169.8	− = S/2001 U2
25. Perdita	0.638	76.417	0.003	0.007	− = S/1986 U10
26. Mab	0.922 9	97.734	0.003	0.134	− = S/2003 U1
27. Cupid	0.617 94	74.392	0.0	−	− = S/2003 U2

Neptune

1. Triton	5.8768541[a]	354.76	0.000016	157.345	0.5232
2. Nereid	360.13619	5513.4	0.7512	27.6	0.039
3. Naiad	0.294396	48.227	0.00033	4.74	626
4. Thalassa	0.311485	50.075	0.00016	0.21	551
5. Despina	0.334655	52.526	0.000139	0.07	466
6. Galatea	0.428745	61.953	0.00012	0.05	261
7. Larissa	0.554654	73.548	0.00139	0.20	143
8. Proteus	1.122315	117.647	0.00044	0.039	28.25
10. Psamathe	9620	49 281	0.2679	124.517	− = S/2003 N1

Unnamed satellites of Neptune (as of present writing)

S/2002 N1	1874	16560.335	0.2597	111.77	−
S/2002 N2	2517	20151	0.17	57	−
S/2002 N3	2747	21365	0.47	43	−
S/2002 N4	9361	48387	0.495	132.6	−

(134340) Pluto[d]

1. Charon	6.38723	19.571	0.000	96.145	−
2. Nix	38.2065	64.48	0.005	96.36	− = S/2005 P1
3. Hydra	24.8562	48.7	0.002	96.18	− = S/2005 P2

(136199) Eris[e]

1. Dysnomia	15.774	37.35	0.0	61 or 142	− = S/2005 Eris 1

[a] Relative to ecliptic.
[b] Also J18 = S/1975 J1.
[c] Variation in periapse (not node) longitude.
[d] Data from Buie et al. (2006) for epoch 1452600.5.
[e] Data from Brown et al. (2006); Brown & Schaller (2007).

We next discuss the properties of the major moons of the solar system, in the context of satellite populations of the giant planets and Pluto. Because no permanent natural satellites are known around Mercury and Venus, and because we have already discussed the Earth's moon extensively, we begin with the moons of Mars.

Table 13.2. Physical properties of solar system natural satellites

No. Name	$M/M_{\rm pl}$	R (km)	$P_{\rm rtn}$ (d)/S[a]	$A_{\rm V}$	V_0	B-V	U-B
Earth							
1. Moon	0.01230	1737.4	S	0.12	−12.74	0.92	0.46
Mars							
1. Phobos	1.65×10^{-8}	13.4, 11.2, 9.2	S	0.07	11.3	0.6	–
2. Deimos	3.71×10^{-9}	7.5, 6.1, 5.2	S	0.08	12.89	0.65	0.18
Jupiter							
1. Io	4.70×10^{-5}	$< 1821.1 >$[b]	S	0.63	5.02	1.17	1.30
2. Europa	2.53×10^{-5}	$< 1562.1 >$[b]	S	0.67	5.29	0.87	0.52
3. Ganymede	7.80×10^{-5}	2631.2	S	0.44	4.61	0.83	0.50
4. Callisto	5.67×10^{-5}	2403.3	S	0.20	5.65	0.86	0.55
5. Amalthea	3.8×10^{-11}	131, 73, 67	S	0.07	14.1	1.50	–
6. Himalia	5.0×10^{-11}	75, 60, 60	0.40	0.03	14.84	0.67	0.30
7. Elara	4×10^{-10}	40	–	0.03	16.77	0.69	0.28
8. Pasiphae	1×10^{-10}	18	–	0.10	17.03	0.63	0.34
9. Sinope	4×10^{-11}	14	0.548	0.05	18.3	0.7	–
10. Lysithea	4×10^{-11}	12	0.533	0.06	18.4	0.7	–
11. Carme	5×10^{-11}	15	0.433	0.06	18.0	0.7	–
12. Ananke	2×10^{-11}	10	0.35	0.06	18.9	0.7	–
13. Leda	3×10^{-12}	5	–	0.07	20.2	0.7	–
14. Thebe	4×10^{-10}	55, 55, 45	S	0.04	15.7	1.3	–
15. Adrastea	1×10^{-11}	13, 10, 8	S	0.05	19.1	–	–
16. Metis	5×10^{-11}	30, 20, 20	S	0.05	17.5	–	–
17. Callirhoe	5×10^{-13}	4.3	–	0.06:	20.8	–	–
18. Themisto	4×10^{-13}	4.5	–	0.06:	21.7	–	–
19. Megaclite	–	2.7	–	0.06	21.5	–	–
20. Taygete	–	2.5	–	0.06	21.9	–	–
21. Chaldene	–	1.9	–	0.06	22.3	–	–
22. Harpalyke	–	2.15	–	0.06	22.2	–	–
23. Kalyke	–	2.5	–	0.06	21.5	–	–
24. Iocaste	–	2.6	–	0.06	21.8	–	–
25. Erinome	–	1.6	–	0.06	22.4	–	–
26. Isonoe	–	1.9	–	0.06	22.5	–	–
27. Praxidike	–	3.4	–	0.06	21.2	–	–
28. Autonoe	–	2.0	–	0.06	22.0	–	–
29. Thyone	–	1.5	–	0.06	22.1	–	–
30. Hermippe	–	2.0	–	0.06	21.8	–	–
31. Aitne	–	1.5	–	0.06	23.2	–	–
32. Eurydome	–	1.25	–	0.06	22.6	–	–
33. Euanthe	–	1.25	–	0.06	23.3	–	–
34. Euporie	–	1.25	–	–	23.0	–	–
35. Orthosie	–	1.0	–	–	23.1	–	–
36. Sponde	–	1.0	–	–	23.0	–	–
37. Kale	–	1.0	–	0.06	22.5	–	–
38. Pasithee	–	0.9	–	–	23.2	–	–
39. Hegemone	–	1.5	–	0.04	22.5	–	–
40. Mneme	–	1.25	–	–	23.1	–	–
41. Aoede	–	2.0	–	0.04	22.4	–	–
42. Thelxinoe	–	1.0	–	–	23.4	–	–
43. Arche	–	1.75	–	0.04	23.0	–	–

44. Kallichore	–	0.5	–	–	23.4	–	–
45. Helike	–	1.5	–	0.04	22.5	–	–
46. Carpo	–	1.75	–	0.04	23.0	–	–
47. Eukelade	–	1.5	–	0.04	22.5	–	–
48. Cyllene	–	1.25	–	–	23.0	–	–
48. Kore	–	0.85	–	–	23.3	–	–

Unnamed satellites of Jupiter (at present writing)

S/2000 J11	–	2.0	–	–	22.4	–	–
S/2003 J2	–	1.0	–	–	23.2	–	–
S/2003 J3	–	0.5	–	–	23.6	–	–
S/2003 J4	–	1.0	–	–	22.7	–	–
S/2003 J5	–	1.75	–	–	22.0	–	–
S/2003 J9	–	0.65	–	–	23.6	–	–
S/2003 J10	–	0.9	–	–	23.6	–	–
S/2003 J12	–	0.65	–	–	23.8	–	–
S/2003 J15	–	0.7	–	–	23.3	–	–
S/2003 J16	–	0.75	–	–	23.0	–	–
S/2003 J17	–	1.0	–	–	23.0	–	–
S/2003 J18	–	1.0	–	–	23.0	–	–
S/2003 J19	–	0.75	–	–	23.5	–	–
S/2003 J23	–	1.0	–	–	23.5	–	–

Saturn

1. Mimas	6.60×10^{-8}	$209, 196, 191$	S	0.5	12.9	–	–
2. Enceladus	1×10^{-7}	$256, 247, 245$	S	1.0	11.7	0.70	0.28
3. Tethys	1.09×10^{-6}	$536, 528, 526$	S	0.9	10.2	0.73	0.30
4. Dione	1.93×10^{-6}	560	S	0.7	10.4	0.71	0.31
5. Rhea	4.06×10^{-6}	764	S	0.7	9.7	0.78	0.38
6. Titan	2.37×10^{-4}	2575	S	0.22	8.28	1.28	0.75
7. Hyperion	4×10^{-8}	$180, 140, 113$	–	0.3	14.19	0.78	0.33
8. Iapetus	2.8×10^{-6}	718	S	0.5^c	11.1	0.72	0.30
9. Phoebe	7×10^{-10}	110	0.4	0.06	16.45	0.70	0.34
10. Janus	3.38×10^{-9}	$97, 95, 77$	S	0.9:	14:	–	–
11. Epimetheus	9.5×10^{-10}	$69, 55, 55$	S	0.8:	15:	–	–
12. Helene	–	$18, 16, 15$	–	0.7:	18:	–	–
13. Telesto	–	$15, 12.5, 7.5$	–	1.0:	18.5:	–	–
14. Calypso	–	$15, 8, 8$	–	1.0:	18.7:	–	–
15. Atlas	–	$18.5, 17.2, 13.5$	–	0.4	19.0	–	–
16. Prometheus	–	$74, 50, 34$	S	0.6	15.8	–	–
17. Pandora	–	$55, 44, 31$	S	0.5	16.4	–	–
18. Pan	–	10	–	0.5:	19.4	–	–
19. Ymir	–	7.5	–	–	22.3	–	–
20. Palliaq	–	10:	–	0.06:	21.7	–	–
21. Tarvos	–	6:	–	0.06:	22.7	–	–
22. Ijiraq	–	5:	–	0.06:	23.1	–	–
23. Suttungr	–	3:	–	–	23.8	–	–
24. Kiviuq	–	7:	–	0.06:	22.7	–	–
25. Mundilfari	–	3:	–	0.06:	24.4	–	–
26. Albiorax	–	10.5	–	–	20.3	–	–
27. Skadi	–	3:	–	0.06:	24.1	–	–
28. Errapio	–	4:	–	0.06:	23.5	–	–
29. Siarnaq	–	16:	–	0.16	20.0	–	–
30. Thrym	–	3:	–	0.06:	24.4	–	–
31. Narvi	–	2.5	–	–	23.9	–	–

(Continued)

Table 13.2. (Continued)

No. Name	M/M_{pl}	R (km)	P_{rtn} (d)/S[a]	A_V	V_0	B-V	U-B
32. Methone	–	1.5	–	–	25.	–	–
33. Pallene	–	2:	–	–	25.	–	–
34. Polydeuces	–	2:	–	–	25.	–	–
35. Daphnis	–	3.5:	–	–	24.	–	–
36. Aeigr	–	3:	–	–	24.4	–	–
37. Bebhionn	–	3:	–	–	24.1	–	–
38. Bergelmir	–	3:	–	–	24.2	–	–
39. Bestla	–	3.5:	–	–	23.8	–	–
40. Farbauti	–	2.5:	–	–	24.7	–	–
41. Fenrir	–	2.0:	–	–	25.0	–	–
42. Fornjot	–	3:	–	–	24.6	–	–
43. Hati	–	3:	–	–	24.4	–	–
44. Hyrokkin	–	4:	–	–	23.5	–	–
45. Kari	–	3.5:	–	–	23.9	–	–
46. Loge	–	3:	–	–	24.6	–	–
47. Skoll	–	3:	–	–	24.5	–	–
48. Surtur	–	3:	–	–	24.8	–	–
Unnamed satellites of Saturn (at present writing)							
S/2004 S7	–	1.5:	–	–	23.9	–	–
S/2004 S12	–	2.5:	–	–	24.8	–	–
S/2004 S13	–	3:	–	–	24.5	–	–
S 2004 S17	–	2:	–	–	25.2	–	–
S/2006 S3	–	3:	–	–	24.9	–	–
S/2006 S4	–	3:	–	–	24.4	–	–
S/2006 S6	–	3:	–	–	24.7	–	–
S/2007 S1	–	3.5	–	–	23.9	–	–
S/2007 S2	–	3:	–	–	24.4	–	–
S/2007 S3	–	2.5:	–	–	24.9	–	–
S/2007 S4	–	0.5:	–	–	26.		
Uranus							
1. Ariel	1.55×10^{-5}	$< 579.0 >$[b]	S	0.35	14.16	0.65	–
2. Umbriel	1.35×10^{-5}	585	S	0.19	14.81	0.68	–
3. Titania	4.06×10^{-5}	789	S	0.28	13.73	0.70	0.28
4. Oberon	3.47×10^{-5}	761	S	0.25	13.94	0.68	0.20
5. Miranda	0.08×10^{-5}	240, 234, 233	S	0.27	16.3	–	–
6. Cordelia	–	13	–	0.07:	24.1	–	–
7. Ophelia	–	15	–	0.07:	23.8	–	–
8. Bianca	–	21	–	0.07:	23.0	–	–
9. Cressida	–	31	–	0.07:	22.2	–	–
10. Desdemona	–	27	–	0.07:	22.5	–	–
11. Juliet	–	42	–	0.07:	21.5	–	–
12. Portia	–	54	–	0.07:	21.0	–	–
13. Rosalind	–	27	–	0.07:	22.5	–	–
14. Belinda	–	33	–	0.07:	22.1	–	–
15. Puck	–	77	–	0.075	20.2	–	–
16. Caliban	–	30:	–	0.07:	22.4	–	–
17. Sycorax	–	60:	–	0.07:	20.9	–	–
18. Prospero	–	16:	–	0.07:	23.7	–	–
19. Setebos	–	15:	–	0.07:	23.8	–	–
20. Stephano	–	11:	–	0.07:	24.6	–	–

21. Trinculo	–	5.5	–	–	25.1	–	–
22. Francisco	–	–	–	–	–	–	–
23. Margaret	–	–	–	–	–	–	–
24. Ferdinand	–	–	–	–	–	–	–
25. Perdita	–	13	–	–	–	–	–
26. Mab	–	6–8	–	–	–	–	–
27. Cupid	–	9	–	–	–	–	–
Neptune							
1. Triton	2.09×10^{-4}	1353	S	0.756	13.47	0.72	0.29
2. Nereid	2×10^{-7}	170	–	0.155	19.7	0.65	–
3. Naiad	–	29:	–	0.06:	24.7	–	–
4. Thalassa	–	40:	–	0.06:	23.8	–	–
5. Despina	–	74	–	0.06	22.6	–	–
6. Galatea	–	79	–	0.06	22.3	–	–
7. Larissa	–	104, 89	S	0.06	22.0	–	–
8. Proteus	–	218, 208, 201	S	0.06	20.3	–	–
10. Psamathe	–	15	–	–	24.9	–	–
Unnamed satellites of Neptune (as of present writing)							
S/2002 N1	–	25	–	–	24.0	–	–
S/2002 N2	–	–	–	–	–	–	–
S/2002 N3	–	–	–	–	–	–	–
S/2002 N4	–	–	–	–	–	–	–
Pluto							
1. Charon	0.125	604 ± 1^c	S	0.5	16.19	0.71	–
2. Nix	–	–	–	–	23.33	0.64	–
3. Hydra	–	–	–	–	23.38	0.91	–
Eris[e]							
1. Dysnomia	–	175 ± 75	–	–	–	–	–

[a] $S \equiv$ synchronous case: $P_{orbit} = P_{rtn}$.
[b] Geometric mean of 1830, 1819, 1815 km for Io (J1); of 1564, 1561, 1562 km for Europa (J2); of 581, 578, 578 km for Ariel (U1) for sub-planet, polar, and equatorial (normal to line of centers) radii, respectively.
[c] Weighted mean of occultation results from Gulbis et al. (2006) and Sicardy et al. (2006);
[d] data from Buie et al. (2006).
[e] from Brown et al. (2006) and: http://www.gps.caltech.edu/~ mbrown/planetlila/

13.1.1 The Moons of Mars

The moons were first described in some detail, not in any scientific publication, but in a work of satirical fiction: *Gulliver's Travels*, by Jonathan Swift, published in 1726. Cited by Abell (1969), among many authors, the text describing the discovery of the astronomers in the enlightened land of Laputa, reads:

... satellites, which revolve about Mars, whereof the innermost is distant from the centre of the primary planet exactly three of the diameters, and the outermost, five;

the former revolves in the space of ten hours, and the latter in twenty one and a half, so that the squares of their periodical times are very near the same proportion with the cubes of their distance from the centre of Mars, which evidently shows them to be governed by the same law of gravitation that influences the other heavenly bodies.

The similarity to Phobos and Deimos would be remarkable (the actual distances are 2.76 and $6.9\Re_{\sigma}$ and periods are 7.65 and 30.3h, for Phobos and Deimos, respectively) if it were not the case that Swift attended the same parish church as the Astronomer Royal (B. Marsden, *private correspondence*). Perhaps Britain's weather prevented the requisite follow-ups that a publication by an eminent astronomer would require, but it appears that at least one other astronomer may have suspected the existence of two satellites for Mars, as had Johannes Kepler (1571–1630) much earlier, although Kepler's conjecture was not based on observation. The moons' discovery is formally credited to Asaph Hall (1829–1907), an astronomer with the US Naval Observatory, in 1877.

The moons are non-spherical, and they are heavily cratered; they are blanketed in regolith, especially Deimos, as Figures 13.1 and 13.2

Fig. 13.1. Phobos, the inner natural satellite of Mars. Its reflection spectra suggest it to be similar to asteroids types C or D (see Chapter 15). The large crater is Stickney, 9.6 km in total width. The orbit is within the synchronous orbital radius of Mars and is subject to decay due to tidal interaction (see Milone & Wilson 2008, Chapter 3.7, and Section 13.2). Phobos is expected to collide with Mars within 10^8 years if it does not break up first and form a ring. The signs of severe tidal stress are already present in striations across its surface. From the Mars Orbital Camera aboard Mars Global Surveyor. NASA/JPL/, Malin Space Science Systems image PIA01331

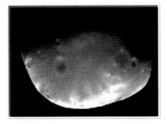

Fig. 13.2. A mosaic of Viking Orbiter images of Deimos, the outer and smaller Martian moon. Note the extent to which its craters are filled with regolith. A NASA/Malin Space Science Systems image

demonstrate. Phobos orbits Mars in so short a period that it is seen to move from west to east in the Martian sky. The densities of these bodies are low, 1.9 g/cm^3 for Phobos and 1.8 for Deimos. The overall impression is that they are captured asteroids of C or D class, with reflection spectra similar to carbonaceous chondrites (refer to Chapter 15), but the mechanism for capture of these objects is unclear. Such asteroids are found in the outer asteroid belt, so they must have been captured. Phobos has a nearly circular orbit today. The capture mechanism for phobos could have been orbital braking and circularization in a short-lived but extensive proto-Mars atmosphere; but Deimos would have had to arrive after this to avoid a collision with Phobos in that moon's initial highly eccentric orbit. Therefore Deimos' capture is the more problematic. Another mystery is why, if it survived from the early solar system, Phobos' orbit is now decaying rapidly: in a relatively short span of time ($< 10^8$ years), it will have either crashed on Mars or disintegrated due to tidal stress. The Russian Phobos mission provided some clues to their origins, but the total picture has yet to be assembled.

13.1.2 The Moons of Jupiter

The satellites of the outer planets are in two main categories according to their orbits: direct and indirect. The direct satellites revolve in the same direction as their planet rotates. The indirect satellites have retrograde revolutions ($i > 90°$) and usually high inclinations. The innermost satellites are connected with the rings of the planet, to be discussed later in this chapter. The outer satellites are almost certainly all captured objects, and the high inclinations and retrograde motions strongly suggest cometary origin.

The Galilean moons constitute a separate group, the major moons of the largest satellite system in the solar system. Discovered by Galileo, these four moons (which he called the Medicean satellites) were apparently not known in antiquity, even though in recent times individuals with exceptional acuity

Fig. 13.3. A composite of Voyager and Galileo images of the Galilean moons of Jupiter in order of decreasing size: Ganymede (the largest moon in the solar system), Callisto, Io, and Europa, from left to right, respectively. North is up for each image, and scaled to 10km/pixel. NASA/JPL/DLR/USGS image PIA00601

have demonstrated that they can see these fifth magnitude objects, and (traditional) knowledge of them has been attributed to certain cultures. As for the physical nature of these moons, the bulk of our data have come from the two Voyager missions, the extended Galileo mission, and the brief more recent visit by Cassini on its way to Saturn. Because their orbital properties have been mentioned before, we briefly describe only their physical properties. We describe them in the usual order, according to proximity to Jupiter, but Figure 13.3 (color plate 8) shows them in arranged order of size.

13.1.2.1 Io This tidally flexed moon must have lost all of its icy volatiles long ago. The remaining volatiles—sulfur and sodium, for instance—constitute major components of the volcanic ejecta and of the surface terrain. This is the most active volcanic body discovered in the solar system, with perhaps a score of active volcanoes at any one time. Pools of lava with 'islands' have been imaged, and lava streams extend for many kilometers across the surface.

There are virtually no impact craters on Io because the surface is so new. Volcanic plumes, analogous to Earth's geysers, may reach as high as 300 km above the surface and their conical sheets may be more than 1000 km across. A direct link between the plumes and active volcanoes has been observed, as, for example, in Figure 13.4 (Plate 9).

In addition to these features, there are also lakes of lava and volcanic flows, as Figure 13.5 (Plate 10) shows. Galileo thermal imagers and ground-based infrared detectors indicate that although most of the thermal anomalies (events where the surface is not at the thermal equilibrium temperature expected from solar heating) are low-temperature events, there are lava temperatures even higher than the 1400 K measured for Earth lava. By itself, at 1 mbar, sulfur boils at 450 K. Therefore, the hotter lava cannot be composed of molten sulfur alone, although impurities can raise the boiling point. The consensus seems to be that molten rock is present, at least in the

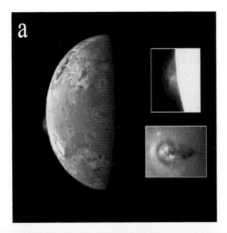

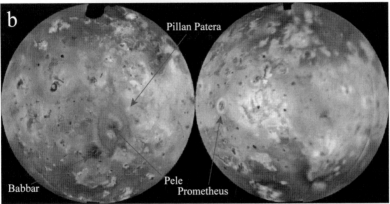

Fig. 13.4. a Io in eruption as seen by Galileo Orbiter. The volcanic plume on the limb originates from Pillan Patera while that at disk center is from Prometheus. Both illustrate the plume and conical sheet of the ejecta. NASA image PIA00703. **b** Annotated US Geological Survey maps of Io, based on Voyager I data. In natural color (seen in Plate 9) and done in equal area production. At the left center of the left figure is the cloven-hoofed ejecta sheet of Pele. Pre-eruption Pillan is to its upper right. Original NASA/JPL/USGS image PIA00318

high-temperature flows (Matson and Blaney 1999). The internal heat flux at Io's surface is $\sim 2\,\mathrm{W/m^2}$.

The atmosphere is primarily plume-fed SO_2 ($\gtrsim 10^{18}$ molecules/cm^2), although SO and NaCl have been detected also (Lellouch 1996; Lellouch et al. 2003). Some of the ejected material finds its way into the orbit; a torus of sulfur and sodium atoms (illustrated in Figure 12.20) has been detected around the orbit. The sulfur has blanketed Jupiter V (Amalthea), giving it an intense red color (see Table 13.2).

Fig. 13.5. From Galileo: a five-color mosaic of the Tvashtar Catena, a chain of calderas on Io. The white spots at left are places where hot lava is emerging from the toes of the flow; the bright white, yellow and reddish patches seen in Plate 10 are false color representations from IR imagery that mark a 60 km long lava stream, now beginning to cool. The diffuse dark region around the main caldera marks deposits from a recent plume eruption. NASA/JPL/University of Arizona image PIA02550

Although there are volcanic mounds on Io, most of the mountains may have been produced by uplift along thrust faults (see Milone & Wilson 2008, Chapter 9.3 for descriptions of the known types of faults). Figure 13.6 illustrates them.

Io has one of the higher densities among the major moons, $\sim 3570 \, \text{kg/m}^3$, attributed to the long-ago depletion of the most volatile ices. Models of the interior include a rocky core of $\sim 500 \, \text{km}$ radius, a mantle of $\sim 800 \, \text{km}$ thickness, a liquid shell (probably molten silicates, tongues of which may heat up the sulfur and salts to produce the eruptive plumes) underlying an asthenosphere and a lithosphere.

13.1.2.2 Europa This icy moon shows evidence of cracks and crustal movement, indirect evidence of fresh deposition on the surface, and a paucity (but not absence) of impact craters that indicate a recent, highly renewed surface. Yet there has been no direct visual evidence of a water mantle or lower portion of the crust beneath the visible icy crust: no geysers. Europa is thought to be mainly rock ($\rho = 2970 \, \text{kg/m}^3$) with a water ice surface. The presence of chaotic terrain, in which the patterning suggests rotation of ice "rafts," extensive (wrinkle) ridges, dark spots (interpretable as upwelling material), and the near absence of craters provide strong indirect evidence of a water layer or, alternatively, of a slushy mantle. Several measurements have confirmed variations in Europa's magnetic field, the most logical explanation

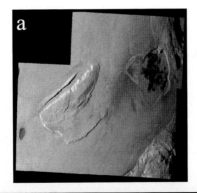

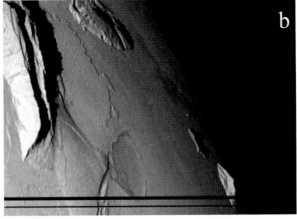

Fig. 13.6. Mountains on Io, seen by Galileo Orbiter. **a** To the left of center is Shamsu Mons, through which a 10-km wide canyon can be seen. NASA/JPL/University of Arizona mosaic image PIA02555. **b** A view of 7-km high Mongibello Mons on the upper left; other mountains are seen even closer to the terminator. NASA/JPL/University of Arizona/ASU image PIA03886

for which may be a briny sea between 5 and 20 km below the surface. The existence of a melt layer must be due to a combination of tidal effects and radioactive heating of the interior, because neither alone should be sufficient to explain the heated layer. Although there are relatively few craters, compared to those on the outer Galilean moons, there are ringed plains, indicating modification and resolidification following impacts.

Features on Europa include:

Linea: relatively smooth linear features, sometimes with linear ridges and grooves within them. They are interpreted as strike–slip faults, e.g., Astypalaea Linea, 290 km in length

Lenticulae: spotted features, darker than surroundings

Chaotic terrain: regions of fractured, rotated blocks of material

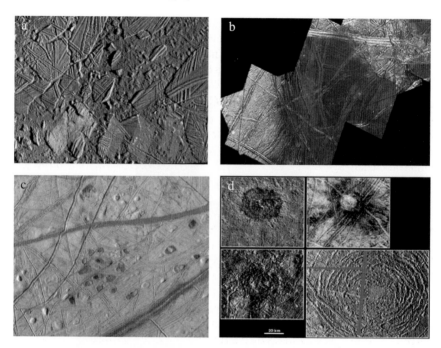

Fig. 13.7. Europan terrain. **a**: Conamara Chaos, where blocks of icy crustal material have been disrupted and refrozen into new consolidations. Long linear ridges and faults as well as arcuate features are characteristic of Europa's surface. Credit: NASA/JPL/Arizona State University. NASA image PIA01403, from Galileo Orbiter's solid-state camera. **b** A mosaic of images of a dark, low area region near the equator. Although few, a cluster of small impact craters suggests that they are secondary craters from a major impact. NASA/JPL image PIA01405 produced by Deutsche Zentrum für Luft und Raumfahrt (DLR). All these images are from the Solid State Camera aboard Galileo. **c** A composite image of a lenticular region on Europa from the Galileo mission. The spots, typically ∼ 10 km in diameter, may mark regions of upwelling material, melting and staining the ice above it, and due to the very low surface temperature of Europa, refreezing. NASA/JPL/University of Arizona/University of Colorado image PIA03878. **d** Craters on Europa. CW from the upper right: Cilix, Tyre, Mannann'an, and Pwyll. Tyre is analogous to the ringed plains of the Moon, Mercury, and Callisto. It may represent a major impact that broke through the hard icy crust into a slushier interior. The part of the structure shown here is 40 km diameter. NASA/JPL image PIA01661, produced by DLR

Figure 13.7 illustrates these features.

Europa's protected ocean makes it a candidate for biotic studies, in company with a number of other venues in the solar system where water exists in abundance, including the interiors of the giant planets, Ganymede, Titan, and on the small Saturnian moon, Enceladus.

13.1.2.3 Ganymede The largest moon of the solar system has a thick, icy crust, with a mixture of mainly two types of terrains: great grooved areas vs. heavily cratered plains. The dark, cratered plains are ancient (3.7–4.0 Gy), as indicated by the high frequency of craters. The grooved terrain is lighter, and more recent (3.1–3.7 Gy). The grooves are 5–10 km wide, and may represent fault-separated blocks of crust. This terrain may represent water ice which penetrated from depth and filled surface cracks and rifts, much like magma on the terrestrial planets. The Galileo Regio is an example of a dark furrowed region, and it, too, is ancient, but not as ancient, judging from crater frequencies, as the ancient cratered plains. New craters are bright, indicating fresh ice exposure. The general appearance of Ganymede is dark. Craters are numerous, their age being indicated, as on the Moon, by rim sharpness and by brightness, with darkness associated with older craters. Principle features (illustrated in Figures 13.8 and 13.9) are:

Sulci: (singular: *sulcus*) similar to the *linea* of Europa
Palimpsests: bright and circular areas, probably sites of impacts
Penepalimpsests (*dome craters*): perhaps an intermediate form between
   craters and palimpsests, suggesting evolutionary connections

The detection of a magnetic field around Ganymede has led to conjecture about a salty ocean mantle beneath the thick crust. Ganymede's moment of inertia, I = 0.311, is indicative of interior diversity and its mean density, 1940 kg/m^3, suggests water ice as well as rock. Many polymorphs of ice, from Ice I to Ice VIII are thought to exist in the deep interior of Ganymede. The latter

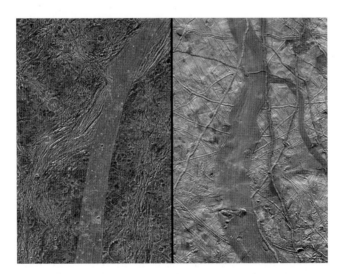

Fig. 13.8. Arbela Sulcus on Ganymede, left, and a similar feature on Europa, right, matched to the same resolution, 133 m/pixel. Tectonic effects are thought to be responsible for both: crustal spreading on Europa and strike–slip faulting on Ganymede. Note, however, the absence of craters on the Europa sulcus indicating a much younger surface. NASA/JPL image PIA02575, produced by Brown University

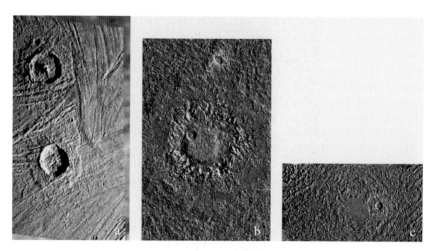

Fig. 13.9. Craters and crater remnants on Ganymede. **a** Craters Gula (top) and Achelous. The latter is ∼35 km diameter and its sharpness, brightness, and relatively undisturbed debris apron, suggest that it is also the younger of the two. **b** Neith, a "dome crater" or "penepalimpsest" with a dome of 45 km diameter, surrounded by concentric roughlands and plain. See Figure 13.11 for a similar feature on Callisto. **c** An image from the Marius Regio of Buto Facula, a palimpsest, ∼45-km in diameter with a smooth center and rings of concentric arcs surrounding it. The morphologies of these features may depend on the energy of the impactor and other circumstances and may not represent a temporal evolutionary sequence, but the penepalimpsest and palimpsest features appear to be far older than the craters in **a**. NASA/JPL images PIA01660, PIA01658, and PIA01659, respectively, produced by Deutsche Zentrum für Luft und Raumfahrt (DLR)

form of ice, for instance, has a density of 1670 kg/m³ and can only be found at pressures above 15 kbar and within a temperature range of 150 to ∼270 K. Ganymede has a tenuous atmosphere.

13.1.2.4 Callisto This darker icy moon is the most heavily cratered of the giant Jovian moons, an indication of a relatively unmodified surface. Its mean density is just slightly less than that of Ganymede. Craters are the dominant feature of Callisto, with age characteristics as on Ganymede. Here, however, cratered plains are not interrupted by extensive grooved and lineated features. The well-known ringed plain, Valhalla, a large ringed basin (Figure 13.10), appears similar to such features as the Caloris Basin on Mercury and Mare Orientale on the Moon. It apparently marks the site of a major impact. Resurfacing is done through impacts, solar wind, and cosmic rays, as on the Moon; younger craters are identified by bright icy surfaces. Figure 13.11 reveals the presence of dome craters on Callisto, indicating a parallel if possibly non-coeval process of crater evolution to that happening on Ganymede.

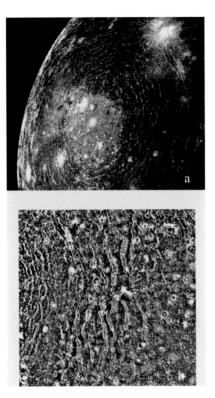

Fig. 13.10. a Valhalla, a great multi-ringed plain on Callisto as seen by Voyager 1. It marks a major impact event, the antiquity of which is indicated by the many smaller craters superimposed upon it. NASA image PIA02277, produced by JPL. **b** The plastic deformations of the concentric rings and craters of various ages are seen in this Voyager 1 detail of Valhalla. NASA/JPL image PIA00484

Models based on data from Voyager and the first two Galileo flybys suggested that Callisto is undifferentiated; but more accurate data from the third Galileo flyby indicate significant although incomplete differentiation into a rock-metal core and a rock-ice mantle, with a moment of inertia coefficient of 0.359±0.005 (Anderson et al. 1998) - see Milone & Wilson (2008, Fig. 5.18). Magnetic anomalies suggest the presence of a salty ocean deep below a cold, dark, and very thick crust of ice.

In three of the four Galilean moons, water ice is the dominant characteristic, but what is meant by the term "ice" depends on where it is in the phase diagram, a plot of the pressure against the temperature. Figure 13.12, after Taylor (1992, Fig. 6.8.1) is the phase diagram of water ice. The density of ice varies with phase, and it is possible for lower density ice to move through overlying material of higher density. On Earth, when salt is the lower density material, and it pushes through overburden, a *diapir* may result in a salt

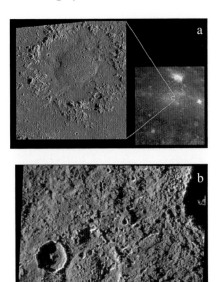

Fig. 13.11. a *(left)* From the Solid State Camera on Galileo, Doh, a dome crater on Callisto, located within the Asgard multi-ring basin visible in the wider image. Note its similarity to the Neith crater on Ganymede (in Figure 13.9). NASA/JPL image PIA01648, produced by Arizona State University. **b** From the same instrument, another dome crater, Har, on which are superimposed a much newer crater of 20 km diameter on the left, many smaller craters, and the secondary craters caused by the ejecta of the larger crater to the upper right, Tindr. NASA/JPL image PIA01054, produced by Arizona State University

dome. Rathburn, Musser, and Squyres (1998) have described the properties that such diapirs must have if they are present on Europa. In particular, the diapirs must be no deeper than the base of the ice layer, whether this be underlain by rock or by liquid, and lay within a few tens of kilometers below the surface. Perhaps the spots in Figure 13.7c are due to such features.

A study by Chuang and Greeley (2000) of material that fell away from impact crater walls indicates that the movement of rock and ice on the surface of Callisto appears to be similar to that of rock glaciers on Earth, but the ice is not as ductile and undergoes brittle deformation. The deposits, individual blocks of which may be several tens of meters high, and up to 1 km across, are seen also at the bases of scarps and cliffs; they have yield strengths similar to dry-rock avalanches on Earth. Nearby impacts appear to trigger some of the slides, but these authors suggest that ice sublimation undermines a near-surface layer and that seismic events generated by impacts trigger them.

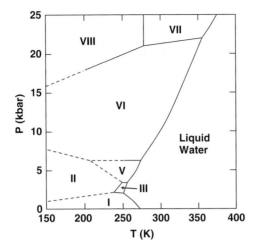

Fig. 13.12. The phase diagram for water ice. Dashed lines indicate estimations/extrapolations. After Hobbs (1974, Table 1.10, Fig. 1.18) and Taylor (1992 Fig. 6.8.1, p. 256). Triple points from Hobbs (Table 1.10). Taylor notes that as Ice VIII changes to Ice I, the ice will undergo an expansion of volume because Ice VIII has a density of $\sim$1670 kg/m^3. This adaptation reprinted with the permission of Cambridge University Press

13.1.3 The Moons of Saturn

Saturn has one giant-class moon and a host of intermediate and small satellites. As in the Jovian system, some of these satellites are associated with the rings; some may provide parent material for them (as Figure 13.13 suggests about Pan). The discovery of propeller-shaped moonlets of $\sim$5 km length and 300 m width, embedded within a small and relatively undisturbed area of the A-ring (Figure 13.14), implies that they are fragmented remains of much larger moons that were the source material for the rings. By extrapolation to the entire ring system, there may be millions of such embedded moonlets.

As also for Jupiter, many satellites have retrograde orbits, most with high inclinations. The orbits of many of these, and some of the moons with direct orbits, are in fact irregular, with major excursions of their orbital elements motions, including eccentricity and inclination.

The intermediate-sized satellites have presented puzzles ever since the Voyager missions revealed fresh ice resurfacing on the faces of several moons (Rhea, Tethys, Dione). Finally, a smoking gun was discovered on Enceladus: geysers were seen (Figures 13.15 and 13.16). As on Europa and Io, tidal interactions coupled with internal heat sources are thought to be responsible.

Mysteries, however, abound, e.g., one side of the small moon Iapetus (Figure 13.17a) has an albedo a factor of ten greater than its other side.

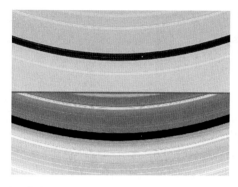

Fig. 13.13. A Cassini Orbiter view of the tiny satellite Pan orbiting in the Encke gap in the Saturnian ring system processed in two ways. Note the trail of ringlet material sharing Pan's orbit in the top panel, and the uneven edges of the adjacent rings in the lower panel. The co-orbital material undergoes horseshoe-shaped orbits relative to Pan, according to Murray (2007), and its source may be impacts on the ring moon by interplanetary material. The uneven ring edges preceding and following Pan are created by Pan's shepherding. NASA/JPL/ESA/Italian Space Agency(ASI), image PIA07528, produced by the Cassini Imaging Team

Figure 13.17b, from Cassini, shows why: bright, heavily frosted areas contrast to dark areas where the frost has sublimated.

The intermediate-sized moons display surface modifications in the form of bright streaks, as Figure 13.18 demonstrates. Occultation data show that Enceladus has a tenuous (and strongly localized) atmosphere (Hansen et al. 2006). Many of these moons show irregular orbits, undergoing variations in eccentricity and inclination, leading to tidal force variation. Hyperion is seen to have such an irregular surface that it more closely resembles a sponge than a moon. Finally, Cassini has detected ring-like debris around Rhea, the first such find around a moon.

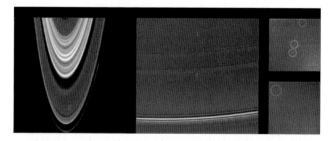

Fig. 13.14. Viewed from the Cassini orbiter, propeller-shaped moons within the A-ring may be fragments of larger moon(s). The moonlets' long axes are tilted by ∼20° offset. NASA/JPL/Space Science Institute, Boulder, image PIA07792

Fig. 13.15. Evidence for plumes of material erupting from the south polar surface of Enceladus, near the features referred to as "tiger stripes" (see Figure 13.16), as caught by Cassini on different dates near "new moon" phase in 2005. a) From image PIA07798, 2005 Feb. 17, at phase angle 153°3, South Pole down; b) from image PIA07758, 2005 Nov. 27, at phase angle 161°4, S. Pole to lower left. The images are reminiscent of the plumes seen on Io, but here, due to water ejected from geysers. Carbon dioxide and methane have also been detected in the plumes. NASA/JPL/Space Science Institute images

13.1.3.1 Titan Discovered by Christiaan Huygens (1629–1695) in 1655, Titan is the largest and brightest of the moons of Saturn; indeed its physical size nearly matches Ganymede, the largest moon of the solar system. It has a nearly circular orbit with a period of 15.95^d, and an inclination to Saturn's equator of $0°35$. It rotates in the same period as it revolves about Saturn, and so is locked in a 1:1 spin–orbit resonance.

Fig. 13.16. The Cassini Orbiter view of Enceladus. Note the varieties of terrain on this icy moon. It undergoes substantial tidal forces from Saturn and Titan and from closer moons. The "tiger stripes" region near the South Pole, the site of newly discovered water plumes, is indicated at the bottom of the image. The moon's icy blue color can be seen in Plate 11. See Porco et al. (2006) for a full discussion. NASA/JPL/Space Science Institute, Boulder, image PIA07800

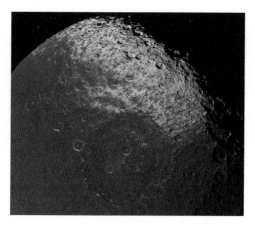

Fig. 13.17a. Cassini reveals the faces of Iapetus: dark terrain on most of the sunlit side and bright terrain at the top (and into the night side). This NASA/ESA/ASI image, taken on December 31, 2004 from a distance of 172,900 km, is a composite of three-passband observations at 338, 568, and 930 nm. NASA/JPL/Space Science Institute, Boulder, image PIA06167

Ground-based radar reflectivity had indicated no large ocean areas to be present, but large lakes were not excluded. The Cassini mission has revealed extensive clouds in Titan's atmosphere and both highland and low-lying areas on the surface. Spectacular images were obtained from the Huygens

Fig. 13.17b. A mosaic of 60 images of Iapetus from the Cassini Orbiter on Sept. 10, 2007 taken at a distance of 73,000 km. Note the heavy frost on this trailing side of Iapetus, and the two huge superposed impact craters. Credits: NASA/JPL/Space Science Institute, Boulder, image PIA 08384

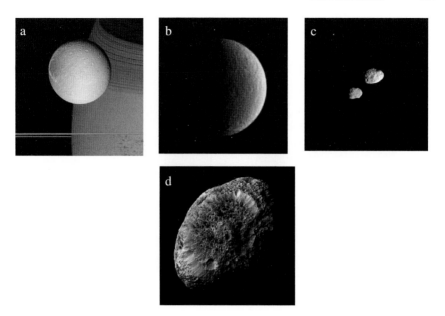

Fig. 13.18. Intermediate-sized moons as seen by Cassini Orbiter. **a** Dione and the rings of Saturn; **b** Rhea seen at phase angle 90°, on Mar. 15, 2005; **c** The co-orbital moons Epimetheus (left) and Janus (right), on Mar. 20, 2006; Janus was in the lower orbit at the time.; and **d** the bizarre surface of Hyperion with high crater frequency in and slumping on the rims of a huge sunken crater. Cassini Orbiter NASA/JPL/Space Science Institute, Boulder images PIA07744, PIA06641, PIA08170, and PIA07740, respectively

probe that was released from the Cassini Orbiter to descend onto Titan's surface. The Huygens probe revealed ground terrain similar to coastal areas on Earth, perhaps shaped not by water but by methane, possibly in solution with ethane. The crustal material may be in the form of a methane *clathrate hydrate*, which methane forms with H_2O at high pressures. The presence of drainage channels, replete with dendritic tributaries and apparent shore lines, lends a decidedly terrestrial appearance to the surface (Figures 13.19 and 13.20). Very dark areas have been observed by Cassini near both poles, and have been attributed to methane lakes filled through seasonal monsoon-like storms involving methane precipitation. As the heated probe landed, it detected sublimed methane gas.

The details on the surface near the probe can be seen in Figure 13.20. A possible cryo-volcano is highlighted in the global false-color image in Figure 13.21 (Plate 12 in color). In addition, bizarre features are seen, such as a series of very dark arcuate lines, that have been likened to cat scratches (perhaps multiple times on a favorite spot on a scratching post!), possibly longitudinal dunes, and long apparent flow patterns. Modification is the rule on Titan's surface and the paucity of impact craters indicates current surfacing. But the absence of methane/ethane oceans then presents a

Fig. 13.19. a and **b** Views of the surface of Titan by the Huygens probe, as it descended into the atmosphere. With conditions near the triple point of methane, the rivers, shoreline, and methane (and/or ethane) clouds visible in these images make sense. **c** A mosaic providing a 360° panoramic view of Titan as seen from Huygens at its landing site. The area over which the probe drifted in image **b** is visible in the upper left part of image **c**. NASA/JPL/ESA/Univ. of Arizona images PIA07236, PIA07231, and PIA07230, respectively.

difficulty, unless Titan is in some sort of ice age, or methane is not as generally abundant on Titan as one would suppose from the observations.

Tobie, Lunine, and Sotin (2006) argue that the bulk density of Titan indicates its mass to be 50–70% silicate material (with densities 3000–4000 kg/m^3) and the rest water ice. A high CH_4 production rate is needed to offset losses from photodissociation by UV photons, but cannot be understood if it arises either from large amounts of primordial material or from cometary impacts. Even in the great bombardment period of the early solar system, cometary delivery of volatiles appears to have been insufficient to be the source of the methane. We describe their solution below, but they note also that ammonia (NH_3) must have been present in large quantities, in agreement with formation and structural models for Titan, which have a homogeneous core of icy rock, overlain by a layer of silicates and a subsurface layer of a water–ammonia mixture.

Some surface features suggest lava flows, which must have been quite viscous to remain molten long enough to produce the lengthy structures seen.

Fig. 13.20. The surface of Titan, as revealed by the Huygens probe. The area resembles a mud flat—with rocks. The bright elongated rock just below the left center of the image is 15 cm across and is 85 cm from the camera. The roundedness of the rocks indicates weathering as an important process. Credits: NASA/JPL/ESA/University of Arizona image PIA07232

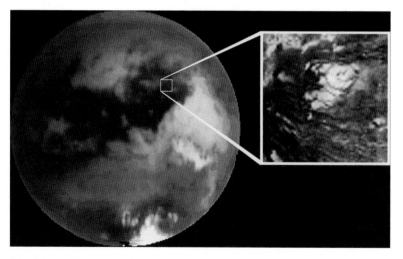

Fig. 13.21. A Titan mosaic of six images, shown in false color in Plate 12, to emphasize the atmospheric haze, areas of light and dark regions; the inset is a 2.3μm image of a possible volcano. Voyager, too, had found haze at 220 km (main layer) and patchy areas from 300 to 350 km. Some of the bright regions are cloud (such as the feature near the South Pole, at the bottom), others represent highly reflecting material on the planet. NASA/JPL/University of Arizona image PIA07965, based on data from the Cassini Orbiter spacecraft

A *eutectic* water–ammonia mixture, one which freezes at much lower temperatures than does water, would satisfy this condition. Ammonia may be decomposed by photolysis, resulting in the overwhelming presence of nitrogen in the atmosphere, and the loss of hydrogen to space. By implication, therefore, ammonia may indeed have been an important primordial component of Titan. The atmospheric composition as deduced through Voyager UVS and IRIS observations is highly altitude dependent, as summarized by Wilson and Atreya (2004). At 1400 km, for example, the abundance of CH_4 has been measured as $20 \pm 2\%$ (Strobel et al. 1992). The atmospheric abundance ranges for CH_4 at $h > 700$ km, those deduced for N_2 and Ar, and those from 100 to 300 km levels for the other constituents, are given in Table 13.3. The water observation is from ISO (Coustenis et al. 1998).

Diacetylene (C_4H_2, 0.001–0.04 ppm), acetonitrile (CH_3CN, 0.001–0.04 ppm), and helium were also detected. Nitrogen is a relatively inert gas, but it may act as a catalyst in reactions involving methane. Given what we have discussed so far, why the hydrocarbons are so abundant on Titan is an interesting question.

Methane was detected in its atmosphere by G. P. Kuiper in the 1940s, and the features in its spectrum are pressure broadened. The atmospheric pressure is 1.5 bar, the highest for any moon in the solar system. The Huygens probe found winds as high as 120 m/s at $h \approx 120$ km, and decreased wind speeds down to the surface, with vertical shears above $h \approx 60$ km. Titan's atmosphere, like that of Venus, superrotates (its body rotation period is the same as its revolution period about Saturn, 15.95 days). Horizontal winds were mainly W to E, like the rotation, with reverses near the surface, where the methane content was $5.0 \pm 1.0\%$ and the relative humidity was $\sim 50\%$. The

Table 13.3. Titan's atmosphere

Constituent		Amount
N_2	(nitrogen)	80–98%
CH_4	(methane)	0.1–20% [2.2% at 850 km]
Ar	(argon)	<1%
CO	(carbon monoxide)	29–52 ppm
H_2	(hydrogen)	[20 ppm]
C_2H_6	(ethane)	1–15 ppm
C_2H_4	(ethylene)	0.01–15 ppm
C_2H_2	(acetylene)	2–5 ppm
C_3H_8	(propane)	0.04–0.7 ppm
CH_3C_2H	(methylacetylene)	0.004–0.5 ppm
HCN	(hydrogen cyanide)	0.05–0.5 ppm
HC_3N	(cyanoacetylene)	$\leq$ 0.002–0.03 ppm
CO_2	(carbon dioxide)	0.001–0.04 ppm
C_2N_2	(cyanogen)	$\leq$ 0.002–0.006 ppm
H_2O	water	4 ppb

high methane abundance on Titan implies that methane is likely outgassing still, because it is broken down by UV photons. Tobie et al. (2006) note that photochemistry would remove the methane on time scales of a few $\times 10^7$ years. For this reason they argue that once the heating of the interior of Titan by tidal forces and the low thermal conductivity and high viscosity of the methane clathrate (compared to water ice) are taken into account, the cooling interval of Titan's interior is lengthened, allowing the clathrate to be dissociated, and resulting in massive episodic methane outgassing on long time scales and, at present, short time scale volcanism.

In any case, the presence of many hydrocarbons in Titan's atmosphere can be understood given the apparent abundance of methane. The expected pathways have been discussed recently by Wilson and Atreya (2004). Some of the methane photolysis pathways are given below, beginning with the five possible ways for the photodissociation of methane high in Titan's atmosphere. There is no consensus on the "branching ratios," i.e., on what fraction of the CH_4 molecules undergoing dissociation follow which pathway that results in other products, but the first of the pathways represented by equations (13.1) through (13.5) seems the most important, especially when photons other than Lyman α photons are considered. Comments are square-bracketed.

$$CH_4 + h\nu \rightarrow CH_3 + H \qquad [CH_3 \equiv \text{methyl radical}] \tag{13.1}$$

$$\rightarrow {}^1CH_2 + H_2 \quad [{}^1CH_2 \equiv \text{ excited singlet state of methylene}] \tag{13.2}$$

$$\rightarrow {}^1CH_2 + 2H \tag{13.3}$$

$$\rightarrow {}^3CH_2 + 2H \quad [{}^3CH_2 \equiv \text{ triplet ground-state of methylene}] \tag{13.4}$$

$$\rightarrow CH + H_2 + H \qquad [CH \equiv \text{methylidyne}] \tag{13.5}$$

Ethylene is produced through:

$$ {}^3CH_2 + CH_3 \rightarrow C_2H_4 + H \text{ and} \tag{13.6}$$

$$CH + CH_4 \rightarrow C_2H_4 + H \tag{13.7}$$

It can be produced also through a chain involving N_2 in an ion exchange reaction:

$$N_2 + h\nu \rightarrow N_2{}^+ + e^- \tag{13.8}$$

$$CH_4 + N_2{}^+ \rightarrow CH_3{}^+ + N_2 + H \tag{13.9}$$

$$CH_4 + CH_3^+ \rightarrow C_2H_5{}^+ + H_2 \qquad [C_2H_5 \equiv \text{ethyl radical}] \tag{13.10}$$

$$HCN + C_2H_5{}^+ \rightarrow H_2CN^+ + C_2H_4 \tag{13.11}$$

concluding with a de-excitation process:

$$e^- + H_2CN^+ \rightarrow HCN + H, \tag{13.12}$$

Equations (13.8) through (13.12) can be summed up in the equation:

$$2CH_4 \rightarrow C_2H_4 + H_2 + 2H \tag{13.13}$$

N^+ and N_2^+ ions are produced through photons with $\lambda < 79.6\,\mathrm{nm}$ and by collisional excitation by electrons.

The comprehensive model produced by Wilson and Atreya (2004), from which this atmospheric chemistry discussion derives, predicts the latter pathway (13.13) of ethylene production to be most important at $h = 1060\,\mathrm{km}$, but the peak production of C_2H_4, according to that model, occurs with (13.6) and (13.7), at $h = 800\,\mathrm{km}$, with branching ratios of 56% and 42%, respectively.

The ethylene is a major source of acetylene. Through other pathways, the methyl and ethyl radicals combine to produce propane in the lower atmosphere. CH radicals help to produce methylacetylene (CH_3C_2H) and propylene (C_3H_6). Acetylene helps to produce benzene (C_6H_6), which may have been detected in the atmosphere (Coustenis et al. 2003). Other potential pathways can produce the other species. For example (from Strobel 1982):

$$C_2H_2 + h\nu \rightarrow C_2H + H \qquad [C_2H \equiv \text{ethynyl radical}] \tag{13.14}$$

$$C_2H + CH_4 \rightarrow C_2H_2 + CH_3 \qquad [C_2H_2 \equiv \text{acetylene}] \tag{13.15}$$

$$CH_3 + CH_3 \rightarrow C_2H_6 \qquad [C_2H_6 \equiv \text{ethane}] \tag{13.16}$$

$$C_2H_6 + C_2H \rightarrow C_2H_2 + C_2H_5 \quad [C_2H_5 \equiv \text{ethyl radical}] \tag{13.17}$$

$$C_2H_5 + CH_3 \rightarrow C_3H_8 \qquad [C_3H_8 \equiv \text{propane}] \tag{13.18}$$

Some alternate pathways are:

Alternative to (13.15) : $C_2H + C_2H_2 \rightarrow C_4H_2 + H$ (13.19)

followed by: $C_4H_2 + h\nu \rightarrow C_4H + H$ (13.20)

Alternative to (13.20): $C_4H_2 + C_2H_2 \rightarrow C_6H_2 + 2H$ (13.21)

$$C_6H_2 + C_2H \rightarrow C_8H_2 + H \tag{13.22}$$

$$C_8H_2 \rightarrow \text{polyacetylenes}$$

where C_6H_2 is triacetylene and C_8H_2 is tetraacetylene. The C_2H_2 and C_2H_4, among other species, can condense, causing mist and rain in the lower atmosphere, as was seen by the Huygens probe. H and H_2 can then escape from the thermosphere of Titan. Lyman α observations of a wide hydrogen torus about Titan's orbit were made by Voyager; CH_4 atoms, the other hydrocarbons, and nitrides are likely sources. At Titan's temperature, 95 K, water on the surface may not sublime, and so probably has been retained by the planet, and it can be produced high in the atmosphere through the reaction:

$$CH_3 + OH \rightleftharpoons H_2O + {}^1CH_2 \qquad (13.23)$$

This concludes our discussion of Titan.

13.1.4 Uranian Moons

The satellites of Uranus are intermediate-sized or small bodies. The first five were named for characters in the Shakespearean comedy, "A Midsummer Night's Dream." The five largest moons were discovered with modest ground-based telescopes. Since then, many more have been added from both space mission imagery and from large ground-based searches. The resurfacing that was seen on the intermediate moons of Saturn is also seen on those of Uranus. The characteristic description of most of the moons is "dark," although resurfacing of some of the icy moons is apparent. The detail of Miranda as revealed by Voyager demonstrated extraordinarily large relief. The "race course" is reminiscent of the sulci of the Galilean moons Europa and Ganymede, but the huge *scissor fault* has no precedent elsewhere (Figure 13.22).

Some of the intermediate-sized Uranian moons, like those of Saturn, appear to have reworked surfaces, probably from the same cause. The moons have

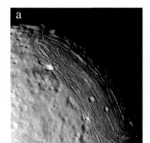 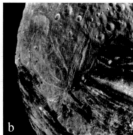 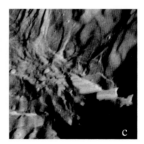

Fig. 13.22. Images of the surface of the Uranian moon Miranda as seen by Voyager 2. **a** The "racetrack," resembling the sulci of Ganymede and Europa. **b** Arcuate features punctuated by large and small craters and highly contrasting terrain. **c** A huge scissor fault, with a progressive scarp, reaching $\sim$15 km above the surface, seen here on the terminator. From the Voyager 2 flyby of Uranus in January, 1986. NASA/JPL images PIA00141, PIA00140, and PIA00044, respectively

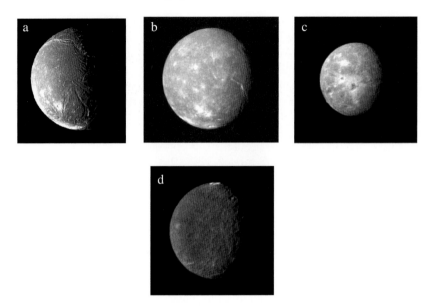

Fig. 13.23. Voyager 2 images of **a** Ariel, **b** Titania, **c** Oberon, and **d** Umbriel. Ariel, Titania, and Oberon have lengthy arcuate valleys and scarps, and bright areas where impacts have excavated fresh ice. Umbriel has limited areas displaying these features, and the overall impression is of an older surface, covered by a very dark regolith. All are heavily cratered. NASA/JPL images PIA01534, PIA01979, PIA00034, and PIA00040, respectively

slightly different albedos (and on some, these vary across the surface), but all are dark, as Table 13.2 indicates. Several of the larger of these Uranian moons are seen in Figure 13.23.

Titania is the largest and most massive. Ariel is the brightest, with a visual albedo, $A_V = 0.35$ (compare to Umbriel, for which $A_V = 0.19$). The darkest of the five largest moons, Umbriel, nevertheless has bright regions, perhaps the sites of more recent impacts which excavated fresh ice. According to de Pater et al. (2006), the small moon Mab shows H_2O ice absorption at 2μm features similar to those of the larger moons, and unlike the smaller moons interior to its orbit.

13.1.5 Neptunian Moons

Little is known of the Neptunian system, but its principal moon, Triton was observed in some detail by Voyager 2. Nereid was discovered in 1949 by Gerard Peter Kuiper (1905–1973); its orbit has a high eccentricity (0.75) and an inclination of 28°. It displays light albedo regions and it has been compared to Phoebe in its physical properties (see Figure 13.24).

 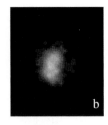

Fig. 13.24. Two small, resurfaced satellites: **a** Phoebe in the Saturnian system as seen by the Cassini Orbiter, NASA/JPL/Space Science Institute, Boulder, image PAI06064; **b** A Voyager 2 view of Nereid in the Neptune system. NASA/JPL image PIA00054

13.1.5.1 Triton Discovered by Lassell in 1846, Triton is the only one of the major moons (under any of the three definitions above) that is moving in a retrograde (CW) orbit around its planet. Yet, its orbital eccentricity is small. Its blue color, high albedo, and visual appearance indicate that it is an icy world, with a measured temperature of 38 ± 4 K. The measured atmospheric pressure is $16 \pm 3\,\mu$bar $= 1.6$ Pa. Voyager 2 determined that the partial pressure of methane at the surface is only 1 nanobar $= 10^{-4}$ Pa. From subsequent ground-based reflection spectroscopy, Brown and Cruikshank (1997) have identified features of N_2, CH_4, CO_2, and CO. The atmosphere is essentially N_2 with traces of CH_4 and upper limits of Ne, A, and CO of less than 1%. N_2 sublimes in sunlight and recondenses in cooler regions, where it may form a translucent glaze coating. It is interesting that in the phase diagram of nitrogen, the triple point, where it can exist as solid, liquid, or gas, is 0.123 bar and 63.15 K. These conditions may be found in Triton's interior. The thermal profile of the atmosphere is such that the temperature decreases up to ∼50 km, and thereafter rises in the thermosphere to ∼100 K above 300 km. Despite its extremely tenuous atmosphere, there are clouds: evidence for, and the extent of, a polar cap layer 4–8 km high can be seen in Figure 13.25.

Features on the planet include a "cantaloupe" region of dimpled and striated areas (see Figure 13.26). There are nitrogen geysers, the plumes of which reach heights of 8 km, and apparently cryovolcanoes, involving melted ice rather than rock. The evidence and possible mechanisms are discussed by McKinnon and Kirk (1999). The surface relief is ∼1 km and this requires a fairly rigid crust. The mean density of Triton is $2075 \pm 19\,\mathrm{kg/m}^3$. These data imply an important role for volatile ices, which, in the outer solar system, serve the same role as water ice does on Earth, and possibly also for water ice, which may serve the same role here as silicate rock does on Earth.

There has been substantial resurfacing. The largest impact crater is Mazomba, 27 km across, and impact craters on Triton are generally sparse. Internal heating must have accompanied Triton's capture from an original parabolic orbit as well as any collisions it may have had before settling into its

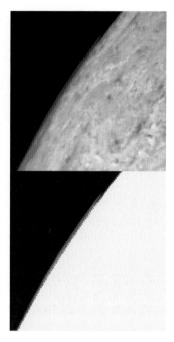

Fig. 13.25. Evidence for clouds in the atmosphere of Triton over its south polar cap provided by Voyager 2. The lower image is of higher contrast to emphasize the extent and height of this layer. NASA/JPL image PIA02203

present-day retrograde but nearly circular orbit. Moreover, tidal dissipation would have heated the moon: today Triton is rotating synchronously.

13.1.6 Pluto–Charon

We discuss Pluto as well as Charon among the moons because the ratio of radii of Pluto to Charon is smaller than that for any major planet to its major moon, making this a kind of binary system, and because in scale Pluto belongs to the population of the major moons — objects that long ago collided with planets, and were either captured by them or expelled from the inner solar system through interactions with the giant planets. It is also an icy denizen of the outer solar system. Its orbital inclination (17°.2) and eccentricity (0.25) are larger than those of any of the major planets, and its perihelion lies within the orbit of Neptune. The two objects cannot collide because Pluto is locked into a 3:2 (orbit:orbit) period resonance with Neptune, a circumstance that makes it a *plutino*; it is the largest known member of this group. Its physical properties suggest Pluto to be a likely member of the Edgeworth–Kuiper Belt of remote, icy bodies, heavily laced with hydrocarbons (refer to Chapters 14

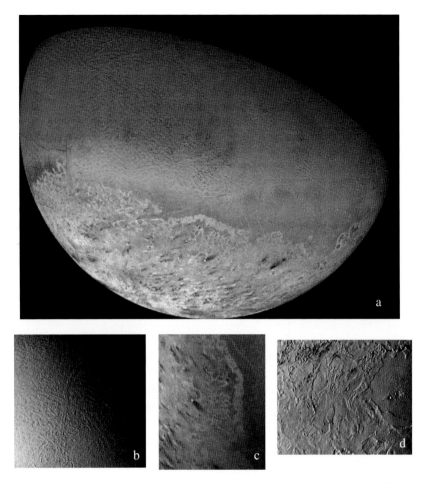

Fig. 13.26. a Triton, the giant moon of Neptune, seen in color mosaic in Plate 13, based on three-passband images recorded by Voyager 2 in its 1989 flyby of Neptune. Notice the blue-green "cantaloupe" terrain at the top, the bright and somewhat pink regions of frost condensation, probably of methane, in a broad area around the South Pole, and the demarcation region between them, in which craters can be seen. In the polar region, dark plumes, likely geysers of nitrogen, are visible. In between are dark spots with bright rims of unknown origin. Produced by the USGS, NASA/JPL image PIA00317. **b** "Cantaloupe" terrain. **c** Plumes from geysers may extend up to 150 km from their vents (see also Plate 14). **d** Ruach Planitia, a possible caldera of a cryovolcano. From left to right, **b–d** are NASA/JPL images PIA01537, PIA02214, and PIA02208, respectively

and 15 for more details). It is one of the largest members of this class known to date, but not the largest.

Historically, Pluto was the last "planet" to be discovered (in 1930 by Clyde Tombaugh (1906–1997), a Lowell Observatory employee engaged in a search for a trans-Neptunian planet undertaken and directed by Percival Lowell

(1853-1916)), and it now has joined a number of asteroids (Ceres and Vesta among others) that were once considered "planets". The IAU has designated Pluto a *dwarf planet*, no longer in the same category as the eight planets. Indeed, at least one other such dwarf object (Eris) is now known to be larger than Pluto. The moon Charon was discovered in 1978 by US Naval Observatory staff members Christy and Harrington while they were carrying out astrometric work on the position of Pluto. This discovery provided the first opportunity to obtain a precise mass for Pluto and the result confirmed what its very small influence on the positions of Uranus and Neptune suggested: not much mass! Moreover, its extreme faintness (~15th magnitude) had strongly suggested a small diameter. Charon was observed to occult a star in 2005, and this provided a precise radius (Table 13.2). Charon's diameter is 1208 km; compared to Pluto's 2300 km, this gives the largest ratio of secondary to primary body diameters in the solar system. Pluto's density is $2029 \pm 32 \, \mathrm{kg/m^3}$, similar to that of Triton ($2060 \, \mathrm{kg/m^3}$). Pluto and Charon are locked into a 1:1 spin:orbit resonance; their revolution and rotations have the same period.

Near perihelion, Pluto has been seen to have an atmosphere of CH_4 and has a surface coating of CH_4 ice. Charon appears to have a water ice coating, but no volatile ices. A determination of the temperatures of Pluto and Charon has revealed that Pluto's measured temperature is 10 K cooler than its equilibrium temperature (i.e., 43 K as opposed to 53 K), whereas Charon's observed temperature is 53 K. This is attributed to the evaporation of methane from Pluto's frosty surface under sunlight, as feeble as it may be, thus cooling the surface.

During the mid-late 1980s the system underwent a series of eclipses. During the last such series, patterns of light and dark across both Pluto and Charon could be determined. Patterns have also been discerned with the Hubble Space Telescope (see Figure 13.27).

One model for Pluto's interior contains 70% rock and 30% ice.

Spectroscopic variations have been seen on both Triton and Pluto, implying that the atmosphere changes with time. Pluto has just passed perihelion, when its equator was more directly exposed to sunlight (most often, as for Uranus, one of its poles is directly exposed to sunlight for very long intervals, a situation that changes very slowly). Therefore both it and Triton undergo seasons, in which differing regions are more directly exposed to sunlight over time. During the Voyager 2 flyby of Triton, the region between −40 and −60° latitudes (where the plumes were observed) was sub-solar. Similar composition and structure is possible for these two objects, although water ice detection on Pluto has been rather elusive. It is possible that it is generally covered by more volatile material.

Dynamically, the Pluto–Charon system is also interesting. The inclination of the rotation and Charon orbital planes are retrograde, with an inclination

Fig. 13.27. Processed images of Pluto taken in June and July, 1994, and (*insets*) raw HST images of Charon, showing contrasting surface features. The scale is 150 km/pixel. North is up in the Pluto images, which show opposite hemispheres. Hubble Space Telescope images, StScI-PR96-09a. (Photojournal image PIA00825). Credits: Alan Stern, Southwest Research Institute; Marc Buie, Lowell Observatory; NASA/ESA. PIA00825

relative to the ecliptic of 99°. Pluto's orbital properties, its possible relation to Triton and other objects in the outer solar system, and its status as a planet are discussed further in Section 13.4.2.

A mission to Pluto and the Kuiper Belt was launched on January 19, 2006 and is enroute to Pluto, at present writing. By means of gravity-assist orbital planning, it is expected not to take more than a century, as it would with a purely Hohmann orbital transfer, but to arrive at the Pluto–Charon system by July 14, 2015, if the mission proceeds as planned. The mission is equipped with a CCD for visual imaging, a near-infrared imaging spectrometer, an ultraviolet imaging spectrometer, and other instruments to analyze particle emissions, dust, and to conduct radio experiments.

13.2 Origins of Ring systems

Ring systems may be the debris of either tidally destroyed moons or of "failed" moons—i.e., material that began to be created through collisions but was unable to accrete into cohesive moon bodies because of tidal disruption.

Tidal instability may be calculated and a critical distance from the planet, the Roche limit, obtained for any satellite of total mass $2m$ and orbital radius r. The tidal force is a differential gravitational force. The magnitude of the acceleration or force per unit mass of this tide-raising force may be expressed as:

$$a_{\Delta g} = \partial/\partial r\,(-GM/r^2) \cdot \Delta r = (2GM/r^3) \cdot \Delta r \qquad (13.24)$$

where M is the pass of the primary, the planet in this case. Considering that the satellite is being pulled apart, we make the assumption that two halves of equal mass are involved; the separation of their centers is then equal to Δr.

To the differential gravitational force we must add the differential centripetal force (or, in the rotating frame, the differential centrifugal force) due to the differential motion of this extended object around the planet.

The centripetal acceleration may be expressed in terms of the angular velocity, $\boldsymbol{\omega} = (\mathrm{v}/r) \cdot \hat{\boldsymbol{\omega}}$:

$$\mathbf{a_c} = \boldsymbol{\omega} \times (\boldsymbol{\omega} \times \mathbf{r}) \qquad (13.25)$$

the magnitude of which, in a circular orbit, becomes:

$$a_c = v^2/r = \omega^2 \cdot r = a_{\text{gravity}} \equiv GM/r^2 \qquad (13.26)$$

so that

$$\omega^2 \cdot \Delta r = GM/r^3 \cdot \Delta r \qquad (13.27)$$

The magnitude of the differential centripetal force per unit mass is therefore:

$$a_{\Delta c} = |\partial/\partial r(\omega \times \omega \times r)| \cdot \Delta r = \omega^2 \cdot \Delta r = GM/r^3 \cdot \Delta r \qquad (13.28)$$

Combining both, we get the magnitude of the total disruptive differential force per unit mass:

$$a_{\Delta f} = a_{\Delta g} + a_{\Delta c} = 3\,GM/r^3 \cdot \Delta r \qquad (13.29)$$

For *stability*, self-gravity must exceed the disruptive force, so, recalling that the satellite mass is $2m$, the force inequality condition becomes:

$$Gm^2/(\Delta \mathrm{r})^2 > 3\,GM(2m)/r^3 \cdot \Delta r \qquad (13.30)$$

or

$$Gm^2/(\Delta \mathrm{r})^3 > 6\,GMm/r^3$$

so that

$$r^3 > 6(M/m) \cdot (\Delta r)^3 \qquad (13.31)$$

With densities $\rho_s = 2m/[(4/3)\pi(\Delta r)^3]$, $\rho_p = M/[(4/3)\pi R_p^3]$, for the satellite and planet, respectively, (13.31) becomes:.

$$r^3 > 6(M/m) \cdot (\Delta r)^3 = 12[\rho_p/\rho_s] \cdot R_p^3 \qquad (13.32)$$

The *critical distance* within which a body cohering exclusively through gravitational attraction will be torn apart by tidal effects is therefore:

$$r_{\text{critical}} = \{12[\rho_p/\rho_s] \cdot R_p^3\}^{\frac{1}{3}} = 2.29\, R_p \cdot [\rho_p/\rho_s]^{\frac{1}{3}} \qquad (13.33)$$

Such a critical distance is usually called a *Roche limit*, after a mid-nineteenth century celestial mechanician named Édouard Roche (1820–1883). With a somewhat more rigorous treatment for a satellite held together purely by gravitational forces, Roche obtained a numerical factor of 2.45 .

One may also calculate the critical density of a satellite, ρ_s, for stability at a given distance from the planet. From (13.32),

$$(\rho_s)_{\text{critical}} > 12[R_p^3/r^3]\,\rho_p \qquad (13.34)$$

This can be used, for example, to study the makeup of the object that gave rise to a particular ring. The demise of the object through tidal disruption can be predicted if

1. It is massive enough to raise a significant tide on its parent planet

2. It is within the synchronous orbit distance

The latter condition arises because in this case, the period of the satellite is smaller than the rotation period of the planet (the synchronous orbital period), so the satellite orbits faster on average than a point on the surface rotates. Tidal friction then produces a bulge on the planet behind the moon, and this bulge retards the moon and causes orbital decay, until the Roche limit (for its density) is reached.

One may use the principle of differential gravitational attraction to study the stability of a satellite orbit around a planet with respect to disruption, by, say, the star of the planetary system. It is also applicable to such cases as the tidal effect on a globular cluster at its pericenter distance to the galactic center, or on one galaxy in proximity to another. Within ring systems, the perturbations by a moon on a ringlet (a name given to a single ring strand in a ring system) can cause the ring to expand outward or to fall back toward the planet, if the moon is revolving faster or slower than the ringlet, respectively). A pair of moons straddling a ring can then "shepherd" the ring, with the inner moon constraining the ring particles from moving into lower orbits and the outer moon constraining particles from moving to higher orbits, as we illustrate below.

13.3 Ring Structures

13.3.1 Jovian Rings

There are basically three ring structures, all discovered in the Voyager 1 flyby. For scale, the radius of Jupiter, $R_J = 71,400$ km.

- Halo ring: $R_{inner} \cong 90,000$ km from Jupiter's center and width of 30,500 km, with thickness $\sim 20,000$ km
- Main ring: $R_{inner} = 122,500$; $\Delta R = 6000$ km; thickness less than 30 km
- Inner Gossamer ring: 128,940–181,000 km

  Associated satellites:

  Metis (J16) with $a = 128,100$ km
  Adrastea (J15), with $a = 128,900$ km
  Amathea (J5), with $a = 181,400$ km

- Outer Gossamer ring: 181,000–225,000 km

The reflection properties of these rings ($A_V \approx 0.05$) suggests $\sim 3\,\mu$m-size particles. Views seen by the Galileo Orbiter while in Jupiter's shadow are reproduced in Figure 13.28. The thickness of the main ring is only 30 km. The greater thickness of the others is explained mainly by the inclinations of the source moons of these rings. Figure 13.29 reproduces NASA graphics showing the orbital locations of the larger ring moons amid the rings.

13.3.2 Saturnian Rings

These were discovered by Galileo in 1610 but identified only as *ansae*, "handles" (as on urns). Their "disappearance" was noted in 1612. Huygens proposed the idea of an equatorial disk in 1655. In 1675, Jean-Dominique Cassini (1748–1845) detected a discontinuity within the rings—now known as *Cassini's Division*, separating Rings A and B. Broad disk instability was demonstrated by Simon de LaPlace, who also showed in 1785 that only narrow

Fig. 13.28. Composite views of Jupiter's rings assembled from Galileo Orbiter images. A brightness enhancement in the upper image permits the cloud of particles constituting the halo ring to be visible. In the lower image, structure in the main ring can be seen. NASA/JPL/Cornell image PIA01622

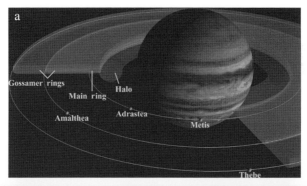

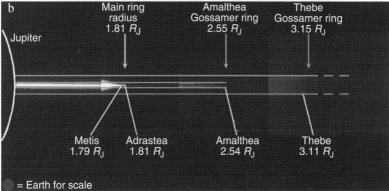

Fig. 13.29. Graphics of the Jovian rings, based on Galileo Orbiter Solid-State Imager data showing the locations of the orbits of small moons associated with the rings (see Table 13.2). The thickness and bluntness of the *Gossamer rings* are due to the orbital inclinations of their sources, the moons Adrastea and Thebe. NASA/JPL images produced by Cornell University: **a** PIA01627 and **b** part of PIA01623, where successive *left* to *right panels* have increasing sensitivity, in order to display the faint outer rings

ringlets were stable because of shearing effects. That the rings were indeed not solid but had to be made of particles was shown by James Clerk Maxwell (1831–1879) in 1857. The measurement of Doppler shift variation across the rings by J. E. Keeler (1857–1900) and W. W. Campbell (1862–1938) in 1895 confirmed Keplerian velocities.

There are four ring systems visible from ground-based telescopes on Earth: A, B, C, D. These are shown in Figure 13.30 (Plate 15). The F, G and E rings are less visible and were discovered by spacecraft. The F and G rings are seen in Figure 13.31. The E ring is so wide that the orbits of Enceladus, Tethys, Dione, and other moons are embedded in it. The inner to outer ring structure is as follows ($R_S = 60{,}268$ km):

- Ring D: 1.11–$1.24 R_S$ ($66{,}900$–$74{,}510$ km);
- Ring C: 1.24–$1.53 R_S$ ($74{,}568$–$92{,}000$ km);

  "Titan ringlet" at $1.29 R_S$ ($77{,}871$ km);
  Maxwell Gap at $1.45 R_S$ ($87{,}491$ km);

- Ring B: 1.53–$1.95 R_S$ ($92{,}000$–$117{,}580$ km; and thickness 0.1–1 km);

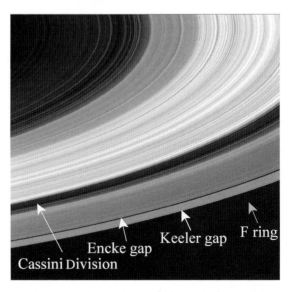

Fig. 13.30. A Cassini view of the rings of Saturn. From the left: the faint D-ring complex of ringlets; the C ring with its narrow Maxwell gap; the very bright B ring, terminated by the Cassini's Division (with faint ringlets); the A ring with the Encke gap and very narrow Keeler gap near its outer edge; and the F ring just beyond the A-ring edge. NASA/JPL/Space Science Institute, Boulder image PIA05421; annotation by E. F. Milone. See also color Plate 15

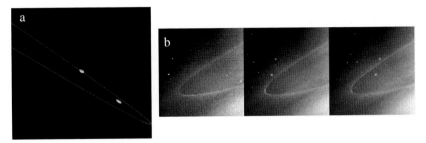

Fig. 13.31. a The F ring of Saturn and the shepherding moons Prometheus (102 km length, 483,500 km distance) on the left and Pandora (84 km long, 459,000 km distance) on the right, as seen by Cassini Orbiter. NASA/JPL/Space Science Institute, Boulder image PIA07653. **b** Cassini's views of the G ring separated by intervals of 45 min, with bright, sharp arcs in an otherwise diffuse ring. The movement of the spacecraft is reflected in the positional shifts in the background stars. NASA/JPL/Space Science Institute, Boulder image PIA07718

Cassini's Division (B to A, 117,500 km) contains ∼20 ringlets;

- Ring A: 2.03–$2.27R_S$ (122,170–136,775 km; thickness 0.1–1 km);

  Encke gap at $2.22R_S$ (133,589 km; 325 km wide);
  Keeler gap at $2.27R_S$ (136,530 km);
  Associated satellites:

    Pan (S18), $a = 133,580$ km;
    Atlas (S15), $a = 137,670$ km;
    Daphnis (S35), $a = 136,530$ km;

- Ring F center: $2.33R_S$ (140,180 km);

  Associated satellites:

    Prometheus (S16), $a = 139,353$ km;
    Pandora (S17), $a = 141,700$ km;

- Ring G: 2.82–$2.90R_S$ (170,000–175,000 km; and thickness 100–1000 km);
- Ring E: 3.00 to ∼$8.0R_S$ (181,000 to ≥483,000 km; and thickness ∼1000 km); see text for associated satellites.

The A_V values for Rings A, B, and C are 0.60, 0.65, and 0.25, respectively; the suggested composition is water ice, at least for the brighter values. Beyond the G ring, the E ring is even more diffuse.

Images from the Cassini Orbiter have spotlighted the actions of the small moons in clumping as well as shepherding the ring particles. Janus (S10)

Fig. 13.32. Cassini discovered Daphnis (∼7 km across) within the Keeler gap, visible here in the center of the image. The Encke gap is visible at the upper right corner and the faint F ring at the lower left corner. This image was obtained on August 1, 2005 when Cassini was 835,000 km from Saturn. A NASA/JPL/Space Science Institute, Boulder image, PIA07584

and Epimetheus (S11) are co-orbital at $a = 151,472$ km; they share the same orbital region, but alternately slow down and speed up each other so that they exchange levels and orbital speeds before they can actually collide. Prometheus and Pandora shepherd the F ring, the outer moon retarding and the lower moon accelerating ring particles so that they stay more or less confined to a narrow, if irregular, ring (Figure 13.31a). Like Pan in the Encke gap, Daphnis (S35) = S2005/S1 plays a similar role in the Keeler gap (Figure 13.32).

Among the larger moons, Enceladus orbits in the E-ring's brightest part and contributes to it. It is perhaps no coincidence that the intermediate moons Dione, Tethys, and possibly Rhea, with evidence of ice flows on their surfaces, are also within the E ring; they too may be contributing to it.

Recently a ring of material around Rhea was announced by the Galileo team. This is the first ring to have been found around a moon.

Finally, we note the presence of "spokes" among Saturn's rings. These had been seen as dark features in the Voyager flybys. Cassini has seen bright spokes (Figure 13.33). The difference is probably the viewing angle: Voyager observed them at low-phase angles and saw little backscattering of the sunlight. Cassini observed the forward scattering component which was large. The ring temperatures (Figure 13.34, Plate 16 in color) provide further clues. We discuss these phenomena further, below.

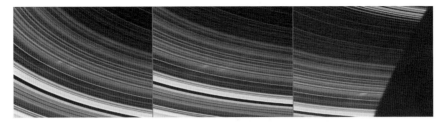

Fig. 13.33. Bright spokes seen in outer B ring by the Cassini Orbiter. They are viewed at a phase angle $\Phi \approx 145°$, while those seen by the Voyagers were viewed at low-phase angles. The spokes are $\sim 3500\,\mathrm{km}$ long and $\sim 100\,\mathrm{km}$ wide. NASA/JPL/Space Science Institute, Boulder images PIA07731

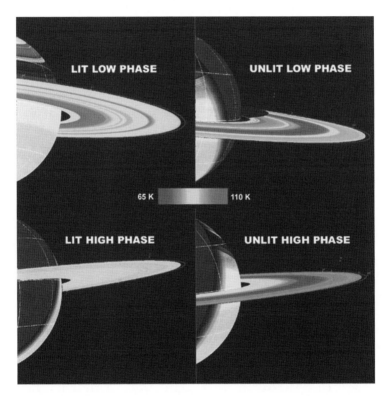

Fig. 13.34. Sun-lit and unlit ring temperatures, measured from low and high-phase angles from Cassini Orbiter. The relative coolness of the B-ring particles is attributable to their high albedo; the darker rings absorb more sunlight. Note the strong thermal emission from the C ring, and the Cassini gap where the ringlets have lower albedo. A NASA/JPL/GSFC graphic, PIA03561, produced by Goddard Space Flight Center. See also color Plate 16

13.3.3 Uranian Rings

Nine rings around Uranus were discovered through the occultation of a star on March 19, 1977 by airborne (Elliot et al. 1977) and ground-based (Millis et al. 1977) teams. The discovery was confirmed by Voyager 2, which found two more rings. They are all extremely dark ($A \approx 0.03$). The widths vary between 1 and 100 km; there is one broad ring (ε) and one diffuse one, 1986 U2R.

There are several groups of Uranian rings: 6, 5, 4; α, β; η, γ, δ, ε; and rings 1986 U2R and 1986 U1R (see Figures 13.35 and 13.36 and Plate 17). Associated moons are Cordelia (U6), at $a = 1.947 R_{\mathrm{U}}$, and Ophelia (U7), at $a = 2.105 R_{\mathrm{U}}$. The individual ring semi-major axes (in units of the radius of Uranus, $R_{\mathrm{U}} = 25,559$ km) and eccentricities are as follows:

- Ring ζ (R/1986 U2) 1.45–1.55R_{U};
- Ring 6: 1.638R_{U}, $e = 0.0014$;
- Ring 5: 1.654R_{U}, $e = 0.0018$;
- Ring 4: 1.667R_{U}, $e = 0.0012$;
- Ring α: 1.751R_{U}, $e = 0.0007$;
- Ring β: 1.788R_{U}, $e = 0.0005$;
- Ring η: 1.847R_{U};
- Ring γ: 1.865R_{U};

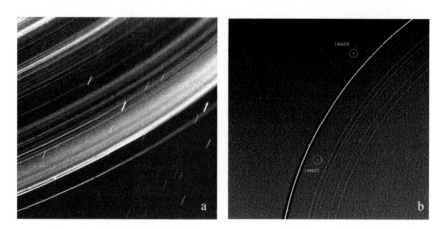

Fig. 13.35. The Uranian ring system as seen by Voyager 2. **a** A 96-sec exposure of the rings. The star images, as well as the blurring of the ringlets on the upper right, reflect the motion of the spacecraft. **b** An annotated high-resolution image showing shepherding satellites. NASA/JPL image PIA00142 and PIA01976, respectively

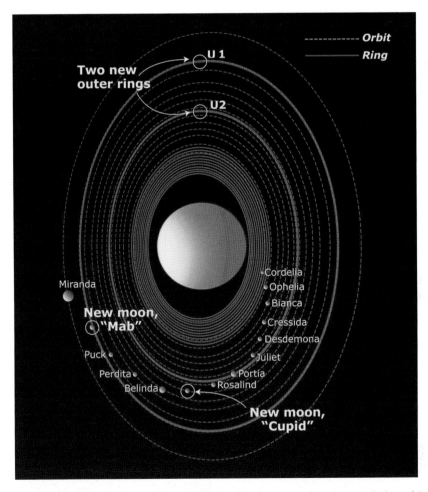

Fig. 13.36. A graphic showing the relationship between the rings and the orbits of associated moons in the Uranian system. Some of the newer (HST) discoveries are indicated. Note the orbit of the satellite Mab, centered on R/2003 U1, here abbreviated to "U1"; the moons Portia and Rosalind flank R/2003 U2, labeled "U2". A NASA graphic, slightly altered by E.F. Milone. Credits: NASA/ESA/A. Feild (STSci). See Plate 17 for a color version.

- Ring δ: $1.891R_U$, $e = 0.0005$;
- Ring λ (R/1986 U1): $1.96R_U$;
- Ring ε: $2.002R_U$, $e = 0.0079$.

  R/2003 U2: $2.59\text{–}2.74R_U$; associated satellites: Portia and Rosalind

  R/2003 U1: $3.37\text{-}4.03R_U$; associated satellite: Puck and Mab

The thickness of the rings is typically ~100 m; the epsilon ring is less than 150 m thick. All have visual albedos ~0.03. Two outer rings were discovered more recently (Showalter and Lissauer, 2006; de Pater, et al., 2006). The ring system is apparently unstable. Individual rings varied in position between the Voyager mission and recent HST and Keck observations; and ring brightnesses changed because of changes in the distribution of dust in the ring plane, and because the Uranian equinox occurred in 2007, producing ring plane crossings and views of the dark side of the rings as seen from the Earth (de Pater et al. 2007). Optically-thin rings brighten as the opening angle decreases to zero (edge-on view) because more particles are visible along each line of sight, whereas optically-thick rings fade because particles are both shaded and obscured by other particles. For this reason, the ϵ ring, dominant to 2006, appeared to be absent in 2007.

13.3.4 Neptunian Rings

Stellar occultations provided evidence of partial rings ("ring arcs"); variable thickness rings were confirmed by Voyager 2. There are five known rings; of these, two are wide and one is quite clumpy. Figure 13.37 shows the extent and clumpiness of the rings and the presence of shepherds. It has been estimated that the amount of material in the Neptunian rings is only ~1% of that in the Uranian ring system. The Neptunian ring structure (in units of $R_N = 24,800$ km) is as follows:

- Galle = 1989N3R: $1.5R_N$ [inner extent of 1989N3R];
                        1.69 (1700 km width); and
                        $2.0R_N$ [outer extent of 1989N3R].
  Associated satellites: Naiad (N3), $a = 1.945R_N$;
                        Thalassa (N4), $a = 2.019R_N$.

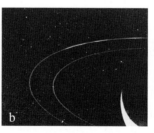

Fig. 13.37. Neptune's rings as seen by Voyager 2. **a** Composite long-exposure images of the rings on either side of Neptune; **b** short exposures showing clumpiness; and **c** the presence of a shepherding satellite outside a faint ring. NASA/JPL images PAI01997, PIA01493, and PIA00053

- Leverrier = 1989N2R: $2.15R_N$ (narrow and unresolved by Voyager 2);
  Associated satellite: Despina (N5), $a = 2.118R_N$.
- Lassell = 1989N4R: 2.14 to $\sim 2.31R_N$ (~ 4000 km width).
- Arago = 1989N5R: $2.31R_N$ (~ 100 km width).
- Adams = 1989N1R: $2.54R_N$ (16 km width); (clumpy: three bright arcs, 4°, 4°, and 10°);
  Associated satellite: Galatea (N6), $a = 2.498R_N$.

13.3.5 Nature and Possible Origins of the Ring Structures

The detailed views returned by the Pioneer, Voyager, Galileo, and Cassini missions to the giant planets demonstrated very complex structure in the rings. The presence of gaps among the rings, of large numbers of thin individual rings, kinked and eccentric rings, and co-orbital and shepherding satellites were largely unexpected!

The gaps can be generally understood as analogous to the Kirkwood gaps created in the asteroid belt by Jupiter, with particular moons taking the role of Jupiter. In one case, the Encke gap in Saturn's A ring is apparently due to the moon *Pan* moving within it. *Atlas* seems to moderate the outer A-ring dynamics.

The extremely thin ringlets can be understood as a consequence of small moons located on either side (the *shepherding moons*), whose accelerations act mutually to confine the material of the ring to a narrow torus between them. Saturn's narrow F ring is such a structure with shepherds *Prometheus* and *Pandora*.

The kinked nature of some of the rings, such as the F ring of Saturn, may be due to a shepherding moon in an eccentric orbit, or involvement of more moons (such as Pan in Figure 13.13) so that shepherding action is performed in a non-uniform way. Perhaps eccentric rings have an explanation in moon eccentricities, but the dynamics of the individual ring particles, through perturbation-induced collisions, certainly add to the non-uniformities. Indeed the propagation of the effects persists around the ring following the passage of the moon.

To the class of unexpected phenomena can be added Neptune's outer ring, which is very clumpy (Earth-based occultation data suggested that the rings were only arcs; Voyager 2 images showed highly variable ring brightness but not the absence of material in any parts of the ring). The explanation for the bright arcs is not yet known, although dust is suspected. Collisions among small moons have been suggested to account for the faint inner rings of the Jovian system. It may also suggest recent breakup of an object due to tidal forces, but collisions probably play a very important role.

The makeup of the rings varies from planet to planet, and from one ring to another. Jupiter's rings are very wide but very tenuous. The bluer outer rings of Saturn and Uranus suggest material of different composition.

In contrast to Saturn's bright rings, those of the other giants are extremely dark, preventing their early visual detection from Earth. Radar observations suggest that most of the mass in Saturn's main rings consists of debris in the range 1 cm to 10 m diameter. Uranus' must be analogous to smoke particles. Dust is nearly ubiquitous in ring systems (with important exceptions, such as the ε ring of Uranus). The "spokes" (radial or nearly radial streaks) in Saturn's rings may be dust which has been charged and lifted out of the ring plane by electrostatic forces, creating shadows on the rings below, as viewed by Voyager (but see Cuzzi et al. 1984, Fig. 25, for a bright, forward-scattered image taken at phase angle 135° from the Voyager 1 mission), but shining in forward scattered light when viewed by Cassini (Figure 13.33).

The intensity of the sunlight I, at any wavelength, λ, scattered by a small, spherical dust particle depends on the intensity it sees, I_0, its radius, R, the index of refraction of the material of which it is composed, m, and the distance from the observer (in this case, the spacecraft), r. It also depends on the scattering function appropriate for the particle, F. The dependencies can be folded into the equation (after Greenberg, 1968, Section 8.5, esp. p. 347):

$$I_r\,(\theta,\lambda) = (\lambda/2\pi)^2\,I_0\,(\lambda)\,F(\theta,\lambda,R,m)\,r^{-2} \qquad (13.35)$$

In the case of photometric observations, the expression must be integrated over the passband in which the observation is made. The function F is defined by Van de Hulst such that

$$F(\theta,\lambda,R,m) = \tfrac{1}{2}\left[i_\parallel\,(\theta) + i_\perp\,(\theta)\right] \qquad (13.36)$$

where $i_\parallel$ and $i_\perp$ refer to the scattering functions for polarized radiation along and perpendicular to the scattering plane. In practice, there is a distribution of sizes of dust grains, $n(R)$, which must be known, in order for the scattering to be computed. This is a difficult problem in the rings, because the radio and radar observations have revealed meter-sized objects in some of the rings, and, as we have seen, the rings tend to be collections of closely spaced ringlets, perturbed and shepherded by moons, and the shapes of the particles may need to be better known.

The propagation of waves due to dynamical interactions has been discussed by Shu (1984), Franklin et al. (1984) and by Dermott (1984), among others. More recently, Fridman and Gorkavyi (1997, especially Section 2.1) make a convincing case that the ringlet structure in the Saturnian rings is largely due to resonance effects produced by the moons: linear wave in Cassini's Division due to Iapetus; extended spiral wave in the B ring due to Janus; a 3:5 damped wave resonance in the A ring due to Mimas; bending waves in the C ring due to resonances with Titan. Moreover, they indicate that one mode of the resonances may also involve vertical oscillations.

The source of dust seen among the ring systems may in part be collisions involving close satellites, and impacts on them by interplanetary material. Fragments chipped off these moons could be accompanied by the ejection of micron or sub-micron-sized particles. The discovery of apparently-fragmented moonlets embedded in a placid region of the Saturnian rings (Figure 13.14) supports the idea as does the existence of ringlets accompanying some satellites, such as the material sharing Pan's orbit in the Encke gap (see Fig. 13.13).

Charged particles may arise from ionization or from electron impact. The magnetic fields of the gas giants sweep through the ring systems, and charged particles will be accelerated as a consequence, leading to periodic phenomena such as spokes. However, the attempts to explain the spokes through a variety of mechanisms all seem to have difficulties (Grün et al. 1984; Mignard 1984).

The dynamics and the history of study of the rings and their phenomena, and their interactions with the moons, can be found in the comprehensive treatment by Fridman and Gorkavyi (1997). However, ring mysteries remain; for example the D and F rings of Saturn, the arcs of the Adams ring of Neptune, and, most dramatically, the rings of Uranus all have exhibited major changes on timescales of years.

13.4 Orbital Stability of the Moons and the Case of Pluto

13.4.1 Satellite Stability

The severe tidal stress that satellites undergo if they are within the Roche limit of their planet would seem to make it less likely that the Moon originated in a rapidly rotating Earth and was sheared off through rotational instability, as was suggested by George H. Darwin (1845–1912). The orbital evolution of the Moon in any case depends on the tidal bulge of the Earth. Since the Earth is rotating more rapidly, the tidal bulge is ahead of the Moon on the Moon-facing side, and is accelerating it so that a_{Moon} is increasing by a few centimeters per year. In future, the Earth will be slowed by tidal friction so that the length of the day will increase to the length of the month; i.e., the Moon will be in a geosynchronous orbit. It has been conjectured that, at that point, braking by solar tides will cause a_{Moon} to decrease slightly, resulting in the Moon orbiting faster than the Earth is rotating. The Earth's tidal bulge then lags behind the Moon, and the Moon will spiral inward, eventually entering the Roche limit to be tidally disrupted. If this is correct, Earth is destined to become a ringed planet.

In addition to the stability with respect to the Roche limit or the synchronous orbit radius, satellites are also subject to the gravitational attraction of the Sun. The possibility that some of the satellites were captured from solar orbits rather than born in situ has been examined through orbital analysis.

Victor Szebehely (1979) used a parameter of the restricted three-body problem to study the stability of satellite orbits that were known at the time. The analysis is based on a theorem of G. W. Hill (1878) that a satellite orbit around the smaller mass of a restricted three-body problem is stable if it is located inside a zero-velocity curve surrounding only this smaller mass (the planet). Hill used this definition of stability to conclude that the Moon's orbit is stable. Szebehely applied the definition to all the planets of the solar system and found critical distances from each of the planets within which satellites would be in "stable" orbits. Several of the outer satellites of Jupiter (J8, J9, J11, and J12) are outside the stable zone, while J7, Phoebe (S9, in a retrograde orbit), and the Moon, are just inside the stability line (Table 13.4).

This suggests that capture origins are therefore possible for these bodies also. Indeed, the currently favored theory for the origin of the Moon is through a collision of the Earth with a larger, perhaps Mars-sized, object, with only crust and mantle materials being ejected as debris from which the Moon subsequently formed.

If Szebehely is correct, most of the major satellites originated in the orbital planes in which we find them today. Presumably the coplanar planetary satellite systems formed through condensation and accretion proceeded similarly to the solar system itself. That these systems are in the rotational plane of the planet implies that the obliquities were determined before the satellites were formed, within disks around the planets.

Table 13.4. Critical satellite orbital radii[a]

Planet	r_{max} (km)	r_{moon}[b] (km)
Mercury	74×10^3	–
Venus	338×10^3	–
Earth	501×10^3	384×10^3 (Moon)
Mars	362×10^3	–
Jupiter	177×10^5	188×10^4 (Callisto)
Saturn	217×10^5	130×10^5 (Phoebe)
Uranus	233×10^5	584×10^3 (Oberon)
Neptune	388×10^5	551×10^5 (Nereid)
Pluto	192×10^5	196×10^2 (Charon)

[a] From Szebehely (1979, Table 1, pp. 177, 179).
[b] r_{moon} is the distance for the large moon farthest from the planet.

13.4.2 Conjectures about Pluto

Prior to the discovery of large numbers of icy objects in the outer solar system, climaxing in the discovery of such an object larger than Pluto that resulted ultimately in its reclassification, Pluto presented an interesting and apparently unique case among the outer planets. It had been argued that Pluto in its present orbit was a product of a catastrophic interaction between a large intruder and an earlier Neptunian satellite system that disrupted that initial system and resulted in the ejection of Pluto. It is not impossible that this scenario did indeed occur, because the origins of the moons (Section 13.4.3) are themselves far from well established, but current thought tends toward a different (Kuiper Belt) origin for Pluto and its moons. In favor of the Neptune collision theory were the circumstances that:

1. Pluto's orbit is so eccentric that perihelion falls within the solar distance of the orbit of Neptune.

2. Pluto's size made it the smallest "planet", (in the pre-2006 usage of this term) smaller than Triton, Neptune's major moon.

3. The orbits of the known moons of the Neptune system are unusual: Triton's orbit is retrograde ($i = 157°$); Nereid has an inclination of $27°6$ and an eccentricity $e = 0.751$; the second largest moon is triaxial in shape and the fourth largest is also non-spherical.

4. Triton and Pluto have similar sizes and densities and they both have icy surfaces on which nitrogen has been identified (the surface of Charon, on the other hand, has water ice on its surface).

Against were:

1. The Neptune–Pluto resonance: at present Neptune and Pluto are orbit–orbit coupled. The ratio of sidereal periods is 3:2, making interaction impossible, at present, anyway. Orbital integrations over millions of years suggest that while Pluto's orbit is chaotic, this basic resonance does not change. Over gigayears, however, the integrations cannot be performed with sufficiently high accuracy to cast further light on the issue.

2. A major interaction within a satellite system is intrinsically unlikely.

3. Pluto was known to have a satellite (at present writing it is known to have three, actually), making it resemble a planet more closely than some type of minor planet. The Pluto–Charon system is even more of a "double-planet" than is the Earth–Moon system. The satellite/planet mass ratio was the largest in the solar system (when, that is, Pluto was considered a planet).

Arguments 2 and 3 were hardly conclusive, however. Unlikely events have happened in the solar system's history; the Moon is thought to have had its origin in a major collision with a Mars-sized body and an early Earth (see Chapter 8). Regarding argument 3, even before Pluto's demotion from planetary status, the reality of asteroid companions was accepted, thanks to the evidence of occultations and of radar and optical imaging from spacecraft (such as Galileo's images of Ida and Dactyl). Even more to the point, three of the four brightest icy objects of the outer solar system that were investigated by Brown et al. (2006) were shown to have companions. So, satellites by themselves are not sufficient for either a planetary definition or a collisional origin. Finally, Venus and Mercury show that it is not a necessary one either.

A preferred current theory is that Pluto has escaped from the Kuiper Belt. *Centaurs* are asteroids that cross the orbit of Saturn and they are thought to be escaped former denizens of the Kuiper Belt, itself the supposed source of short-period comets. The comet-like asteroid Chiron (more recent discoveries have shown these objects to be not uncommon) is also thought to have originated there. Prior to the discovery of Eris (Brown et al. 2005), Pluto could have been considered the largest known trans-Neptunian object, and the largest known asteroid. This theory would not help to explain the strange condition of the Neptunian system, but there is no necessary reason why it should. A number of icy, trans-Neptunian objects have been discovered to have the 3:2 orbital resonance with Neptune. Indeed, it appears that Pluto is merely the largest object in this class; appropriately, the members of this class are referred to as *plutinos*.

2003 UB313 (Eris), with an estimated albedo of 0.60, and infrared flux implying a surface temperature of 23 K, appears to have a diameter of 2400 ± 100 km, compared to Pluto's 2250 km; this situation required the IAU to define what a planet is and is not, which it attempted to do at the General Assembly meeting in Prague in 2006. Originally, a committee created to deal with this issue recommended an expansion of the definition to include both Pluto, Eris and a number of other objects; but this was rejected by the assembly, and, instead a less inclusive definition was adopted that named eight planets (Mercury through Neptune) and designated smaller but still round objects (that could pull themselves together through self- gravity) such as Ceres, Pluto, and Eris as "dwarf planets."

The strange orbit of Triton and the fact that it resembles Pluto in some ways have continued to intrigue astronomers. Scenarios of a proto-Triton interaction with a proto-Pluto have been created by Dormand and Woolfson (1989), Farinella et al. (1979), and Harrington and Van Flandern (1979). A review of these works is provided by Woolfson (1999). The most recent collision argument is that Triton needed to have grazed Pluto in order for Charon to have been formed.

The discovery of two more moons about Pluto (Weaver et al. 2005) adds complication but does not decrease the likelihood of an impact origin for

Charon. Stern et al. (2006) note that the specific angular momentum of Charon is so high that the Pluto-Charon pair was likely created in a collision and the proximity of the two smaller moons argues for their creation at the same time. Where such a collision took place, however, is not certain.

13.5 Origins of the Moons

One may conclude that the present moons of the solar system represent two very different populations (with possible examples given in parenthesis):

1. Small bodies (asteroids; Phobos and Deimos; the outer moons of the giant planets), including, for present purposes, the icy bodies of the outer solar system, and cometary objects

2. Primordial/reconstituted moons (the major moons of the solar system)

Primordial large moons that are primarily rocky no longer exist as isolated bodies in the inner solar system except, arguably, for Mercury and maybe Mars. During the early stages of the solar system they careened about and collided with each other or with the larger planets or were ejected; today most of the remaining bodies of this population are preserved in the potential wells of the major planets, like the sabre-toothed tigers in the tar pits beneath Los Angeles.

Some, perhaps most, of the larger moons revolve in the equatorial planes of their primaries. This is not really convincing evidence that they originated in disks about the planets because the tendency is for (close) orbits to become coplanar and circular with time. Thus, cases where there are departures from this scenario, as for Triton, or our Moon, become very interesting. Clearly the solar system continues to be a dynamic place, the complexity of which we are just beginning to understand.

Next we will discuss another class of small bodies, the comets. In Chapter 15, we discuss their cousins, the myriad asteroids of the solar system, including the icy denizens of the outer solar system, to which Pluto and some of the icy moons of the outer planets may be closely related.

References

Abell, G. 1969. *Exploration of the Universe*, 2nd Ed. (New York: Holt, Rinehart and Winston) (First Ed. 1964).

Anderson, J. D., Schubert, G., Jacobson, R. A., Lau, E. L., Moore, W. B., and Sjogren, W. L. 1998. "Distribution of Rock, Metals, and Ices in Callisto," *Science*, **280**, 1573–1576.

Brown, R. H., and Cruikshank, D. P. 1997, "Determination of the Composition and State of Icy Surfaces in the Outer Solar System", *Annual Review of Earth and Planetary Sciences*, **25**, 243–277.

Brown, M. E. and Schaller, E. L. 2007. "The Mass of the Dwarf Planet Eris," *Science*, 316. 1585.

Brown, M. E., van Dam, M. A., Bouchez, A. H., Le Mignant, D., Campbell, R. D., Chin, J. C. Y., Conrad, A., Hartman, S. K., Johansson, E. M., Lafon, R. E., Rabinowitz, D. L., Stomski, Jr, P. J., Summers, D. M., Trujillo, C. A., and Wizinowich, P. L. 2006. "Satellites of the Largest Kuiper Belt Objects," *Astrophysical Journal*, **639**, L43–L46.

Buie, M. W., Grundy, W., Young, E. F., Young, L. A., and Stern, S. A. 2006. "Orbits and Photometry of Pluto's satellites: Charon, S/2005 P1 and S/2005 P2," *Astronomical Journal*, **132**, 290–298.

Burns, J. A., ed. 1977. *Planetary Satellites* (Tucson, AZ: University of Arizona Press).

Chuang, F. C. and Greeley, R. 2000. "Large Mass Movements on Callisto," *Journal of Geophysical Research*, **105**, 20,227–20,244.

Coustenis, A., Salama, A., Lellouch, E., Encrenaz, Th., Bjoraker, G. L., Samuelson, R. E., De Graauw, Th., Feuchtgruber, H., and Kessler, M. F. 1998. "Evidence for water vapor in Titan's atmosphere from ISO/SWS data," *Astronomy and Astrophysics*, 336, L85.

Coustenis, A., Salama, A., Schulz, B., Ott, S., Lellouch, E., Encrenaz, Th., Gautier, D., Feuchtgruber, H. 2003. "Titan's atmosphere from ISO mid-infrared spectroscopy," *Icarus*, **161**, 383–403.

Cuzzi, J. N., Lissauer, J. J., Esposito, L. W., Holberg, J. B., Marouf, E. A., Tyler, G. L., and Boischot, A. 1984. "Saturn's Rings: Properties and Processes" in *Planetary Rings*, eds R. Greenberg and A. Brahic (Tucson, AZ: University of Arizona Press), pp. 73–199.

Dermott, S. F. 1984. "Dynamics of Narrow Rings," in *Planetary Rings*, eds R. Greenberg and A. Brahic (Tucson, AZ: University of Arizona Press), pp. 589–637.

Dormand, J. R. and Woolfson, M. M. 1989. *The Origin of the Solar System: The Capture Theory* (Chichester: Ellis Horwood).

Elliot, J. L., Dunham, E. W., and Mink, D. J. 1977. "The Rings of Uranus," *Nature*, **267**, 328–330.

Farinella, P., Milani, A., Nobili, A. M., Valsacchi, G. B. 1979. "Tidal Evolution and the Pluto–Charon System," *Moon and the Planets*, 20, 415–421.

Franklin, F., Lecar, M., and Wiesel, W. 1984. "Ring Particle Dynamics in Resonances" in *Planetary Rings*, eds R. Greenberg and A. Brahic (Tucson, AZ: University of Arizona Press), pp. 562–588.

Fridman, A. M. and Gorkavyi, N. N. 1997. *Physics of Planetary Rings: Celestial Mechanics of Continuous Media* (Berlin: Spinger-Verlag) (tr. D. ter Haar).

Greenberg, J. M. 1968. "Interstellar Grains," Chapter 9 in *Nebulae and Interstellar Matter*, eds B. M. Middlehurst and L. H. Aller (Chicago, IL: University of Chicago Press).

Grün, E., Morfill, G. E., and Mendis, D. A. 1984. "Dust-Magnetosphere Interactions," in *Planetary Rings*, eds R. Greenberg and A. Brahic (Tucson, AZ: University of Arizona Press), pp. 275–366.

Color Plates

Plate 1. A composite image of Jupiter in three passbands, as viewed from the Cassini Orbiter on October 8, 2000, at a distance of 77.6 million km from Jupiter. Note the bright zones, dark belts, Great Red Spot, white ovals and brown barges. NASA/JPL/CICLOPS/University of Arizona image PIA02821

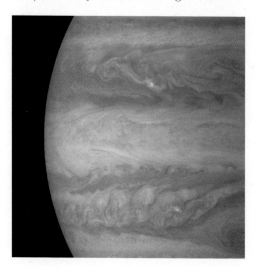

Plate 2. Structure in the equatorial zone and belts of Jupiter, as viewed by the Cassini Orbiter near its closest approach. Note the blue-grey clouds. The picture is a modified composite of images in three passbands: 728, 752, and 889 nm, in bands of methane and the continuum. The Great Red Spot is seen at the lower right. Bright white points may be high-altitude lightning. NASA/JPL/CICLOPS/University of Arizona image PIA02877

Plate 3. NASA/JPL/California Institute of Technology visualization PIA01192 of the structure of the atmosphere of Jupiter near the equator, as revealed by the Galileo Probe. The empty layer is clear dry air; above it is haze and below are thick clouds. The *blue area* marks a region of dry, descending air, while the *whitish region to the right* is ascending, moist air. The *blue–green–red colors* indicate enhanced reflection/emission in the 889, 727, and 756 nm passbands, respectively

Plate 4. Saturnian aurorae seen simultaneously at both poles, captured in UV light with the STIS instrument onboard the Hubble Space Telescope. Note the auroral oval, similar to that seen on Earth. Note also the similarity in alignment of both magnetic and rotational poles in Saturn. During the impacts of the components of the Comet Shoemaker–Levy 9, aurorae were triggered at both magnetic poles of Jupiter. Credits: HST/ESA/NASA/JPL/J.T. Trauger. Courtesy, John Trauger

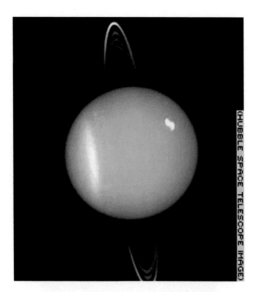

Plate 5. Uranus as seen by Voyager 2, showing backscattered rings, polar clouds, and a relatively high cloud, probably of methane, the red absorption of which gives rise to the greenish blue hue of the planet's atmosphere. Credits: NASA/JPL

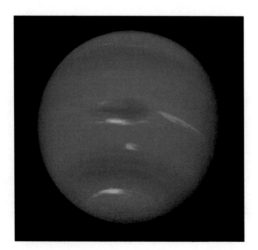

Plate 6. A Voyager 2 image of Neptune showing the main discernible features of its atmosphere: The Great Dark Spot (GDS) at disk center, bright streaks of cirrus, and Spot D2, below the GDS. As on the greater giants, belts and zones are present, if subdued due to overlaying haze. NASA/JPL image PIA02245

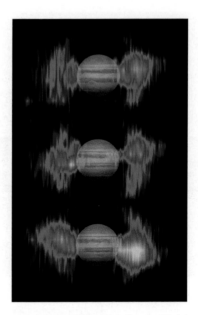

Plate 7. The inner magnetosphere of Jupiter is revealed by the radiation belts at three moments in the 10 h rotation period of Jupiter. The particle density is coded so that *light colors* indicate higher density. They were observed by Cassini during its flyby of Jupiter en route to Saturn. NASA/JPL image PIA03478

Plate 8. A composite of Voyager and Galileo images of the Galilean moons of Jupiter in order of decreasing size: Ganymede (the largest moon in the solar system), Callisto, Io, and Europa, from left to right, respectively. North is up for each image, and scaled to 10 km/pixel. NASA/JPL/DLR/USGS image PIA00601

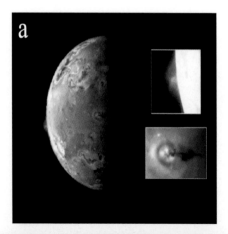

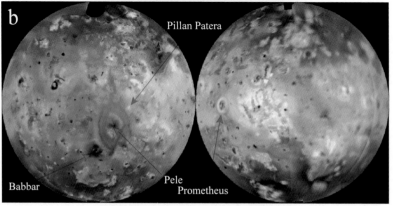

Plate 9. a Io in eruption as seen by Galileo Orbiter. The volcanic plume on the limb originates from Pillan Patera while that at disk center is from Prometheus. Both illustrate the plume and conical sheet of the ejecta. NASA image PIA00703. **b** Annotated US Geological Survey maps of Io, based on Voyager I data. In natural color and done in equal area production. At the left center of the left figure is the cloven-hoofed ejecta sheet of Pele. Pre-eruption Pillan is to its upper right. Original NASA/JPL/USGS image PIA00318

Plate 10. From Galileo: a five-color mosaic of the Tvashtar Catena, a chain of calderas on Io. The white spots at left are places where hot lava is emerging from the toes of the flow; the bright white, yellow and reddish patches are false color representations from IR imagery that mark a 60 km long lava stream, now beginning to cool. The diffuse dark region around the main caldera marks deposits from a recent plume eruption. NASA/JPL/University of Arizona image PIA02550

Plate 11. The Cassini Orbiter view of Enceladus. Note the varieties of terrain on this icy moon. It undergoes substantial tidal forces from Saturn and Titan and from closer moons. The "tiger stripes" region near the South Pole, the site of newly discovered water plumes, is indicated at the bottom of the image. See Porco et al. (2006) for a full discussion. NASA/JPL/Space Science Institute, Boulder, image PIA07800

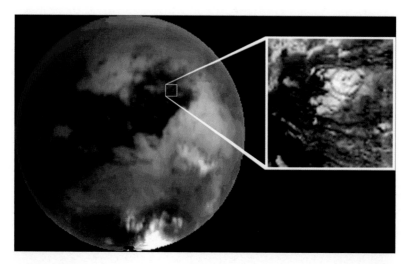

Plate 12. A Titan mosaic of six images, shown in false color, to emphasize the atmospheric haze, areas of light and dark regions; the inset is a 2.3 μm image of a possible volcano. Voyager, too, had found haze at 220 km (main layer) and patchy areas from 300 to 350 km. Some of the bright regions are cloud (such as the feature near the South Pole, at the bottom), others represent highly reflecting material on the planet. NASA/JPL/University of Arizona image PIA07965, based on data from the Cassini Orbiter spacecraft

Plate 13. Triton, the giant moon of Neptune, seen in color mosaic, based on three-passband images recorded by Voyager 2 in its 1989 flyby of Neptune. Notice the blue-green "cantaloupe" terrain at the top, the bright and somewhat pink regions of frost condensation, probably of methane, in a broad area around the South Pole, and the demarcation region between them, in which craters can be seen. In the polar region, dark plumes, likely geysers of nitrogen, are visible. In between are dark spots with bright rims of unknown origin. Produced by the USGS, NASA/JPL image PIA00317

Plate 14. A false-color composite image of Triton's surface as seen by Voyager 2 on Aug. 25, 1989. Plumes from geysers may extend up to 150 km from their vents. NASA/JPL image PIA02214

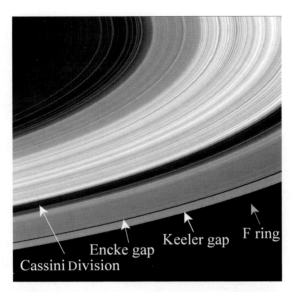

Plate 15. A Cassini view of the rings of Saturn. From the left: the faint D-ring complex of ringlets; the C ring with its narrow Maxwell gap; the very bright B ring, terminated by the Cassini's Division (with faint ringlets); the A ring and Encke gap near its outer edge; and the F ring just beyond the Keeler gap, just visible here, near the A-ring edge. NASA/JPL/Space Science Institute, Boulder image PIA05421; annotation by E. F. Milone

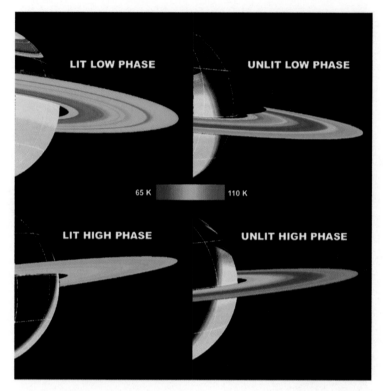

Plate 16. Sun-lit and unlit ring temperatures, measured from low and high-phase angles from Cassini Orbiter. The relative coolness of the B-ring particles is attributable to their high albedo; the darker rings absorb more sunlight. Note the strong thermal emission from the C ring, and the Cassini gap where the ringlets have lower albedo. A NASA/JPL/GSFC graphic, PIA03561, produced by Goddard Space Flight Center

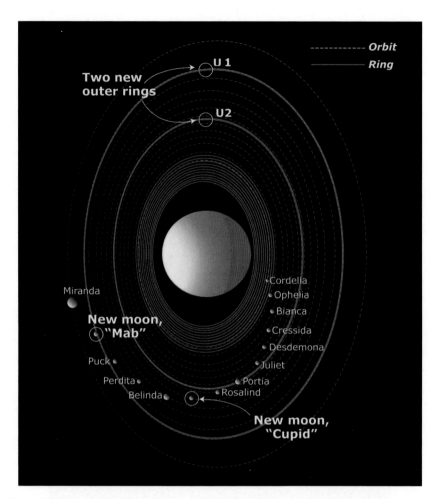

Plate 17. A graphic showing the relationship between the rings and the orbits of associated moons in the Uranian system. Some of the newer discoveries are indicated. Note the orbit of the satellite Mab, centered on R/2003 U1, here abbreviated to "U1"; the moons Portia and Rosalind flank R/2003 U2, labeled "U2". A NASA graphic, slightly altered by E.F. Milone. Credits: NASA/ESA/A. Feild (STSci)

Plate 18. C/1995 O1 (Hale–Bopp) as it appeared on April 1, 1997. Note the two types of tails: a thick amorphous dust tail and a blue ion tail. Image by John Mirtle (from his Ektachrome photography)

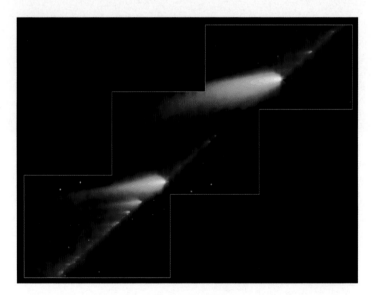

Plate 19. An infrared false-color mosaic of 44 fragments (of at least 65 components known at that time) and sun-warmed IR-emitting dust of P73/Schwassmann–Wachmann 3 on May 4–6, near its 2006 perihelion passage, as viewed from the Spitzer Space Telescope. Altogether, 45 fragments are visible in this image. Credits: W. T. Reach/Caltech/JPL/NASA. Image PIA08452. Courtesy, William T. Reach

Plate 20. A small slice of the Allende meteorite, viewed from both sides, with each side photographed under slightly different lighting conditions. Note the fusion crust, the round chondrules, the surrounding matrix, and the white calcium–aluminum-rich inclusions (CAI)

Plate 21. Fragments of the Zagami DSR achondrite Shergottite fall near Zagami, Nigeria, October 3, 1962. Noble gas analysis suggests a Martian source

Lunar Meteorite MET 01210

Maximum dimension is 4 centimeters.   NASA

Plate 22. Lunar meteorite MET 01210. A basaltic regolith breccia, it was collected in 2001 in the Meteorite Hills, Antarctica, and described in the Antarctic Meteorite Newletter, 27, No. 1, Feb. 2004. It is composed mainly of very-low-titanium crust, which can be seen at the right. It is basically a collection of stone and mineral clasts or fragments in a glassy matrix. NASA photo, curator.jsc.nasa.gov/antmet/amn/amn.cfm

Plate 23. A fragment of the Henbury iron meteorite find from the Northern Territory, Australia, 1931, according to the Bethany Sciences Company. It is classified as a medium octahedrite (IIIAB), with bandwidth 0.9 mm

Plate 24. A fragment of the Sikhote-Alin meteorite fall in eastern Siberia, February 12, 1947. It is classified a Type IIAB iron, coarsest octahedrite, with bandwidth of 9 mm. The circumstances of this spectacular meteoritic fall are described by Fessenkov (1955)

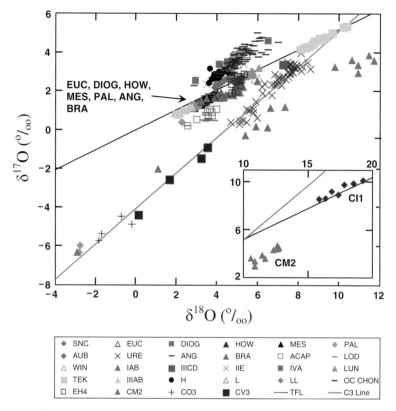

Plate 25. Fractional oxygen isotopes of various meteorites, the Earth–Moon, and two lines: one approximating the theoretical mixing line for anhydrous minerals and the other terrestrial fractionation line TFL. This sort of evidence points to heterogeneous origins in the early solar nebula and proto-planetary disk. The ratios of oxygen isotopes plotted are from analyses by Clayton & Mayeda (1984; 1996), and Clayton et al. (1991). The terrestrial fractionation line has a slope 0.52; the "C3 line" is a linear fit to the combined CO3 and CV3 data with slope 0.918 and zero point −4.036; similarly, the carbonaceous chondrite chondrules lay along a line with a slope ∼1. These lines typify the distributions of the meteorites, and suggests that exchanges between at least two reservoirs (Clayton, 1993; Wasson, 1985, esp. pp. 68–74, 215–217), involving gas and dust components of the solar nebula, had occurred. The heating source is not yet agreed on. Achondrites: ACAP = Acapulcoites; ANG = Angrites; AUB = Aubrites; BRA = Brachinites; DIOG = Diogenites; EUC = Eucrites; HOW = Howardites; LOD = Lodranites; LUN = Lunar; SNC = Martian (Shergottites, Nakhlites Chassigny, Orthopyroxene); URE = Ureilites; WIN = Winonaites. Stony-irons: MES = Mesosiderites; PAL = Palasites Chondrites: OC CHON = Ordinary chondrite chondrules. TEK = tektites (Earth) For more chondrite group identifications see Table 15.2 (recall that numbers indicate petrologic classes), and for the irons', see Table 15.5

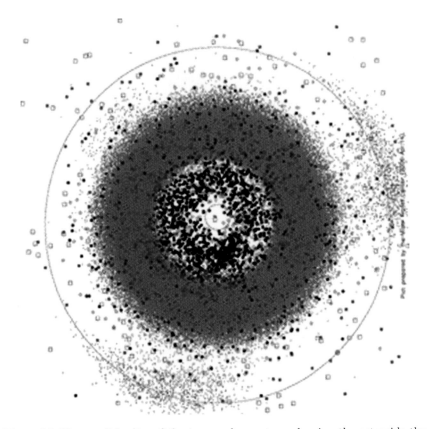

Plate 26. The small bodies of the inner solar system, showing the asteroids that cross the orbits of the Earth and of Mars, the "main belt" asteroids (in *green*) between the orbits of Mars and Jupiter, and the Trojan asteroids clustering along Jupiter's orbit (*the outer circle*) and centered on stations preceding and following Jupiter by 60°. The Hilda group of asteroids (Section 15.7.3 and Figure 15.10) lies between the Trojans and the main belt asteroids, and also at the upper left, opposite Jupiter, which is indicated by a *filled circle* at the lower right. Courtesy Minor Planet Center, SAO, Cambridge, MA

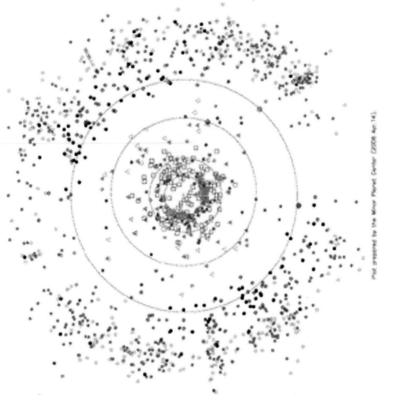

Plot prepared by the Minor Planet Center (2006 Apr 14).

Plate 27. The small bodies of the outer solar system, from the Jovian orbit to beyond Neptune. *Filled* and *unfilled squares* indicate numbered periodic comets and all other comets, respectively. *Triangles* represent Centaurs. *Black circles* near the orbit of Neptune are Plutinos (the largest of which is Pluto, near 3 o'clock, and just outside Neptune's orbit, the *outer blue circle*). Neptune is seen at ∼1 o'clock. The outermost *filled* and *unfilled red circles* are main Edgeworth–Kuiper Belt and scatter objects, respectively. Courtesy Minor Planet Center, SAO, Cambridge, MA

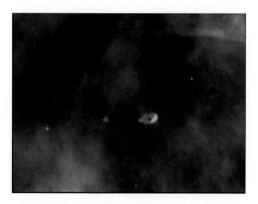

Plate 28. Possible proplyds in Orion, as imaged with the HST Wide Field and Planetary Camera 2 on December 29, 1993 (O'Dell and Wen 1994, Figure 3c on plate 7). 183-405 is the dark object on the right, seen in silhouette. Courtesy, C. R. O'Dell, Vanderbilt University, NASA, and ESA. Reproduced by permission of the AAS

Gulbis, A. A. S., Elliot, J. L., Person, M. J., Adams, E. R., Babcock, B. A., Emilio, M., Gangestad, J. W., Kern, S. D., Kramer, E. A., Osip, D. J., Pasachoff, J. M., Souza, S. P., and Tuvikene, T. 2006. "Charon's Radius and Atmospheric Constraints from Observations of a Stellar Occultation," *Nature*, **439**, 48–51.

Hansen, C. J., Esposito, L., Stewart, A. I. F., Colwell, J. Hendrix, A., Pryor, W., Shemansky, D., and West, R. 2006. "Enceladus' Water Plume," *Science*, **311**, 1422–1425.

Harrington, R. S. and Van Flandern, T. C. 1979. "The Satellites of Neptune and the Origin of Pluto," *Icarus*, **39**, 131–136.

Hill, G. W. 1878. "Researches in the Lunar Theory," *American Journal of Mathematics*, **1**, 5–26, 129–147, and 245–260.

Hobbs, P. V. 1974. *Ice Physics* (Oxford: Clarendon Press)

Johnson, T. V. 1990. "The Galilean Satellites" in *The New Solar System*, 3rd Ed., eds J. K. Beatty and A. Chaikin (Cambridge: University Press), pp. 171–188.

Lellouch, E. 1996. "Io's Atmosphere: Not Yet Understood," *Icarus*, **124**, 1–21.

Lellouch, E., Paubert, G., Moses, J. I., Schneider, N.M., and Strobel, D.F. 2003. "Volcanically-Emitted Sodium Chloride as a Source for Io's Neutral Clouds and Plasma Torus," *Nature*, **421**, 45–47.

Matson, D. L. and Blaney, D. L. 1999. "Io" in *Encyclopedia of the Solar System*, eds P. R. Weissman, L.-A. McFadden, and T. V. Johnson (San Diego, CA: Academic Press), pp. 357–376.

McKinnon, W. B. and Kirk, R. L. 1999. "Triton" in *Encyclopedia of the Solar System*, eds P. R. Weissman, L.-A. McFadden, and T. V. Johnson (San Diego, CA: Academic Press), pp. 405–434.

Mignard, F. 1984. "Effects of Radiation Forces on Dust Particles in Planetary Rings" in *Planetary Rings*, eds R. Greenberg and A. Brahic (Tucson, AZ: University of Arizona Press), pp. 333–366.

Millis, R. L., Wasserman, L.H., and Birch, P. V. 1977. "Detection of Rings Around Uranus," *Nature*, **267**, 330–331.

Milone, E. F. and Wilson, W. J. F. 2008. Solar System *Astrophysics: Background Science and the Inner Solar System.* (New York: Springer).

Murray, C. D. 2007. "Saturn's Dynamical Rings," *Physics Today*, **60**, No. 8, 74–75.

Pappalardo, R. T., Reynolds, S. J., and Greeley, R. 2006. "Extensional tilt blocks on Miranda: Evidence for an Upwelling Origin of Arden Corona," *Journal of Geophysical Research*, **102**, No. E6, 13,369–13,379.

de Pater, I., Hammel, H. B., Gibbard, S. G., and Showalter, M. R. 2006. "New Dust Belts of Uranus: One Ring, Two Ring, Red Ring, Blue Ring," *Science*, **312**, 92–94.

de Pater, I., Hammel, H. B., Showalter, M. R., van Dam, M. A. 2007. "The Dark Side of the Rings of Uranus," Science, **317**, 1888–1890.

Porco, C. C. et al. (25 authors) 2006. "Cassini Observes the Active South Pole of Enceladus," *Science*, **311**, 1393–1400.

Rathbone, J. A., Musser Jr., G. S., and Squyres, S. W. 1998. "Ice Diapirs on Europa: Implications for Liquid Water," *Geophysical Research Letters*, 25, no. **22**, 4157–4160.

Shu, F. 1984. "Waves in Planetary Rings," in *Planetary Rings*, eds R. Greenberg and A. Brahic (Tucson, AZ: University of Arizona Press), pp. 513–561.

Showalter, M. R. and Lissauer, J. J. 2006. "The Second Ring-Moon System of Uranus: Discovery and Dynamics," *Science*, **311**, 973–977.

Sicardy, B. et al. (45 authors) 2006. "Charon's Size and an Upper Limit on its Atmosphere from a Stellar Occultation," *Nature*, **439**, 52–54.

Stern, S. A., Mutchler, M. J., Weaver, H. A., and Steffl, A. J. 2006. A Giant Impact Origin for Pluto's Small Moons and Satellite Multiplicity in the Kuiper Belt. *Nature*, **439**, 946–948.

Strobel, D. F. 1982. "Chemistry and Evolution of Titan's Atmosphere," *Planetary and Space Science*, **30**, No. 8, 839–848.

Strobel, D. F., Summers, M. E., and Zhu, X. 1992, "Titan's Upper Atmosphere: Structure and Ultraviolet Emissions," *Icarus*, **100**, 512–526.

Szebehely, V. 1979. "A General Study of Satellite Stability" in *Natural and Artificial Satellite Motion*, eds. Nacozy and S. Ferraz-Mello (Austin, TX: University of Texas Press), pp. 175–180.

Taylor, S. R. 1992. *Solar System Evolution: A New Perspective.* (Cambridge: Cambridge University Press)

Tobie, G., Lunine, J. I., and Sotin, C. 2006. "Episodic Outgassing as the Origin of Atmospheric Methane on Titan," *Nature*, **440**, 61–64.

Weaver, H. A. et al. 2005. IAU Circular 8625, 1.

Weaver, H. A. Stern, S. A., Mutchler, M. J., Steffl, A. J., Buie, M. W., Merline, W. J., Spencer, J. R., Young, E. F., and Young, L. A. 2006. "Discovery of Two New Satellites of Pluto," *Nature*, **439**, 943–945.

Wilson, E. H. and Atreya, S. K. 2004. "Current State of Modeling the Photochemistry of Titan's Mutually Dependent Atmosphere and Ionosphere," *Journal of Geophysical Research*, **109**, 1–39, E06002, doi:10.1029/2003JE002181.

Woolfson, M. M. 1999. "The Neptune–Triton–Pluto System," *Monthly Notices of the Royal Astronomical Society*, **304**, 195–198.

Challenges

[13.1] Plot the densities of the major moons (say of size 100 km and up) against their distances from the planet for each of the giant planets. Place on the plot the synchronous orbit limit for each planet. From the results of this plot, what can you deduce about the origins of the moons of the outer planets? The synchronous radius has importance for accelerating satellites into higher and lower orbits. Illustrate how this happens. Make a copy of your plot for Q. [13.2]. [Hint: spreadsheets are very useful for this type of work.]

[13.2] Calculate the Roche limits for density values in the range 1000–3000 kg/m^3 and place these along with the observed limits for the ring systems on a copy of your plot for Q. [13.1]. Now what can you deduce about the origin of the *ring* systems?

[13.3] In an earlier chapter the Galilean satellites were shown to have orbital resonances among them. Examine closely the periods of the other giant planet moons for similar effects.

[13.4] Organize the satellites of the outer planets in groups as suggested (a) by orbital characteristics and (b) by physical characteristics. Comment on the differences and similarities between the two groups and what this suggests for the origins of the moons in each group.

[13.5] Consider the situation and conclusions of Section 13.2 for the case of a retrograde satellite.

[13.6] The heights of the eruptive plumes on Io vary from 60 to 400 km. Given the mass and radius of Io from Table 13.2, compute the ejection velocities required to reach these heights, and compare them to the escape velocity.

[13.7] Compute the escape velocity of Titan, and the molecular weight of a molecule that can be retained over billions of years, assuming no major or prolonged temperature increases. Discuss the retention of NH_3 and CH_4 on Titan, and of volatile gases generally from the objects in the outer solar system.

14. Comets and Meteors

14.1 Comets in History

Because of the effects comets have had on our intellectual history, and on the development of astronomy, in this chapter we delve a bit more into past perceptions than we have done in most of the earlier chapters.

14.1.1 Early History

Comets, a name derived from "hairy stars" (in the Greek, $\alpha\sigma\tau\eta\rho$ $\kappa o\mu\acute{\eta}\tau\eta\varsigma$, aster' come'tes), have been studied, feared, and admired throughout history, partly because they commanded attention. To many they represented a sign or a message, and given human history such as it has been, comets often did precede some momentous event. The early Christian theologian Origen ($\sim$184–254), in his discussion of the Star of Bethlehem (*Contra Celsus*, Book 1, Ch. LIX; excerpted in Kelley and Milone 2005, p.134), comments,

It has been observed that, on the occurrence of great events, and of mighty changes in terrestrial things, such stars are wont to appear, indicating either the removal of dynasties or the breaking out of wars, or the happening of such circumstances as may cause commotions upon the earth. But we have read in the Treatise on Comets by Chaeremon the Stoic [Alexandria, 1st century], that on some occasions also, when good was to happen, [comets] made their appearance; and he gives an account of such instances. [Brackets added by present authors.]

Records of probable comet observations go back to at least the seventh c. BC; here, we provide only the briefest summary of other historical commentaries on comets:

- The records of the "Wên Hsien Thung Khao" or "Historical Investigation of Public Affairs" by Ma Tuan-Lin and a supplement list 372 comets observed in the period 613 BC–1621 AD. The list includes several observations of Halley's comet, for the dates: 240 BC, 87 BC, 11 BC, 66 AD, and all subsequent returns to the end of the chronicle; records of 467 and 163 BC also may refer to this comet.

- The Pythagoreans in the 6th c. BC considered comets to be planets which appeared infrequently.
- Anaxagoras of Clazomenae (5th c. BC) and Democritus of Abdera (5th-4th c. BC) thought comets were conjunctions of planets.
- Ephorus of Cyme (4th c. BC) observed the comet of 371 BC to split into two stars.
- Apollonius of Myndus (4th c. BC) commented that comets were distinct heavenly bodies with orbits.
- Aristotle (4th c. BC) stated that comets were:

  - Outside the ecliptic, therefore they were *not* planets;
  - Also, therefore, they could not be conjunctions of planets;
  - They were *not* coalescences of stars; but were
  - Atmospheric phenomena (possibly omens of droughts or high winds).

- Posidonius (1st–2nd c. BC) also thought they were atmospheric phenomena. He saw one during a solar eclipse and argued that sometimes comets were invisible because they were lost in the Sun's glare.
- Seneca (1st c. AD) reviewed previous writings.
- Pliny the Elder (1st c. AD, who died studying the eruption of Vesuvius that buried Pompeii) cataloged comets.

Other writings on comets were contributed by:

- The Venerable Bede (672–735)
- Thomas Aquinas (1225–1274)
- Roger Bacon (1214–1294)

In a scientifically more active era, the fifteenth century, the pace began to pick up:

- Paolo Toscanelli (1397–1482), later adviser to Columbus, made positional measurements of several comets relative to stars.
- Georg von Purbach (1423–1461) attempted to measure the distance to a comet.
- Johann Müller, also known as Johannes de Monte Regio or Regiomontanus (1436–1476), student of Purbach, observed and studied the structure of comets.

These activities marked advances; on the other hand, a contemporary figure, Matthew of Avila, expounded that comets caused evil, because "their hot, putrid vapours contaminated the air ."

Fortunately, progress seems to have continued. In ~1550 the mathematician Jerome Cardan (Girolamo Cardano, 1501–1576) described a comet as "a globe in the sky illuminated by the Sun, the rays of which, shining through the comet, give the appearance of a beard or tail." More science was to come.

14.1.2 Tycho Brahe and the Comet of 1577

Tycho Brahe (1546–1601) had determined the "nova" of 1572 (*SN 1572*) to be located somewhere beyond the Moon, and thus not to be a comet or meteor (which were thought to be formed below the orbit of the Moon). The discovery of a truly "new" star clearly outside the terrestrial regions damaged the credibility of the Aristotelian immutability of the heavens. In 1577, Brahe turned his attention to the great comet of that year (C/1577 V1) and determined its parallax to be within his observational error, set mainly by his measuring device. His instrument was a cross-staff which produced a mean measurement error of ~4 arc-min. This established an upper limit to the parallax, and therefore a lower limit to the comet's distance:

$$r = R_\oplus / \sin p \geq R_\oplus / 0.0011636 = 859 R_\oplus = 5.48 \times 10^6 \, \text{km}$$

This minimal distance is ~14× the Moon's distance or $60 R_\oplus$. This result demonstrates that Aristotle's theory of comets is incorrect, because comets, or least the Comet of 1577, could not be in the Earth's atmosphere without producing a large, measurable parallax.

The imperial court physician and astronomer Thaddeus Hagecius at Prague is also said to have attempted a measurement of the parallax. The separation between the two sites was 600 km, and the differential parallax was found to be $\lesssim 2'$ implying a distance of at least 1.03×10^6 km, again, further than the Moon.

Brahe concluded that at the time of observation the comet lay between the Moon and Venus. He suggested further that the orbit was "oval", possibly the first informed argument that non-circular motion occurred in the heavens.

The parallax observations were confirmed by, among others, Michael Maestlin (or Mästlin, 1550–1631), who found that the comet's distance varied between 155 and $1495 R_\oplus$. The latter result meant that part of the comet's orbit was within the celestial sphere associated with Venus. The ancient view was that the planets traveled on celestial spheres, and various later proponents in the European and Islamic worlds held this view also. The observations of the comet of 1577 indicated that the crystalline sphere of Venus had been penetrated.

Tycho's instruments improved greatly in the years following and by 1585, when another comet appeared, the achievable precision of his measurements was ~1 arc-min. This comet, too, was demonstrated to be more distant than the Moon. This was a further refutation of ancient ideas, and this was made public, for the first time, apparently, by Christoph Rothmann (1560–1600) in a book published in 1585, regarding the comet observed in that year. Thus the credibility of the celestial spheres had been shattered forever by the end of the sixteenth century. See Gingerich and Westman (1988) for a delightful account of the personalities of that era, the discoveries, and the reception of those discoveries.

Returning to the Comet of 1577, we note that the linear dimensions of the comet measured 8' for the head and 22° and 2°.5 for the length and width of the tail, respectively. Brahe concluded that the comet was huge and porous, permitting sunlight to filter through it. Attacks were made on these interpretations by, among others, Galileo Galilei (1564–1642). In a curious kind of defense of Aristotle, Galileo argued that comets could be "optical illusions;" he suggested that vapors arising from the Earth's atmosphere could reflect sunlight in such a way that they would be visible once outside the Earth's shadow. This may seem odd in retrospect, but Copernicus, too, had thought that comets were a terrestrial phenomenon. The attacks were countered by Kepler (in *Tychonis Brahei Dani Hyperaspistes*), who defended the accuracy of Brahe's observational results.

14.1.3 Later Historical Studies

Johannes Kepler (1571–1630) studied the comets of 1607 and 1618 and published the results in *de Cometis*, in 1619. He thought comets moved along straight lines but with irregular speed. The description is comprehensible when one considers that the paths of comets traveling in highly eccentric ellipses or parabolas when seen on edge could resemble straight lines.

In the seventeenth century a number of others carried out comet studies as well; for example:

- Johannes Hevel (1611–1689) made the first comprehensive survey in 1654 and suggested that comets originated in the atmospheres of Jupiter and Saturn;

- Giovanni Borelli (1608–1679) thought that cometary orbits were parabolic, as did

- Georg S. Dörffel (1643–1688), who argued that the orbit of the comet of 1680, that was observed to make a rapid turn near the Sun and move back in the same direction it had come, was a narrow parabola with a very short perihelion distance;

- Edmond Halley (1656–1742) was the first to determine that comets could be periodic and the first to successfully predict a return date, 1758;

- Isaac Newton (1642–1727) discussed the elliptical nature of cometary orbits (as determined by Halley) and noted the high eccentricities of comets. In the *Principia*, III, 1686 (Cajori tr. 1962, pp. 497, 498), he stated that

"I am out in my judgement, if they are not a sort of planets revolving in orbits returning into themselves with a continual motion; ..."
and proposed that
"...their orbits will be so near to parabolas that parabolas can be used for them without sensible error."

Some of Newton's other points were that:

– comets shine by reflected sunlight

– tails arise from comets' atmospheres (following Kepler)

– the mutual gravitational effects of the large numbers of comets will perturb their orbits, so that elements will change from one apparition to the next

– a "new star" could be produced by the infall of a comet into an "old star." (Newton cites Tycho's supernova of 1572 as an example; cf., Cajori tr., 1962, pp. 540–542)

- Alexis-Claude Clairaut (1713–1765) performed the first perturbation calculations and predicted a return date of Halley's comet perihelion passage of April 15, 1759;

- Johann Franz Encke (1791–1865) was the second to predict the return of a comet—and to determine the periodicity of the shortest-period comet, that of: January, 1786; November, 1795; October, 1805; November, 1818. He also noted that the period was decreasing: $\Delta P/\Delta t < 0$. This implies that the semi-major axis is decreasing, so the comet is essentially spiraling into the Sun. From Kepler's third law,

$$P^2 = (M_\odot + m)a^3 = Ma^3 \tag{14.1}$$

where P is the sidereal period in years, M the total mass, m the comet's mass, and a the semi-major axis in astronomical units. Differencing and then dividing through by (14.1), one finds:

$$2\Delta P/P = \Delta m/M + 3\Delta a/a \tag{14.2}$$

Although a comet loses mass continually and hemorrhages material as it rounds perihelion, the first term on the RHS is still negligible due to the solar mass in the denominator; hence,

$$\Delta a/a = (2/3)\Delta P/P \tag{14.3}$$

Encke attributed the period decrease he determined for the comet that bears his name to a resisting medium or to a belt of meteoritic particles. Other deviations in the orbit he attributed to the heating effects of sunlight on the comet. The effect on the orbit is determined by the direction as well as the speed of the comet's rotation.

- John Herschel (1792–1871) observed that Biela's Comet of 1772 split into two in 1846 and that a luminous bridge joined the two parts, each of which subsequently developed tails. The twin comets returned in 1852, but have not been seen since.

The idea that non-gravitational forces act on comets was probably first advanced by Friedrich Wilhelm Bessel (1784–1846), but, as noted above, Encke, and, much later, Brian Marsden, applied it effectively so that the effects could be quantified.

For more comprehensive views of the development of ideas about comets, see Brandt and Chapman (1981), Bailey et al. (1990), or Yeomans (1991).

14.2 Comet Designations

In ancient Mesopotamia, the name for a comet was *sallamu*. Ephorus of Cyme (405–330 BC) observed a comet split in two in 371 BC. According to Ronan and Needham (1981, p. 208), ancient Chinese names for comets were "brush-stars," "long stars," and "candle-flame stars." Several of the objects referred to in the Chinese annals as "sweeping stars" are comets; and it is possible that the large sizes, and blue colors attributed to some, refer to appearances of comets or at least to some phenomenon other than novae. Objects reported for 76 BC ("candle star"), 48 BC ("blue-white", "star as big as a melon") and 5 BC ("sweeping star") fall into this category. Some are reported as being as "big" as the Sun, or some other object. Presumably, in a phrase such as "as big as the Sun," angular size is intended. However, the context of comments about size in the Chinese annals usually makes it clear that brightness and not a true angular size is being described; for example, the *Sung-shih* and *T'ung-k'ao* annals report for that from July 28 to August 6, 1203, a star with a blue-white color, "no flame or tail," as "big" as Saturn, stayed at *Wei* (Scorpius).

In modern usage, a comet is always named for its discoverer(s), but additional designations are given to it to indicate the order of appearance or detection and the nature of the comet. Thus, for example, the designation 1P refers to the first comet discovered for which an accurate orbit was computed (Comet Halley), and the fact that it is a periodic comet (see below for letter designations).

Formerly comets had two types of year designations. One indicated the year of discovery with a lower case letter indicating the order of discovery in that year; another usage gave the year of perihelion passage followed by a Roman numeral indicating the sequence within the year. A year-type designation is still used for the discovery or recovery date, but now it contains a code for every half month (A = Jan 1–15; B = Jan 16–31, etc.) followed by a sequence number in Arabic numerals. Thus, for example, *1P/1982 U1* or *2P/1822 L1*.

The naked eye appearance or *apparition* of Halley's Comet in 1985–1986 (Figure 14.1) was preceded by its detection as a very faint object in 1982; its former year designations were 1982i and 1986 III; its current provisional designation is 1982 U1.

A third modern designation indicates if the comet is periodic (*P*), i.e., has appeared at two (or more) perihelia, or not (*C*), if it is defunct (*D*), i.e., has failed to return or is known no longer to exist, is later found to be an asteroid (*A*), or, if it is not possible to compute an orbit either because the data are inadequate or because the comet is one for which a meaningful orbit cannot be computed (*X*), for whatever reason. Thus, for example, *D/1993 F2-K*, designates one of the now defunct components ("K") of the disintegrated comet formerly known as Shoemaker-Levy 9. The distinction between the asteroids and comets is likely to be increasingly blurred as the icy bodies of the outer solar system are investigated in detail; the objects in the Edgeworth–Kuiper Belt (see below and Chapter 15) are, in fact, thought to be an ultimate source of short-period comets. The term "short-period" has traditionally been applied to comets with periods of 200 years or less.

The current preferred designation for Halley's Comet is *1P/Halley*.

Fig. 14.1. Halley's Comet (1P/Halley) during the January 1986 apparition, a relatively unfavorable one for visibility in the northern hemisphere. Photo by E. F. Milone.

A description that applies to a particular appearance would be, for example, *Comet 1P/1982 U1*, or *Comet 1P/-239 K1*; and, as we note, the designation for the comet per se, would be comet *1P/Halley*. A number of short-period comets can be observed around the orbit, and thus no longer receive an appearance or recovery designation (for example, comet *2P/1822 L1* was a provisional designation for comet *2P/Encke*), and it has been observable continuously since discovery.

14.3 Cometary Orbits

There are two types of comets distinguished by their orbits:

- Periodic comets (i.e., $P \lesssim 200^y$)
- Parabolic comets or nearly parabolic, sometimes called long-period comets

There were 2221 comets in the Marsden and Williams (2005) catalog of cometary orbits. Of these, 341 comets had periods of less than 200 y; of the 341 comets, 203 had been seen at least twice, 138 were (to that date) single-apparition comets. Seven short-period comets had become "defunct" in the sense that they had not been detected at recent predicted returns.

Marsden and Williams (1997) state that the average short-period comet has the properties: $P = 7$ y; $q = 1.5$ au; $i = 13°$.

They note further that there are twice as many short-period comets with longitudes of perihelion in the semi-circle centered on Jupiter's perihelion than in the opposite semi-circle. Jupiter not only dominates the short-period comets, it sometimes captures them into its own satellite family. A recent example was Comet Shoemaker-Levy 9 (D/1993 F2), that collided with Jupiter in 1994. Some have argued that most comets with periods up to 20 y, should be designated "Jupiter-family" comets. In fact, however, there are families of comets connected to each of the giant planets, but dynamical modeling has shown these comets as capable of migrating inward, past the outer giants.

Several families of comets approach the Sun very closely, each family likely being the fragments of a single comet. The *Kreutz sungrazing group*, with perihelia less than 0.02 au (a few solar radii), is by far the largest: the discovery of the 1000th member, C/2006 P7 (SOHO), was announced by ESA on August 10, 2006. Although the family was first recognized by Heinrich Carl Friedrich Kreutz (1854-1907) in the nineteenth century, all but about thirty have been discovered in SOHO coronographs since 1996.

Three other sungrazing families have been recognized in SOHO images: the *Meyer group* with perihelia 0.03 - 0.04 au, the *Marsden group* with perihelia near 0.05 au, and the *Kracht group*, believed to be related to the Marsden group, with perihelia only slightly smaller than those of the Marsden group. The 2005 General Catalogue lists 59, 23, and 24 members, respectively, in these three families.

Of the parabolic or near parabolic orbit comets in the General Catalogue (2005), about 200 had eccentricities greater than one, but only two of these had eccentricities greater than 1.01:

Comet Spacewatch (C/1997 P2), with $e = 1.028407$
Comet Bowell (C/1980 E1) with $e = 1.057322$

Both are in direct orbits with relatively small inclinations ($14°.5$ and $1°.7$), and with arguments of perihelia, $\omega = 25°.4$ and $135°.1$, respectively.

Plots of the cometary distribution against eccentricity and against inclination show the rarity of hyperbolic speeds. Therefore, these objects are bound to the solar system, i.e., they share the common motion of the Sun orbiting the galactic center. There are no comets known with $e \gg 1$, although some speculate that such highly eccentric comets may arise at a frequency as high as 1/century.

It is not difficult to see how a comet may acquire a parabolic or even mildly hyperbolic orbit, however: at great distances from the major planets, the net gravitational field felt by a comet is produced by the entire planetary system; as it approaches, it may be accelerated by local encounters, the effects of which may accumulate until the orbital speed equals or even slightly exceeds the solar escape velocity while in the inner solar system. This is a natural form of the slingshot-assists designed by space probe engineers. Most likely it will not escape from the solar system into the galaxy, but it is certainly possible. This possibility for our solar system suggests the possibility of extrasolar comets. Similarly, a comet may lose energy through encounters, and wind up in a tighter orbit, or enthralled to a gas giant. We reconsider the origin of comets again in Section 14.7.

The techniques for computing cometary obits are given by Danby (1988), Moulton (1914), Murray and Dermott (1999), among others. A particular concern for calculating orbits or ephemerides for many comets is the large eccentricity, which limits the methods that can be employed.

Comets have small masses, so their effects on planets are relatively minor, while the effects of the planets on comets are considerable. One may apply the equations of a restricted three-body system (defined in Milone & Wilson, 2008, Chapters 3.3 and 3.6), in which a body of negligible mass (the comet) moves in the combined gravitational field of two massive bodies (Sun and planet) that follow circular orbits about each other, to explore this. Consider Figure 14.2, after Danby (1988, Fig. 8.1), where $\mu = m_2$ and $1 - \mu = m_1$, in a new unit of mass such that $m_1 + m_2 = 1$, and with $m_1 > m_2$ so that $\mu < 0.5$.

Let the unit of distance be $x_2 - x_1$. The orbital motion of m_2 is equivalent to the rotation of this coordinate system's x-axis about the origin, O. The

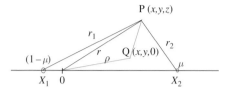

Fig. 14.2. A coordinate system rotating about O, the origin placed at the center of mass of two massive bodies, $\mu = m_2$ and $1 - \mu = m_1$. The y-axis is positive upward from the origin and in the plane of the page, the z-axis is positive out of the page; it is the rotation axis. In the restricted three-body system, m_1 and m_2 are massive bodies in circular orbits and the object at P has so negligible a mass, m, that it has no effect on the other two objects. Q is the projection of P onto the (x,y) plane, a distance ρ from the origin. Illustration adapted from Danby (1988, Fig. 8.1) with permission.

angular velocity of rotation, Ω, may be written (Danby 1988, p. 253; see also (3.7) and (3.10) of Milone & Wilson 2008, noting that, for a central force and circular orbits, there is no θ component and r = constant),

$$\Omega = \kappa\{[(1 - \mu) + \mu]/[x_2 - x_1]^3\}^{1/2} \qquad (14.4)$$

where the constant $\kappa = \sqrt{G}$ includes units of time. Changing the time unit such that $\kappa = 1$ (and therefore G = 1) yields the result that $\Omega = 1$ (see Danby (1962, pp. 138–141; 1988, pp. 146–149).

The distance of m at P from the two finite masses is given by

$$r_1{}^2 = (x - x_1)^2 + (y - 0)^2 + (z - 0)^2$$

and

$$r_2{}^2 = (x - x_2)^2 + (y - 0)^2 + (z - 0)^2 \qquad (14.5)$$

In this rotating system, the square of the speed, v', of m at P is given by

$$v'^2 = \dot{x}^2 + \dot{y}^2 + \dot{z}^2 \qquad (14.6)$$

Recall from (10.58) and (10.59), that the acceleration of the object at point Q in a rotating frame is

$$\mathbf{a} = \mathbf{a}' + \mathbf{\Omega} \times \mathbf{v}' + \mathbf{\Omega} \times (\mathbf{v}' + \mathbf{\Omega} \times \mathbf{r}) = \mathbf{a}' + 2\mathbf{\Omega} \times \mathbf{v}' + \mathbf{\Omega} \times (\mathbf{\Omega} \times \mathbf{r}) \quad (14.7)$$

where $\mathbf{a}$ is acceleration and $\mathbf{v}$, velocity, and the primed and non-primed quantities relate to motion as seen in and outside the rotating frame, respectively. Also, from the exposition of the two-body solution in Milone and

Wilson, 2008, Chapter 3.1, the dot product of the velocity and the acceleration is

$$\mathbf{v} \cdot \mathbf{a} = -\mathbf{v} \cdot \mathbf{r}/r^3 = v_r/r^2 \tag{14.8}$$

where we have substituted $\mathbf{v}$ for $d\mathbf{r}/dt$, $\mathbf{a}$ for $d^2\mathbf{r}/dt^2$, and v_r for the radial component of velocity, dr/dt. Recall also that $m_1 + m_2 = 1$ and $G = 1$. In a rotating coordinate system (see Chapter 10.3.1), this becomes,

$$\mathbf{v} \cdot \mathbf{a} = \mathbf{v} \cdot [\mathbf{a}' + 2\Omega \hat{\mathbf{z}} \times \mathbf{v}' - \Omega^2 \mathbf{r}] = v_r/r^2 \tag{14.9}$$

From (3.16), the two-body energy equation is

$$v^2 = (2/r) - h^2/r^2 + 2E \tag{14.10}$$

where E is the total energy, the sum of the kinetic and potential energy. This energy integral is modified in the rotating frame of the restricted three-body case to *Jacobi's integral* (Danby 1988, p. 254), namely,

$$v'^2 = x^2 + y^2 + 2(1 - \mu)/r_1 + 2\mu/r_2 - C \tag{14.11}$$

Note that $\boldsymbol{\Omega} = \Omega\hat{\mathbf{z}} = \hat{\mathbf{z}}$, because $\Omega = 1$ here, so that the motion of m in the rotating reference frame is:

$$\mathbf{v}' \equiv d\mathbf{r}'/dt = (\mathbf{v} \equiv d\mathbf{r}/dt) - \hat{\mathbf{z}} \times \mathbf{r} \tag{14.12}$$

In Section 10.3.1, we assumed for exposition purposes that the point under consideration was in the plane of rotation. In fact, this is not essential, because $\mathbf{r} = \boldsymbol{\rho} + z\hat{\mathbf{z}}$, so that $\hat{\mathbf{z}} \times \mathbf{r} = \hat{\mathbf{z}} \times \boldsymbol{\rho}$. We find, after substituting this in (14.12), taking the self-product, and rearranging,

$$
\begin{aligned}
v'^2 &= v^2 - 2\mathbf{v} \cdot (\hat{\mathbf{z}} \times \boldsymbol{\rho}) + \rho^2 \\
&= v^2 - 2\mathbf{v} \cdot (\hat{\mathbf{z}} \times \mathbf{r}) + \rho^2 \\
&= v^2 - 2\hat{\mathbf{z}} \cdot (\mathbf{r} \times \mathbf{v}) + x^2 + y^2
\end{aligned}
\tag{14.13}
$$

With the last equation of (14.13), Jacobi's integral, (14.11), becomes:

$$v^2 - 2\hat{\mathbf{z}} \cdot (\mathbf{r} \times \mathbf{v}) = 2(1 - \mu)/r_1 + 2\mu/r_2 - C \tag{14.14}$$

Suppose now that a comet is the object at P, the Sun is at x_1, and a giant planet is at x_2 in Figure 14.2. Jupiter is the most massive gas giant, so from

Table 12.1, $\mu \lesssim 10^{-3}$. Given the position and velocity of the comet at any instant, some of the elements are readily determinable. From the vis-viva equation [Milone & Wilson, 2008, (3.37)], we obtain the semi-major axis, a:

$$v^2 = 2/r - 1/a \qquad (14.15)$$

From the dot product of the $\hat{z}$ vector and the angular momentum, $\mathbf{h}$ [see Milone & Wilson, 2008, (3.11a)], actually the *specific angular momentum*—the angular momentum per unit mass, and from [Milone & Wilson, 2008, (3.24)], we have:

$$\hat{z} \cdot (\mathbf{r} \times \mathbf{v}) = \hat{z} \cdot \mathbf{h} = [a(1 - e^2)]^{1/2} \cos \iota \qquad (14.16)$$

where e and ι are the eccentricity and the inclination, respectively. These relationships may be useful in determining whether an observed comet has been observed in a previous apparition. Substitution of (14.15) and (14.16) into (14.14) yields the relation,

$$(1/a) + 2[a(1 - e^2)]^{1/2} \cos \iota = 2/r - 2(1 - \mu)/r_1 - 2\mu/r_2 + C \qquad (14.17)$$

What is interesting about (14.17) is that even though the comet's orbital elements may have changed due to the perturbation of a giant planet, C is a constant. If we assume that $r = r_1$ and judge that μ is small enough to be ignorable, we may make the further simplification,

$$a^{-1} + 2[a(1 - e^2)]^{1/2} \cos \iota \approx C \qquad (14.18)$$

Having computed C for a previous comet, we can test if a comet seen at another apparition is the same, by computing its C as well and comparing them. The criterion for identifying the similitude of the two comets, namely, the validity of the equality

$$a_1{}^{-1} + 2[a_1(1 - e_1{}^2)]^{1/2} \cos \iota_1 = a_2{}^{-1} + 2[a_2(1 - e_2{}^2)]^{1/2} \cos \iota_2 \qquad (14.19)$$

is known as *Tisserand's criterion*, after the French celestial mechanician, François Félix Tisserand (1845–1896). Equation (14.18) or an equivalent expression, has been referred to as the *Tisserand invariant* or *Tisserand parameter*. Recently, more uses have been found for this remarkable relation. It has been used to study the scattering of dust grains in various parts of the solar system (Gor'kavi et al. 1997), as well as to speed up the planning of space probe trajectories (Strange and Longuski, 2002). The criterion has also been used to distinguish different groups of short-period comets (see Section 14.7).

From time to time, comets approaching a major planet are perturbed so badly that their orbital elements are greatly changed. Marsden (1963) investigated a number of these "lost" comets, and by integrating the orbits over time and taking into account the perturbations to which they were subjected during close approaches, was able to predict future apparitions. In this way, Comet 9P/Tempel 1 was recovered in 1972, and, on reexamination, was found as a very faint image on a plate taken by E. Roemer in June, 1967, as Marsden had predicted. It has librated about a 2:1 resonance with Jupiter since the last major perturbations by Jupiter in 1941 and 1953.

Thanks to its recovery, 9P/Tempel 1 was the successfully targeted site of the "Deep Impact" mission in 2005; as a consequence it has become the best imaged comet, at least at high resolution, and one of the best studied to date (A'Hearn et al., 2005; Sunshine, et al., 2005).

14.4 Typical and Historically Important Comets

In addition to Halley's Comet and the historic comet of 1577 noted in Table 14.1, some others of historic interest are indicated below.

The shortest-period periodic comets with more than one appearance are:

- Comet 2P/Encke ($P = 3\overset{y}{.}3$), seen every revolution since 1822
- Comet 107P/Wilson–Harrington ($P = 4\overset{y}{.}30$), seen six times up to 1997
- Comet 26P/Grigg–Skjellerup ($P = 5\overset{y}{.}11$), seen 18 times up to 1997;
- Comet 79P/du Toit–Hartley ($P = 5\overset{y}{.}21$), seen three times up to 1997.

Other examples of short-period comets are:

- Comet 29P/Schwassmann–Wachmann 1 ($P = 14\overset{y}{.}9$), seen six times up to 1997
- Comet 55P/Temple–Tuttle ($P = 33\overset{y}{.}2$), seen five times since 1366
- Comet 95P/Chiron (= 1977 UB) ($P = 50\overset{y}{.}7$), once classified as an asteroid (now classified a Centaur)—one that grew a tail!
- Comet 13P/Olbers ($P = 69\overset{y}{.}6$), seen in 1815, 1887, and 1956
- Comet 109P/Swift–Tuttle ($P = 135^{y}$), seen five times since 69 BC (See Figure 14.3)
- Comet 9P/Tempel 1 ($P = 5\overset{y}{.}50$), seen twice before it was lost in 1879, and continually since its 1967 recovery.

The orbital parameters of selected comets can be found in Table 14.1, excerpted from Marsden and Williams (2005) (see also Figure 14.4).

Table 14.1. Selected comets

Comet name (or other designation)	T_{per}	P_{sid} (y)	q (au)	e	i (deg)	Ω (deg)	ω (deg)
C/1577 V1	1577.83	$>10^3$	0.178	1.0	104.8	31	256
3D/Biela[a] (1772 E1, 1832 S1)	1832.62	6.65	0.879	0.751	13.2	251	222
6P/d'Arrest (1678 R1; E:02/02/15)	2002.60	6.53	1.353	0.613	19.5	140	177
2P/Encke[b] (1786 B1; E:03/12/27)	2004.00	3.30	0.338	0.847	11.8	335	186
21P/Giacobini–Zinner (1900 Y1; E:05/07/09)	2005.50	6.62	1.038	0.706	31.8	195	173
1P/Halley[c] (-239 K1, E:86/02/19)	1986.11	76.0	0.587	0.967	162.2	59	112
C/Ikeya–Seki[d] (1965 S1-A, -B)	1965.80	880.	0.008	1.000	141.9	347	69
C/1973 E1[e] (Kohoutek); E:73/12/24	1973.99	...	0.142	1.000	14.3	258	38
13P/Olbers (1815 E1, 1956 A1)	1956.46	69.6	1.179	0.930	44.6	86	65
12P/Pons–Brooks (1812 O1)	1954.39	70.9	0.774	0.955	74.2	256	199
29P/1902 E1 (Schwassmann–Wachmann-1, 1927 V1; E:41/07/04)	2004.52	14.7	5.724	0.044	9.4	312	49
D/Shoemaker–Levy 9[f] (1993 F2)	1994.24	17.8	5.380	0.210	5.9	221	355
109P/Swift–Tuttle[g] (-68 Q1, 1992 S2)	1992.95	135.	0.964	0.960	113.4	139	153
55P/Tempel–Tuttle (1366 U1, 1965 IV)	1998.16	33.2	0.977	0.906	162.4	235	172
C/Hale-Bopp (1995 O1)[h]	1997.25	$\geq 10^3$	0.914	0.995	89.4	282	131

[a] Discovered by M. Biela in 1826; later identified with comets of 1772 and 1806, it was seen again in 1832. John Herschel (1792–1871) saw it split in two in 1846; both parts returned in 1852 but not thereafter. A meteor stream occupies the orbit.

[b] Méchain made the first recorded observation in January, 1786. The comet was observed by Caroline Herschel in October, 1805. The first computed elements and prediction of return were by Johann Encke (1791–1865) for 1822. Till 2003, there were 59 recorded appearances, the largest number of known appearances of any comet. It is not visible to the naked eye.

[c] This is the oldest known extant periodic comet. There have been 30 recorded appearances: 240, 164, 87, and 12 BC, and AD 66, 141, 218, 295, 374, 451, 530, 607, 684, 760, 837, 912, 989, 1066, 1145, 1222, 1301, 1378, 1456, 1531, 1607, 1682, 1759, 1835, 1910, and 1986. This is probably the comet depicted in the Bayeaux tapestry and is the most likely model for the comet in Giotto's *Adoration of the Magi*.

[d] A long "short-period" comet. The eccentricity is slightly smaller than 1.000, but is rounded off.

[e] An alleged "virgin comet," its early brightness far from the Sun led some to predict that it would be one of the brightest comets ever seen. Its subsequent dimness suggests that its likely origin was in the Kuiper Belt, not in the Oort Cloud, as initially supposed, so that its material had been reprocessed to some degree. C/1973 E1 was studied from SkyLab. Its eccentricity was computed to be 1.000008.

[f] This comet was independently discovered by three groups on photographic images in 1993; it was captured by Jupiter after tidal interaction probably in July 1992 into a satellite orbit; the interaction broke up the comet into 21 major fragments which collided with Jupiter in July, 1994. This comet was not visible to the naked eye, but is historically important as an example of the explosive effects of such collisions. The elements of Chodos & Yeomans (1994) are given for a middle segment (K).

[g] There have been five recorded appearances: 69 BC, and AD 188, 1737, 1862, and 1992 (see Figure 14.3).

[h] See Figure 14.4 and color plate 18.

Fig. 14.3. Comet 109P/Swift Tuttle as imaged by John Mirtle, November 14, 1992. This was only the fifth apparition at which this comet has been observed in recorded history.

Fig. 14.4. C/1995 O1 (Hale–Bopp) as it appeared on April 1, 1997. Note the two types of tails: a thick amorphous dust tail and a blue (in Ektachrome photography) ion tail, seen in color Plate 18. Image by John Mirtle.

The date of the last perihelion (to present writing) is indicated along with the other elements, presented here for representative purposes only. The earliest provisional designation is given when there are other names. The epoch for the elements (in the format: E:YYMMDD), also enclosed in parentheses but given only for some cases, is typically close to the time of perihelion passage, but usually not exactly on it. Subject to perturbations, small-mass bodies such as comets may undergo large changes in orbital elements, therefore for calculation purposes, the epoch to which the elements refer is an important datum. Other tabulated quantities are: the period, P, perihelion distance in astronomical units, q, argument of perihelion, ω, longitude of the ascending node, Ω, and the inclination, ι. An inclination greater than 90° implies retrograde orbital motion (CW motion as view from above the North ecliptic pole). The elements are "osculating," in the sense that they agree closely with (literally, "kiss") the orbit near the epoch, but depart from the true elements

Table 14.2. Comet 1P/Halley perihelion passages

Date			Associated event
240 BC	Mar.	25	
164	Nov.	13	
87	Aug.	06	
12	Oct.	11	Candidate for Star of Bethlehem
66 AD	Jan.	26	
141	Mar.	22	
218	May	18	
295	Apr.	20	
374	Feb.	16	
451	Jun.	28	
530	Sep.	27	
607	Mar.	15	
684	Oct.	03	Depicted in Nüremberg Chronicles
760	May	21	
837	Feb.	28	
912	Jul.	19	
989	Sep.	06	
1066	Mar.	21	Woven into the Bayeaux Tapestry
1145	Apr.	19	
1222	Sep.	29	
1301	Oct.	26	Possibly depicted by Giotto in the *Adoration of the Magi*
1378	Nov.	11	
1456	Jun.	10	Size measured by Toscanelli; distance sought by Purbach
1531	Aug.	26	Depiction by Apian
1607	Oct.	28	Studied by Kepler
1682	Sep.	15	
1759	Mar.	13	Predicted by Halley
1835	Nov.	16	Birth of Mark Twain (Samuel Clemens)
1910	Apr.	20	Mark Twain's death
1986	Feb.	09	

with time due to the perturbations to which the comet is subjected. These perturbations may be non-gravitational as well as gravitational; we discuss them further in later sections.

A fuller discussion of cometary orbits and designations as well as references to data sources can be found in the latest edition of the Catalogue of Cometary Orbits (at current writing: Marsden and Williams, 16th edition, 2005; the Minor Planet Center, Smithsonian Astrophysical Observatory, Cambridge, MA.).

Aspects of some of the more recent developments in cometary dynamics can be found in Fernández (1999).

1P/Halley is the best-known and one of the best-studied comets. At its last apparition, there were 7469 astrometric observations alone made of it. It is one of two extant, bright, short-period comets (the other being Comet 109P/Swift–Tuttle), and its apparitions have had impacts on human history. Mark Twain said that he came in with it and would go out with it; he did. The earliest firm date we have for a perihelion passage is 239 BC; the 30 known returns are listed in Table 14.2.

14.5 Cometary Structure

Far from the Sun, the comet shines by light reflected from the comet surface, later (when the comet is closer to the Sun) to be known as the *nucleus*. The observed brightness thus depends both on solar and terrestrial distances:

$$\ell \propto (1/r_\oplus^2)(1/r_\odot^2) \tag{14.20}$$

As it gets closer to the Sun, however, the equation changes to:

$$\ell \propto (1/r_\oplus^2)(1/r_\odot^n) \tag{14.21}$$

where $n \approx 4.2 \pm 1.5$, and is variable both from comet to comet and over time for the same comet.

As it approaches the Sun, the nucleus is warmed by the Sun. Local, sub-solar heating warms the ices and causes sublimation. The gases may explosively escape from warm pockets, bringing out dust with them.

The dust is acted upon by sunlight and may be driven away by radiation pressure. A repulsive force 2–3× gravity will drive out the dust; a broad, curved tail will result from the net accelerations on each dust particle as it travels away from the moving nucleus.

The solar UV and x-ray photons ionize the liberated gas molecules and the solar wind then drives the ions away at higher speeds than the photons drive away the dust. A repulsive force 20–30× gravity causes a fast ejection and a straight tail.

Consequently, the basic components of comets near perihelion are:

- *Nucleus* of dust and ices, ~0.2:-100:: km (where the colons imply uncertainty) in diameter. The nucleus of Comet 19P/Tempel 1 has a mean radius of 3.0 ± 0.1 km.
- *Coma* of evaporated gases and dust, typically ~10^4–10^5 km diameter
- *Hydrogen envelope*, typically ~10^7 km across
- Dust and ion *tails*, typically 10^8 km in length near perihelion

The maximum size of a comet nucleus is unknown at present. If the icy bodies beyond the orbit of Neptune and those that have migrated inward turn out to have the same composition and structure (see below), the limit may be of order 10^3 km. The *coma* is a parabolic envelope of dust particles; in fact a series of overlapping comae (or comas) are expected for particles of different sizes, the smallest being closest to the nucleus; closest of all is the icy halo of ~10 km separation from the nucleus, and the most extensive dust halo reaching to 10^4 km or more.

The hydrogen envelope or corona contains the shock front (magnetopause) separating the solar wind and the cometary magnetosphere, typically at ~10^6 km from the nucleus. The presence of such a cloud was predicted by Biermann (1968).

Far from the Sun, the comet presents a heavily cratered, non-spherical surface with ice and dust. We discuss cometary composition in the next section, but thanks to the imagery provided by space missions, we have a good idea of the appearance of a few comets. Figure 14.5 shows several of them.

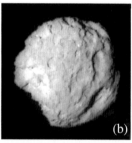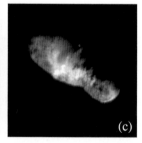

Fig. 14.5. Images of comets. **(a)** 1P/Halley from the *Giotto* Mission. Its size is ~$15 \times 7 \times 7$ km. ESA image; **(b)** 81P/Wild 2, ~5 km across, imaged by the *Stardust* Mission spacecraft Navigation Camera on January 2, 2004. NASA/JPL-Caltech image PIA06285; **(c)** 19P/Borrelly, 8 km-long, imaged at a distance of 3417 km, during the *Deep Space 1* mission, NASA/JPL image PIA03500. See also Figure 14.10.

14.6 Cometary Composition

Comets have been observed since the 1910 perihelion passage of Halley to exhibit spectral emission of hydrocarbons and their radicals, especially CN and C_2. The classic model for a cometary nucleus was suggested by Fred L. Whipple: the "dirty iceball". The model explains several properties of comets:

- Asteroidal appearance at large solar distances
- Extensive gas production near the Sun
- Dust tail development with approach to the Sun

Photochemistry of the subliming gases can occur immediately or in the tail as the molecules and dust separate from the nucleus and enter first the coma and then the tail. The density appears high enough near the nucleus to involve chemical reactions.

The identified parent molecules in the nucleus are H_2O, HCN, and CH_3CN.

The coma shows evidence of: CO, CN, C_2, C_3, CH, NH, NH_2, OH, O, C, and CS.

The ion tail shows features of: H_2O^+, CO^+, N_2^+, CH^+, CO_2^+, and OH^+.

In 1P/Halley, it was observed that different vents showed different rates of NH_3 and H_2CO relative to H_2O venting. This implies a heterogeneous structure.

The gas composition in Halley was found to be (Taylor 1992, p.125):

H_2O: 80%;   CO: 10%; CO_2: 3%; CH_4: 2%;
NH_3: < 1.5%; HCN: 0.1%.

Table 14.3. Comets' hydrocarbons composition relative to water ice[a]

Comet	CO	CH_3OH	CH_4	C_2H_6	C_2H_2	HCN
C/Hyakutake[b]	14.9 ± 1.9	1.7-2.0	0.70 ± 0.08	0.62 ± 0.07	0.16 ± 0.08	0.18 ± 0.04
Hale-Bopp[b]	12.4 ± 0.4	2.1	1.45 ± 0.16	0.56 ± 0.07	0.16 ± 0.08	0.18 ± 0.04
Ikeya-Zhang[b]	4.7 ± 0.8	2.5 ± 0.5	0.51 ± 0.06	0.62 ± 0.13	0.18 ± 0.05	0.18 ± 0.05
1P/Halley[b]	3.5	1.7 ± 0.4	<1	0.4	0.3	0.2
Lee[b]	1.8 ± 0.2	2.1 ± 0.5	1.45 ± 0.18	0.67 ± 0.07	0.27 ± 0.03	0.29 ± 0.02
C/1999 S4	0.9 ± 0.3	<0.15	0.18 ± 0.06	0.11 ± 0.02	<0.12	0.10 ± 0.03
Tempel 1[b,c]	...	1.3 ± 0.2	...	0.19 ± 0.04	...	0.18 ± 0.06
Tempel 1[b,d]	4.3 ± 1.2	1.0 ± 0.2	0.54 ± 0.30	0.19 ± 0.04	0.13 ± 0.04	0.21 ± 0.03

[a] Water ice abundance = 100.
[b] See Figures 14.4–14.10 for images of these comets.
[c] Pre-impact values only; comet showed slight activity level prior to impact.
[d] Post-impact values only; rapid fall-off in intensity of lines implies no ongoing jet.

The relative mix of these gases differs from that in the interstellar medium, indicating that some processing has occurred. The abundances of measured hydrocarbons in Halley and other recent comets are shown in Table 14.3, taken from Mumma et al. (2005). Relative abundances for carbon monoxide (CO), methanol (CH_3OH), methane (CH_4), ethane (C_2H_6), acetylene (C_2H_2), and hydrogen cyanide (HCN) are shown. Some data are rounded off. The authors note that for 9P/Tempel 1, the number of molecules of water increased after the impact by about a factor of 2 (Figures 14.6–14.10).

The element composition in 1P/Halley, as given in Taylor (1992, Table 3.10.2), shows that, compared to the Sun, Halley is strongly depleted in H; slightly enriched in C, O, Na, Si, S, Ti, and Co; slightly deficient in Al, K, Cr, Mn, Fe, and Ni; and about the same in N, Mg, and Ca.

Although volatiles have been observed previously, the actual detection of water ice *on the surface* through reflectance spectroscopy of a comet was first achieved by the Deep Impact mission on 9P/Tempel 1 (Sunshine, et al. 2005).

Various processes which can affect the nucleus are:

- Impacts, which heat, cause "farming", ejection of matter and secondary cratering
- Micrometeorites, which add to the regolith
- Irradiation of the crust by cosmic rays
- Solar heating, usually ineffective at aphelia, where most of the time is spent

These processes can lead to several conditions on the nucleus:

- Refractory veneer on the crust
- Hydration reactions
- Loss of volatiles through "volcanoes"
- Formation of organic compounds
- Shock effects

During its flyby of 1P/Halley, the Vega spacecraft measured the dust properties: No calcium-aluminum-rich inclusions (CAI), the presence of which is characteristic of carbonaceous chondrites (a type of undifferentiated meteorites—see Chapter 15.2.2), were seen. The dust had three main components (Grün and Jessberger, 1990):

1. CHON particles (for the elements that dominate)
2. Silicate particles

Fig. 14.6. C/1996 B2 (Hyakutake). Note the disconnects in the ion tail, indicative of the effects of the solar wind plasma on the ionized particles created by photolysis in sunlight. Photo by John Mirtle, March 24, 1996.

Fig. 14.7. Comet C/1999 H1 (Lee). Note the sharp "beak," or spikey sunward ion tail. This particular "antitail" is probably not an optical illusion, produced solely by geometric projection, but is seen sometimes in first-time (sometimes referred to as "virgin") comets, that alone may have the icy volatiles in sufficient amount and location necessary to produce sizable and rapid ejection responses to sunlight. Perhaps a slow rotation aids the effect. Such a prominent antitail was seen also in C/1956 R1 (Arend–Roland) in 1957. Image obtained by John Mirtle, August 11, 1999.

Fig. 14.8. Comet 153P/Ikeya–Zhang. Note the tight region of streamers feeding the ion tail, but contained within the broad dust tail. This would seem to present difficulties for the theory that the solar wind sweeps around the head of the comet to produce the ion tail and constrain the streamers. Photo by John Mirtle. See also the photo of Comet 23P/Brorsen–Metcalfe, Figure 14.9, where the streamers extend out to a considerable distance from the nucleus.

Fig. 14.9. Comet 23P/Brorsen–Metcalfe. Note the clean features of the ion tail, indicating smooth flow from the solar wind during the time interval of propagation of the ion tail components. Photography by John Mirtle.

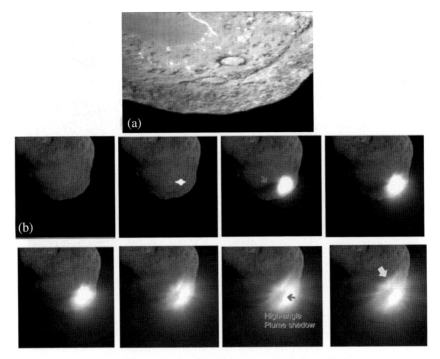

Fig. 14.10. 9P/Tempel 1, from the *Deep Impact* mission. (a) Pre-impact image showing not only heavily cratered areas but also smooth areas and scarps, as bright icy areas gleam in sunlight. NASA/JPL-CalTech/Univ. of Maryland image PIA02135 (b) The impact, rays, spreading fireball, shadows. Credits: NASA/JPL-CalTech/U of Md., M. F. A'Hearn et al. (2005).

3. mixtures of the two, similar to carbonaceous chondrites but enriched in CHON elements

The net results indicate a resemblance to Brownlee particles, captured by high-altitude aircraft.

The overall composition resembles CI-type meteorites (Flynn, 2002); the ices in particular could be primordial, but the Fe/Si and Mg/Si ratios are the lowest in the solar system. If CO is a true parent molecule in 1P/Halley, as it has now been seen to be in other comets, then the nucleus must have a temperature ~ 25K when unheated. For 1P/Halley and other comets, the carbon isotopes ratio, $^{12}C/^{13}C \approx 90$ (Wyckoff et al. 2000) (although individual grains vary from 1 to 5000)! This compares to ~ 89 for the solar system and 72–94 for the interstellar medium at 10 Kpc from the galactic center (Milam et al. 2006).

The impact of the Deep Impact mission's 370 kg space probe on the nucleus of Comet Tempel 1 (A'Hearn et al., 2005) caused a clear increase in the brightness of the comet that was noted by ground-based observers. The resulting dust cloud was observed at the Frederick C. Gillette (Gemini North) Telescope by Harker et al. (2005). In infrared spectra between 8 and 13 μm they observed in the ejecta a broad silicate emission feature which they attributed to amorphous pyroxene, amorphous olivine, and magnesium-rich crystalline olivine. They also saw features of amorphous carbon grains. The addition of orthopyroxene did not improve their fitting. Modeling the rate of cooling and flux dispersal, they estimated that 10^4–10^5 kg of dust had been ejected; none had been observed prior to impact. Sugita et al. (2005) came to a similar conclusion about silicate grains from data taken with the Subaru telescope. Figure 14.10 shows the comet and the fireball created by the impact.

Boice and Huebner (1999) estimated the composition of comets to be 85% H_2O, 4% CO, 3% CO_2, 2% H_2CO (formaldehyde), 2% CH_3OH, 1% N_2, and 3% H_2S, HCN, NH_3, CH_4, CS, and other hydrocarbons, combined. They also noted that the make up and composition is not uniform for all comets. The ratio of gas to dust varies from comet to comet, with some comets having very little dust (see, for example, Figure 14.11). However, IR measurements

Fig. 14.11. C/2004 Q2 (Machholz), with sharply defined streamers but very faint dust tail at 5 o'clock, obscured by twilight. The Pleiades star cluster is in the upper left. Photo by John Mirtle.

seem to suggest greater uniformity than do those made in the visible part of the spectrum. The CO/H_2O ratio varies as well, so that the prominent CO^+ ion tail seen in comets C/1961 R1 (Humason) and C/1908 R1 (Morehouse) are weak in other comets.

In most comets, the emission features of C_2 and C_3 are 20–50% stronger than those of CN but in some comets they are deficient by up to a factor 20. Finally, oxides tend to be more abundant in comets other than those in the short-period comet group. The greater oxide composition is closer to interstellar values and suggests that the comets in a disk near the ecliptic plane were formed in a more chemically reducing environment. As we discuss in the next section, kinematic evidence, too, points to a different place of origin for the longer period comets compared to the short-period variety. Once near the Sun, photolytic reactions can produce the observed species. The chemical reactions in comets are reviewed by Colangeli et al. (2004).

14.7 Origins of Comets

If comets were truly interstellar in origin, at very large distances they would have positive velocities relative to the Sun, and as a consequence, would appear to have strongly hyperbolic paths. This not being the case, an overwhelming majority appear to be bona fide members of the solar system.

The number of comets in the direction of the apex of the Sun's motion[1] is approximately the same as from other directions; therefore, there is little evidence that any significant number of these comets has come from outside the solar system. The perturbations on comets by the giant planets, and especially Neptune, suffice to increase the speeds at which some comets approach the inner solar system so that they appear to have just slightly more energy than 0.

The distribution of the inclinations of the comets with parabolic and nearly parabolic orbits shows them to be roughly isotropically distributed. Thus, they probably originate in a spherical or spheroidal region around the Sun. This region is called the *Oort cloud*, after the astronomer who first suggested it, Jan Oort (1900–1992). If there were a cloud of comets at $\sim 10^5$ au from the Sun, to satisfy the observed frequency of comets in the inner solar system, $\sim 10/y$, from the ratio of the order of the angular size of the solar system to

---

[1] Sometimes called the *solar apex*, the direction in space toward which the Sun is moving with respect to the motions of nearby stars; it is roughly at $\alpha = 18^h$ and $\delta = 30°$; see Milone & Wilson, 2008, Chapter 4.2.

that of the entire sphere (4π), the probability of a comet moving toward the inner solar system is

$$\Omega/(4\pi) \approx 10/N \tag{14.22}$$

where Ω is the solid angle taken up by the planetary orbits, say a region 10 au across, and N is the total number of comets in the cloud. For a comet coming from high-ecliptic latitudes, then, the chance is

$$\Omega/(4\pi) = (10/10^5)^2 = 10^{-8}$$

so that, if all comets had the same probability, the population of the Oort cloud would be $N \approx 10^9$. However, the probability depends also on the ecliptic latitude, so that in the ecliptic plane, the inner solar system presents a very narrow target region and the probability is much less, perhaps $\leq 10^{-9}$.

Estimates of comet numbers in the Oort cloud have been as high as $\sim 10^{12}$, and its radius is estimated at $\sim 40,000$ au.

Although Stagg and Bailey (1989) argued for the existence of the original, large-scale Oort cloud alone for cometary origin, albeit one with some central concentration, Bailey and Stagg (1990), other work has suggested three sources of comets:

1. The outer Oort cloud, 2–5×10^4 au, spherical, 1–2×10^{12} comets

2. The inner Oort cloud, 2–20×10^3 au, flattened, 2×10^{12}–10^{13} comets

3. The Edgeworth-Kuiper Belt (or Cloud), highly flattened, 30–50 au, with mostly relatively circular, low-inclination orbits

The orbital properties of the short-period comets suggest, in fact, two groups of these comets: those with periods less than 20 y and with prograde orbits that lie as close to the ecliptic plane as the asteroids, and a smaller group with higher inclinations and often retrograde orbits, and periods that may be less than 20 y, but as great as 200 y. Most of the former have been referred to as *Jupiter-family* comets. The latter, sometimes called the *Halley group*, have been thought to be from the Oort cloud and the Jupiter family from the *Edgeworth–Kuiper Belt* (referred to also as the *Kuiper–Edgeworth*, or simply the *Kuiper Belt*), or possibly the inner Oort cloud, if it exists. Recent work suggests that the direct source of the Jupiter-family comets and maybe a significant fraction of the Halley group also is the *scattered disk*, a population of higher-eccentricity, higher-inclination objects, fed by migrants from both the classical Edgeworth–Kuiper Belt and from the Uranus–Neptune region. The dwarf planet Eris is thought to be a member of this scattered population of icy objects.

Fernández (1999, pp. 550–551) shows that the *Tisserand invariant* [written as one side of (14.17), but with a substitution involving the perihelion distance

$q = a(1 - e)]$, T, can be used to distinguish the Jupiter-family (with $\sim 2 <$ $T < 3.0$) comets from the one-pass comets (sometimes called "non-periodic" or "long-period" comets) and Halley group comets ($T < 2$). This, Fernández argues, makes the Oort cloud a less likely site of origin for the short-period comets. Note, though, that the Tisserand criterion and "invariant," as useful as they are, can only be approximate. The solar system, after all, is an n-body system, and the restricted three-body condition (Sun–Jupiter–comet) is approximate only.

A body of what were formerly considered to be asteroids, the *Centaurs*, has been found at distances of ≥ 10 au. It has been conjectured that they are "escaped" members of the Edgeworth–Kuiper Belt, a collection of objects with semi-major axes extending over a range of several tens of aus. Although these objects will be discussed in more detail in Chapter 15, we note here that there are of the order of 100 of these objects known thus far.

Identified as an "asteroid–comet transition object," and by some as a Centaur despite its small q value, 944 *Hidalgo* has a reflectance spectrum described as that of a "dead comet" (Beatty and Chaikin 1990, p. 196). It has a mean distance, $a = 5.85$ au, eccentricity, $e = 0.655$, and inclination to the ecliptic plane, $i = 42°.36$, and thus moves between the orbits of Mars and Saturn. It appears to be an object that has migrated inward.

The first Centaur recognized as such is 2060 *Chiron*, designated 95P/1977 UB following the detection of a persistent coma and sometimes tail from this object by Meech and Belton (1990). Chiron has orbital elements: $a \approx 13.6$ au, $e = 0.38$, and $i = 6°.95$. This orbit ranges from within that of Saturn to nearly the orbit of Uranus, and is subject to chaotic variation. From the duration of an occultation of a star, this object appears to have a diameter ~ 160 km or more. If it is a true comet, Chiron has the largest nucleus detected thus far.

Cruikshank et al. (1998) obtained reflectance spectra between 0.4 and 2.4 μm of the Centaur 5145 *Pholus* that revealed a surface much more character-istic of comets than of asteroids, with water vapor absorption bands (1.5 and 2.1 μm), and possibly one due to methanol, CH_3OH; the spectrum was modeled with two types of terrain: amorphous carbon (carbon black) covering $\sim 62 \pm 5\%$ of the surface; and a second type (over 38% of the surface) made up of 55% olivine, 15% water ice, 15% methanol, and 15% organic solids (tholins). They did not achieve better fitting with NH_3 included. The investigators suggest that the active components in the model are principal constituents of a comet nucleus and that Pholus is basically a comet that has never been active.

Such research provides the critical clues needed to understand the outer solar system. It reveals clearly that comets are icy bodies with insufficient gravity to retain their volatiles as the latter sublime when in close proximity to the Sun. Therefore, the dividing line between a comet and some similarly constructed icy object may be decided by its escape velocity and its perihelion distance.

We have traced the origin of comets to the Oort cloud and to the Edgeworth–Kuiper Belt, but we have not discussed how comets get to the planets' orbital region, other than through random motion.

Perturbations by passing stars or by tidal effects of the galaxy or of inter-actions with massive gas/dust clouds have been suggested as triggers for comets in the Oort cloud itself. The probability discussion about how often a perturbed comet will enter the realm of the major planets still holds. The existence of a spherical region has itself been argued to have developed as a result of randomizing perturbations by passing stars (Fernández 1999, p. 555), or gas/dust clouds.

For objects in the Edgeworth–Kuiper Belt, simulations suggest that if they come closer than $\sim$40 au, they will be perturbed into chaotic behavior and will eventually migrate through the zones of the major planets into that of Jupiter, thus accounting for the Jupiter family of short-period comets. The origin of the scattered population we will take up in Chapter 15.

14.8 Cometary Demise

Comets are indeed delicate. The Minor Planet Center in a recent posting lists 23 that were observed to have split and 21 that have faded out or disintegrated completely since 1975 alone. Boehnhardt (2004) provides a list and suggests that multiple splittings may evolve into families of comets, citing the Kreutz sun-grazing comets as examples, although most such fragments have short lives. His Fig. 2 suggests that very few survive more than 1000 days.

The mechanism of splitting is likely tidal action. D/Shoemaker–Levy 9 disintegrated following a close approach to Jupiter, was perturbed into a Jovian orbit, and eventually all 21 fragments collided with Jupiter. 73P/Schwassmann–Wachmann 3 started to break up 16 days before its 1995 perihelion passage, accompanied by strong OH emission, increased dust ejecta, and a brightness increase of four magnitudes (Sekanina 2005). On the same pass, it fragmented further. An infrared mosaic (Figure 14.12; color Plate 19) from the Spitzer Space Telescope shows 45 of the 65 of more fragments and a long dust trail connecting them during the 2006 apparition.

A comet loses mass every time it rounds perihelion, hence it has a finite lifetime when near the Sun. The attrition rate depends critically on the perihelion distance and on the period. A typical comet with perihelion distance, $q \gtrsim 0.6$ au has sufficient material for $\sim 10^3$ perihelion passes. In the

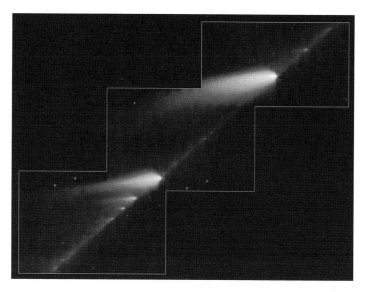

Fig. 14.12. An infrared false-color mosaic of 44 fragments (of at least 65 components known at that time) and sun-warmed IR-emitting dust of P73/Schwassmann–Wachmann 3 on May 4–6, near its 2006 perihelion passage, as viewed from the Spitzer Space Telescope. Altogether, 45 fragments are visible in this image. Credits: W. T. Reach/Caltech/JPL/NASA. Image PIA08452. Courtesy, William T. Reach. See Plate 19.

case of Halley's Comet, for example, the ESA spacecraft *Giotto* measured emission indicating a gas loss rate of $\sim$20 t/s and a dust loss rate of 3–10 t/s. Therefore, in 100 days centered on perihelion, 1.8×10^8 t of gas and $\sim$4 $\times 10^7$ t of dust are lost. The mass of the comet from perturbations on the spacecraft was determined as 10^{11} t. Thus, we see that

$$\tau \approx 10^{11}/2 \times 10^8 = 500 \, \text{passes}$$

reasonably close to prediction.

The dust material becomes entrapped in planetary gravitational fields or orbits the Sun until it is either driven out into the solar system or spirals into the Sun. Large clouds of very fine dust were seen in the vicinity of Halley during its 1986 perihelion passage. This dust moves in associated orbits with the comet and spreads out along the orbit with time. The ascending and descending nodes of the comet thus become the intersections of the Earth with the cometary debris.

We next discuss the dust entering the atmosphere, in the form of cometary debris, viz., meteor showers; then we discuss meteors generally, and, finally, micrometeorites in the upper atmosphere of Earth.

14.9 Meteor Showers

A list of prominent meteor showers can be found in Table 14.4, taken from Kelley and Milone (2005). For the most part, these showers represent the interaction between cometary dust tails and the Earth. Occasionally, larger fragments, resulting in *bolides* (bright, often exploding, meteors—

Table 14.4. Selected meteor showers

Name	UT date	Δt	Radiant (2000)		Associated comet or asteroid
Quadrantids[a]	Jan. 4.0	1^d	15^{h}30^m	+51°	
Lyrids[b]	Apr. 23.7	3	18 16	+35	1861 I = C/Thatcher (1861 G1)
η Aquarids[c]	May. 5.4	10	22 27	+00	1P/Halley
S δ Aquarids	Jul. 29.2	12	22 36	-17	
N δ Aquarids	Aug. 12	5	21 24	-05	
Perseids[d]	Aug. 12.9	20	03 04	+58	109P/Swift–Tuttle
Draconids	Oct. 10	1	17 40	+54	21P/Giacobini–Zinner
Orionids[e]	Oct. 22.5	7	06 20	+15	1P/Halley
S Taurids	Nov. 3.5	30	03 32	+14	2P/Encke
N Taurids	Nov. 13.5	30	03 55	+23	
Leonids[f]	Nov. 18.0	6	10 08	+22	55P/Tempel–Tuttle (1366 U1, 1997 E1)
Andromedids[g]	Nov. 20	21	00 52	+55	3D/Biela (1772 E1; 1852 III)
Geminids[h]	Dec. 14.8	8	07 32	+32	3200 Phaethon
Ursids	Dec. 23.0	5	14 28	+76	8P/Tuttle

[a] From an old constellation named *Quadrans muralis*; the radiant is in Bootes. Typical modern rate: ~80/h.

[b] The last major shower was in 1803; Chinese records list showers on March 23, 687 BC ("stars flew") and March 26 15 BC ("stars fell like a shower"); a Korean record lists an April 3, 1136 shower. Typical modern rate: 15/h.

[c] The earliest of five possible showers in Chinese records occurred on April 8, 401. Typical modern rate: 60/h.

[d] Many records of Perseids are found in Asian sources. The earliest firm identification is that of a shower on July 17, 36 AD ("more than 100 small stars flew"). Typical modern rate: 95/h.

[e] Widely variable in rates from year to year ($\sim 4\times$). Chinese records indicate major displays on September 23, 585 AD ("many stars chased each other"), September 25, 930, and several dates in the fifteenth, and one in the seventeenth century. Typical modern rate: 30/h.

[f] The period of comet 1866 I = Tempel-Tuttle was 33^{y}176. Every 33 y, with some disappointing displays such as in 1933, Leonids have given spectacular displays, as in 1966. Chinese, Korean, and Japanese annals record many such displays starting on October 15 and 16, 931 AD. Several clumpings in the orbit allow for showers in adjacent years also.

[g] Also called the *Bielids. Meteor storms* were seen in 1872 and 1885 with rates ~75,000 meteors/h. Little activity has been seen since, but the nodes of the orbit undergo rapid regression.

[h] 3200 Phaethon is, at least at present, considered an asteroid. Typical modern rate for the Geminids: $\sim$ 90/h.

see Section 14.10) are visible. For more observational detail, consult Kronk (1988).

14.10 Meteors

14.10.1 Basic Meteor Phenomena and Circumstances

The term *meteor* derives from the Greek, $\mu\eta\tau\eta\rho\varnothing\nu$, meteron, "thing of the air." It is a luminous phenomenon, an incandescence caused primarily by ram-pressure heating of an object as it traverses the upper atmosphere. Ram pressure is the pressure exerted on a fast-moving object because it compresses the air in front of it. In the case of a meteor, the extremely rapid compression heats the air, and part of this heat is transferred to the object itself. Prior to entry, i.e., in space, the object is called a *meteoroid*.

The estimated frequency of meteors is $\sim 8 \times 10^9$/day of which 25×10^6 are bright enough to be seen. The conditions for visibility are:

- A distance from the observer of less than ~ 200 km
- Altitude 80–95 km (only the brightest last long enough to be seen lower than 80 km)

The very brightest are called *bolides* or *fireballs*, which occur at a rate of 10^4–10^5/day. These meteors sometimes explode in the air and can be heard.

One estimate of the total meteoric mass accumulated by the Earth in one day is 10^3 t (Dubin 1955).

There are two types of meteors:

1. *Sporadic meteors*—individual objects
2. *Shower meteors*—swarms of objects, mostly from comets, appearing from distinct *radiants* in the sky.

Spectra indicate that 99% of the light is from gases glowing at temperatures $\sim 10^3$ K. The species seen are: Ca II, Fe II, Al, N, Na, Mg, Mn, Si, and H.

14.10.2 Meteor Heating and Incandescence

The visibility of the meteor as it moves through the atmosphere indicates that its kinetic energy is partly converted into radiant energy. The underlying principles and some of the data can be found in Hawkins (1964) and in the papers presented by Kaiser (1955). Here, we develop some of these principles from basic physics.

Model the meteoroid as a spherical object of mass m, radius R, and cross-section $A = \pi R^2$, traveling at speed v through air of density ρ_{air}. In time Δt it travels a distance $\Delta \ell = v \Delta t$, and encounters air of volume $V_{\text{air}} = A \Delta \ell$ and mass $m_{\text{air}} = \rho_{\text{air}} \, V_{\text{air}}$. In the upper part of the trail most of the air flows freely around the meteoroid and is pushed sideways to roughly speed v by the passage of the meteoroid. Lower in the trail, where the air is denser, most of the air in volume V_{air} is swept up and compressed into a cap traveling at the same speed as the meteoroid. The process reaches a quasi-steady state where the rate of new air being swept up equals the rate of old air streaming off the side (quasi-steady because the air density increases with time as seen by the meteoroid).

We now derive equations for the drag force on the meteoroid, its acceleration, and its rate of mass loss by ablation. The work-kinetic energy theorem states that the work, W, done on any object by all forces acting on it equals the change in its kinetic energy, K. Then, treating the air as the system and the meteoroid as an external agent doing work on the system, we have

$$\frac{W_{\text{air}}}{\Delta t} = \frac{\Delta K_{\text{air}}}{\Delta t} \quad \text{or} \quad \frac{dW_{\text{air}}}{dt} = \frac{dK_{\text{air}}}{dt} \qquad (14.23)$$

where the first of equations (14.23) equates the average rate at which work, W_{air}, is done on the air by the meteoroid over any time interval Δt to the average rate of change of kinetic energy of the air over the same time interval, and the second equates the instantaneous rate at which the meteoroid does work on the air at time t to the derivative of the kinetic energy at that time. W_{air} and dW_{air} are macroscopic and microscopic *amounts* of work done, whereas ΔK_{air} and dK_{air} are macroscopic and microscopic *changes* in kinetic energy.

Because the process is quasi-steady state, $\Delta K_{\text{air}}/\Delta t$ and dK_{air}/dt are essentially equal. Then if the meteoroid accelerates all of the air in volume V_{air} from $v_0 = 0$ to $v_{\text{f}} = v$ in time Δt, the rate of change of the kinetic energy of the air is

$$\frac{dK_{\text{air}}}{dt} = \frac{\Delta K_{\text{air}}}{\Delta t} = \frac{\frac{1}{2} m_{\text{air}} v^2 - 0}{\Delta t} = \frac{\frac{1}{2} \rho_{\text{air}} A \Delta \ell v^2}{\Delta t} = \frac{1}{2} \pi R^2 \rho_{\text{air}} v^3 \qquad (14.24)$$

where A and $\Delta \ell$ are defined above.

The meteoroid does work W_{air} on the air and, conversely, the air does work W_{M} on the meteoroid. Make the postulate that these quantities may be equal, or they may not be equal (a postulate that seems unassailable!). Then the two quantities are related by

$$\frac{dW_M}{dt} = \Gamma \frac{dW_{air}}{dt} \tag{14.25}$$

where at this point Γ is simply the appropriate ratio of one to the other.

The work done on the meteoroid by the air may be regarded as being caused by a drag force, F_d, exerted on the meteoroid by the air. Then the rate at which the air does work on the meteoroid equals the power, P_d, delivered to the meteoroid by the drag force. Power is force times velocity, so

$$\frac{dW_M}{dt} = P_d = F_d v. \tag{14.26}$$

Substituting (14.26), (14.23), and (14.24) into (14.25), we obtain the drag force equation,

$$F_d = \tfrac{1}{2}\Gamma \pi R^2 \rho_{air} v^2. \tag{14.27}$$

Thus, Γ is the drag coefficient. From (14.25) and Newton's third law, one might expect that in many circumstances $\Gamma = 1$. This is the case where the air flows freely around the meteoroid in the upper part of its trail. Lower in the atmosphere, where a cap of air forms over the oncoming face of the meteoroid, $\Gamma \to 0.5$; peak brightness occurs at intermediate values of Γ.

The drag equation giving the acceleration of a meteoroid of mass m may be obtained from (14.27) using $F_d = m\, dv/dt$, and replacing m by $(4/3)\pi r^3 \delta$, where δ is the meteoroid's density:

$$\frac{dv}{dt} = \frac{3\Gamma \rho_{air} v^2}{8R\delta} \tag{14.28}$$

Some fraction, Λ, of the work that the air does on the meteoroid enters the meteoroid as heat, Q, and causes ablation through melting and vaporization. Then, from (14.25),

$$\frac{dQ}{dt} = \Lambda \frac{dW_M}{dt} = \Lambda \Gamma \frac{dW_{air}}{dt} \tag{14.29}$$

Define ζ = heat of ablation, the amount of heat required to ablate a unit mass of meteoroid, measured from the undisturbed (interplanetary) state of the material. For silica (SiO_2), $\zeta = 8 \times 10^{10}$ erg/g $= 8 \times 10^6$ J/kg. Then

$$\frac{dQ}{dt} = \zeta \frac{dm}{dt} \tag{14.30}$$

Using (14.23), (14.24), and (14.30) in (14.29), we obtain the ablation equation,

$$\zeta \frac{dm}{dt} = \tfrac{1}{2} \Lambda \Gamma \pi R^2 \rho_{air} v^3. \tag{14.31}$$

The meteor emits light as a result of ablation, excitation and ionization of its atoms by collisions with air molecules, as suggested by Levin (1955),

$$I = \tfrac{1}{2}(dm/dt)v^2\,\tau \tag{14.32}$$

where I is the luminous energy radiated per second, and the *luminous efficiency*, the fraction of energy of the vapor trail atoms that goes into radiation is

$$\tau \approx 2.0 \times 10^{-10}\,v \tag{14.33}$$

where v is expressed in cm/s. Speeds of meteors vary from $\sim$12 km/s for "slow" meteors to $\sim$72 km/s for "fast" ones. Typically, $\tau \approx 10^{-4}$ for $v \approx$ 12 km/s, and $\tau \approx 10^{-3}$ for $v \approx$ 72 km/s. The ablation rate,

$$dm/dt = (4/9)(m/H)v \cos z \tag{14.34}$$

where the scale height $H = kT/(mg) = 6.5 \times 10^5$ cm, m is in g, and z is the path angle.

Hawkins (1964, p. 7) relates the intensity to the visual magnitude through the relation:

$$V \approx 24.3 - 2.5 \log I \tag{14.35}$$

where the constant presumably includes a correction from bolometric (i.e., across all wavelengths) to visual brightness. A distance of 100 km from the observer is assumed.

The meteor may be detected by radar. Its visibility at radio wavelengths depends on the number of ions produced. The evaporation rate of the meteoric atoms can be written as

$$<m>^{-1}\,dm/dt \tag{14.36}$$

where $<m>$ is the mean mass of the atoms of the meteor. The rate of ionizations (the number produced per second) is:

$$N_i = (\beta/<m>)\,dm/dt \tag{14.37}$$

where β is the probability of ionization, which, in turn, depends on the velocity:

$$\beta \approx 2.0 \times 10^{-26} v^{3.4} \qquad (14.38)$$

where again v is expressed in cm/s. For "slow" meteors, $\beta \approx 10^{-5}$ and for "fast" meteors, $\beta \approx 5 \times 10^{-5}$.

The number of electrons produced per unit length is then given by:

$$N_q = (1/v)N_i \qquad (14.39)$$

The relative energy that goes into heat, light, and ionization, Hawkins (1964, p. 19) gives as 10^5:10:1 for "slow" meteors, and 10^5:10^2:5 for "fast" meteors, respectively.

The brightness will not stay constant, but is found to vary according to an empirical pattern. For bright meteors, I depends very much on the air density. If $\rho_{I\,\text{max}}$ is the atmospheric density at which the meteor achieves maximum brightness, then the relative brightness at any other density is:

$$I/I_{\text{max}} = (9/4)(\rho/\rho_{I\,\text{max}})\{1 - [(\rho/(3\rho_{I\,\text{max}})]^2\} \qquad (14.40)$$

However, faint meteors do not follow this prescription as closely. Fragmentation is thought to be responsible. Break-up pressure is about 2×10^4 dynes/cm^2 = 2000 N/m^2.

Given n fragments of a total mass of m and radius R, each fragment will have average mass m/n and radius $r \approx R/n^{1/3}$. In this case, the drag equation (14.28) becomes

$$dv/dt = 3\Gamma \rho v^2 n^{1/3}/(8R\delta) \qquad (14.41)$$

For the fragmented case, the vaporization rate, (14.31), becomes, for each fragment:

$$\zeta(dm/dt)_i = \tfrac{1}{2}\Lambda\,\Gamma\,\pi\,\rho\,R_i{}^2\,v_i{}^3 \qquad (14.42)$$

and for all fragments:

$$\zeta\,dm/dt = \tfrac{1}{2}\Lambda\,\Gamma\,\pi\,\rho\,R^2 v^3 n^{1/3} \qquad (14.43)$$

This means that a fragmented meteor decelerates more quickly and evaporates faster than an unfragmented meteor, resulting in an initial burst of light, followed by fading, higher in the atmosphere.

The brightness of a meteor also depends on the angle at which it descends. If z is the angle between the meteor trail and the vertical, the V magnitude becomes

$$V = 24.3 - 2.5 \, \log(6.84 \times 10^{-17} \, \mathrm{m} \, \mathrm{v}^4 \, \cos z) \qquad (14.44)$$

With $m = 1 \, \mathrm{gm}$, speed $v = 30 \, \mathrm{km/s} = 3 \times 10^6 \, \mathrm{cm/s}$, and with $z = 45°$, $V = +0.3$. At $v = 72 \, \mathrm{km/s}$, the brightness increases to -3.5, brighter than all the objects in the sky but the Sun, Moon, and sometimes Venus.

The length and duration of a meteor depend on its speed. Hawkins estimates the height range (we will call it Δh) over which a meteor can stay visible (brighter than $V = +5$, say) as 35 km; hence the path length will be given by

$$\Delta \ell \approx \Delta h / \cos z = 35 \, \sec z \, \mathrm{km} \qquad (14.45)$$

The duration of visibility, Δt, then becomes,

$$\Delta t = \Delta \ell / v = 35 \, \sec z / v \qquad (14.46)$$

Therefore, a meteor with $v = 30 \, \mathrm{km/s}$, $z = 45°$, and peak magnitude $V \approx 0$ will remain visible ($V \gtrsim 5$) over a path range of $\Delta \ell = 49 \, \mathrm{km}$, and will be seen for $\Delta t \approx 1.6 \, \mathrm{s}$.

Shower meteors greatly increase in number for an interval of time when the Earth intersects the orbit of the debris; sporadic meteors show a more gradual variation, involving time of day and year, and location on the Earth. The number of meteors brighter than ~ 0 magnitude decreases with increasing brightness. Hawkins (1964, p. 10) cites data from the American Meteor Society of the average influx rates of very bright meteors. These are reproduced in Table 14.5.

N is the number per hour, per square kilometer of meteors brighter than or equal to magnitude m. A meteor of -15 magnitude would be $10\times$ brighter than the full moon, but a meteor this bright or brighter occurs only once over a square kilometer every $\sim 770,000 \, \mathrm{h}$, on average. This is once per 32,000 days, or once in $\sim 88 \, \mathrm{y}$, effectively, once in a lifetime.

Table 14.5. Bolide[a] rates

m	-5	-6	-7	-8	-9	-10	-15
N (hr^{-1} km^{-2})	1.3×10^{-2}	5.1×10^{-3}	2.0×10^{-3}	8.1×10^{-4}	3.2×10^{-4}	1.3×10^{-4}	1.3×10^{-6}

[a] A bolide is a meteor that exceeds the brightness of any planet at any time; this means a magnitude of ~ -5 or brighter. "Fireball" is an equivalent term.

Hawkins also gives the distribution of all meteors with photographic and visual magnitudes brighter than m_p and m_v, respectively. The compilation is for meteors that are at or brighter than the naked eye limit, $\sim +5$. He summarizes the tables in the following relations, where N is again expressed in numbers per km^2 per hour:

$$\log N = 0.538m_p - 4.34$$
$$\log N = 0.538m_v - 5.17$$

(14.47)

Because the brightness as a function of mass is relatively understood, the distribution (number of meteors per year per km^2 of mass equal to or greater than m) can be expressed in terms of the meteor mass (Hawkins 1964, p. 68):

$$\log N = 0.4 - 1.34 \log m \qquad (14.48)$$

14.11 Micrometeorites

These are mainly spheroidal particles with diameters between 25 and 500μm (Rietmeijer, 2002). From data obtained from the Long Duration Exposure Facility satellite, an estimated $4 \pm 2 \times 10^4$ metric tons of micrometeoroids are encountered by the Earth each year (Hutchison 2004, p. 3). The rates of accretion are:

$1\text{m}^{-2}\text{d}^{-1}$ for $10\,\mu\text{m}$-sized particles
$1\text{m}^{-2}\,\text{y}^{-1}$ for $100\,\mu\text{m}$-sized particles

Brownlee particles are scooped up by stratospheric collectors aboard aircraft or high-altitude balloons. They are of two types:

1. CP (for chondritic porous). These have:

   – Mixed composition
   – Resemblance to chondrules of chondrite meteorites (possibly a chondrule precursor)
   – Mainly pyroxene composition
   – Metal deficiency

2. CS (for chondritic smooth). These have:

   – Resemblance to CI meteorite matrix

  – Clusters of 0.3 μm spheres

  – Hydrated silicates

  – Probable origin in CI, CM-type meteorites.

Both types show strong enhancement of deuterium (similar to giant molecular clouds).

Particles of size $> 100\,\mu m$ form spherules and contain magnetite.

The origin of the dust is thus the debris of comets and asteroids, the latter component produced by the "asteroid mill." The dust is a distant component of the solar corona (the *F-corona*). Near Earth, the density is $\sim 10^{-10}/m^3$.

The dust is seen phenomenologically in the *zodiacal light*, a conical distribution of material aligned along the ecliptic, before sunrise in the eastern sky or after sunset in the western sky. This is best visible in Spring at dusk, and in Fall at dawn, in the Northern hemisphere, when the ecliptic achieves its highest angle in the sky relative to the celestial equator (see Figure 14.13).

It can also be seen, but only in exceptionally dark and high-altitude sites, in the *Gegenschein*, or "counterglow" opposite the Sun in the sky.

Rietmeijer (2002), citing studies of micrometeorites from a wide range of locations (from ocean sediments to the stratosphere) concludes that although cometary dust is more abundant in the zodiacal cloud, the ratio of asteroidal to cometary dust on Earth is ~4:1, due to the smaller tensile strength of the latter.

Some sources suggest that there is an interstellar component as well.

Next we discuss the fate of this dust.

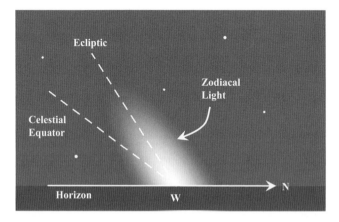

Fig. 14.13. Zodiacal light can be seen optimally in evening twilight in the Northern hemisphere spring, and from other times and places when the ecliptic has a large angle with respect to the horizon.

14.12 Dust Destinies

Basically, there are three possible fates for these little particles:

- To be swept up by a planet or other body
- To be blown away from the Sun by radiation pressure or by the solar wind
- To spiral into the Sun via the *Poynting–Robertson effect*

14.12.1 Radiation Pressure

The pressure due to the momentum of solar photons can be written:

$$P = \Im_\odot/c \; [\text{J/m}^3] \tag{14.49}$$

where $\Im(r)$ is the radiative flux from the Sun at distance r, and is expressed in units of W/m^2. As per [Milone & Wilson, 2008, (6.15)], (14.49) may be rewritten:

$$P = \mathcal{L}_\odot/(4\pi \, cr^2) \tag{14.50}$$

where $\mathcal{L}_\odot$ is the solar luminosity (expressed in W), so that the force on the dust particle due to radiation pressure is:

$$F_{\text{RP}} = PA = [\mathcal{L}_\odot/(4\pi \, cr^2)][\pi R^2] \tag{14.51}$$

where πR^2 is the cross-sectional area and R is the radius of the dust particle. This can be compared to the gravitational force:

$$F_{\text{G}} = GM_\odot m/r^2 = [GM_\odot/r^2]\rho \, (4\pi/3)R^3 \tag{14.52}$$

where the density, ρ, of the particle has been substituted. The ratio of these two forces will indicate which way the particle moves:

$$F_{\text{RP}}/F_{\text{G}} = [\mathcal{L}_\odot/(4\pi GM_\odot)][(4c/3)\rho R]^{-1}$$

$$= [(3\mathcal{L}_\odot)/(16\pi cGM_\odot)](\rho \, R)^{-1} \tag{14.53}$$

$$= 5.8 \times 10^{-4}(\rho R)^{-1} \tag{14.54}$$

when the SI values $\mathcal{L}_\odot = 3.9 \times 10^{26}$ W, $M_\odot = 2.0 \times 10^{30}$ kg, $c = 3.0 \times 10^8$ m/s are substituted.

The forces balance when this ratio is unity, or when the product, $\rho R = 5.8 \times 10^{-4}$. When this product is smaller, the radiation pressure dominates.

Thus, for $\rho = 1000\,\mathrm{kg/m^3}$, $R_{\mathrm{crit}} = 6 \times 10^{-7}\,\mathrm{m} = 0.6\,\mu\mathrm{m}$, so that smaller particles of this density are pushed away from the Sun.

For much denser particles, say $\rho = 6000\,\mathrm{kg/m^3}$, the critical size is smaller: $R_{\mathrm{crit}} = 1 \times 10^{-7}\,\mathrm{m} = 0.1\,\mu\mathrm{m}$.

When gravity dominates, the particle will spiral in. This is why: (14.51) and (14.52) are radial forces, and if these were the only forces acting on the particle then circular orbits would be stable at a reduced speed (compared to orbital speed without radiation pressure) of $v = \sqrt{\{[GM_\odot - 3\mathcal{L}_\odot/(16\pi cR\rho)]/r\}}$. However, even if the transverse motion is at circular velocity initially, aberration will result in a component of the radiation pressure opposing this velocity. The very small acceleration causes the speed to drop below the circular value, causing the dust to spiral into the Sun.

The timescale is given by:

$$t = 7 \times 10^5\ \rho R r^2\ \text{years} \tag{14.55}$$

where ρ is in $\mathrm{kg/m^3}$, R is in m, and (in this expression only), r is in au. Thus, for $R = 1\,\mu\mathrm{m}$, $\rho = 1000\,\mathrm{kg/m^3}$,

$$r = 1.0\,\mathrm{au},\ t = 700\,\mathrm{y}$$
$$2.8\,\mathrm{au},\ t = 5500\,\mathrm{y}$$
$$10\,\mathrm{au},\ t = 70,000\,\mathrm{y}$$

and, for $R = 1\,\mu\mathrm{m}$, $\rho = 3000\,\mathrm{kg/m^3}$,

$$r = 1\,\mathrm{au},\quad t = 2100\,\mathrm{y}$$

and the other times scale up by a factor of 3 also.

For still denser and/or larger particles, the lifetime increases in proportion. Thus, asteroid dust tends to stay around the original orbit longer (and therefore accumulate to a higher concentration) than lighter cometary material.

An improvement to the approximation represented by (14.51) can be obtained if diffraction is included for very small particles (this has the effect of making the effective cross-section of the particle greater than its geometrical area). In addition, the absorptivity $(=1-A)$, the fraction of photons that encounter the particle that are absorbed, should be taken into account. These two effects tend to work against each other, but the former effect is important only for very small particles. Folding the two effects into a quantity Q for radiation pressure coefficient, if gravity is ignorable, the net force is then given by

$$\boldsymbol{F}_{\mathrm{RP}} = [L_\odot/(4\pi cr^2)](Q\pi R^2)[(1 - 2v_{\mathrm{r}}/c)\hat{\mathbf{r}} - (v_\theta/c)\hat{\Theta}] \tag{14.56}$$

where the velocity terms in the right square bracket represent the "Poynting–Robertson drag." What we now call the Poynting–Robertson effect was first suggested by Poynting (1903) and applied to the solar system quantitatively by Robertson (1937). Recent detailed discussions of the state of knowledge about and the fate of the dust can be found in Williams (2002) and in other papers in Murad and Williams (2002) and in Duncan and Lissauer (1999).

Great progress in understanding comets has occurred, but much is yet to be learned, especially if recent reports of detections of main asteroid belt comets (e.g., Hsieh et al., 2004) are verified.

This concludes our discussion of cometary phenomena. Next we shall examine the denizens of the Edgeworth–Kuiper Belt and their generally rockier cousins, the main-belt asteroids. First we turn our attention to the study of meteorites, then to the asteroids, and finally to the likely origin for the solar system in light of what we know both about our planetary system and those of other stars.

References

A'Hearn, M. F. et al. (31 co-authors) 2005. "Deep Impact: Excavating Comet Tempel 1," *Science*, **310**, 258–264.

Bailey, M. E., and Stagg, C. R. 1990. "The Origin of Short-Period Comets," *Icarus*, **86**, 2–8.

Bailey, M. E., Clube, S. V. M., and Napier, W. M. 1990. *The Origin of Comets* (Oxford: Pergamon Press).

Beatty., J. K. and Chaikin, A. 1990. *The New Solar System*, 3rd Ed. (Cambridge, MA: Sky Publishing Corp. and Cambridge, UK: University Press).

Biermann, L. 1968. "On the Emission of Atomic Hydrogen in Comets." JILA Report No. 93 (Boulder, CO: Joint Institute for Laboratory Astrophysics).

Boehnhardt, H. 2004. "Split Comets," in *Comets II*, eds M. C. Festou, H. U. Keller, and H. A. Weaver (Tucson, AZ: University of Arizona Press), pp. 301–316.

Boice, D. and Huebner, W. 1999. "Physics and Chemistry of Comets," in *Encyclopedia of the Solar System*, eds P. R. Wessman, L.-A. McFadden, and T. V. Johnson (San Diego, CA: Academic Press), pp. 519–536.

Brandt, J. C. and Chapman, R. D. 1981. *Introduction to Comets*, (Cambridge, UK: University Press).

Cajori, F., tr., *Sir Isaac Newton's Mathematical Principles of Natural Philosophy and his System of the World*, revision of A. Motte's 1729 translation. Reprinting, 1962 (Berkeley, CA: University of California Press).

Colangeli, L., Brucato, J. R., Bar-Nun, A., Hudson, R. L., and Moore, M. H. 2004. "Laboratory Experiments on Cometary Materials," in *Comets II*, eds M. C. Festou, H. U. Keller, and H. A. Weaver (Tucson: University of Arizona Press), pp. 695–717.

Cruikshank, D. et al. (12 co-authors). 1998. "The Composition of Centaur 5145 Pholus," *Icarus*, **5**, 389–407.

Danby, J. M. A. 1962. *Fundamentals of Celestial Mechanics*, 1st Ed., (New York: The Macmillan Company).

Danby, J. M. A. 1988. *Fundamentals of Celestial Mechanics*, 2nd Ed., (Richmond, VA: Willmann-Bell).

Dubin, M. 1955. "Meteor Ionization in the E-region", in *Meteors*, ed. T. R. Kaiser (London: Pergamon Press), pp. 111–118.

Duncan, M. J., and Lissauer, J. J. 1999. "Solar System Dynamics", in *Encyclopedia of the Solar System*, eds P. R. Wessman, L.-A. McFadden, and T. V. Johnson (San Diego, CA: Academic Press), pp. 809–824.

Fernández, J. A. 1999. "Cometary Dynamics", in *Encyclopedia of the Solar System*, eds P. R. Wessman, L.-A. McFadden, and T. V. Johnson (San Diego: Academic Press), pp. 537–556.

Flynn, G. J. 2002. "Extraterrestrial Dust in the Near-Earth Environment," in *Meteors in the Earth's Atmosphere*, eds Murad, E. and Williams, I. P. (Cambridge: Cambridge University Press).

Gingerich, O. and Westman, R. S. 1988, "The Wittich Connection: Conflict and Priority in Late Sixteenth-Century Cosmology," *Transactions of the American Philosophical Society*, **78**, Part 7.

Gor'kavi, N. N., Ozernoy, L. M., and Mather, J. C. 1997. "A New Approach to Dynamical Evolution of Interplanetary Dust," *Astrophysical Journal*, **474**, 496–502.

Grün, E. and Jessberger, E. K. 1990. "Dust," in *Physics and Chemistry of Comets*, ed. Huebner, W. F. (Berlin: Springer-Verlag), pp. 113–176.

Harker, D. E., Woodward, C. E., and Wooden, D. H. 2005. "The Dust Grains from 9P/Tempel 1 Before and After Encounter with Deep Impact," *Science*, **310**, 278–280.

Hawkins, G. S. 1964. *Meteors, Comets, and Meteorites* (New York: McGraw Hill).

Hsieh, H. H., Jewett, D. C., and Femandez, Y. R. 2004. "The Strange Case of 133P/Elst-Pizarro: A Comet Among the Asteroids," *Astronomical Journal*, **127**, 2997–3017.

Hutchison, R. 2004. *Meteorites* (Cambridge: Cambridge University Press).

Kaiser, T. R., ed. 1955. *Meteors*. Special Supplement (Vol. II) to the Journal of Atmospheric and Terrestrial Physics (London: Pergamon Press).

Kelley, D. H. and Milone, E. F. 2005. *Exploring Ancient Skies: An Encyclopedic Survey of Archaeoastronomy* (New York: Springer-Verlag).

Kreutz, H. C. F. 1888. "Untersuchungen über das System der Cometen 1843I, 1880I and 1882II, I. Theil," *Publ. Sternwarte Kiel*, No. 3.

Kronk, G. W. (1988). Meteor Showers: A Descriptive Catalog (Hillside, NJ; Aldershot, Hants, UK: Enslow Publishers, Inc.)

Levin, B. J. 1955. "Physical Theory of Meteors and the Study of the Structure of the Complex of Meteor Bodies," in *Meteors*, ed. T. R. Kaiser (London: Pergamon Press), pp. 131–143.

Marsden, B. G. 1963. "On the Orbits of Some Long Lost Comets," *Astronomical Journal*, **68**, 795–801.

Marsden, B. G. and Williams, G. V. 1997. *Catalogue of Cometary Orbits 1997*, 12th Ed. (Cambridge, MA: Minor Planet Center).

Marsden, B. G., and Williams, G. V. 2005. *Catalogue of Cometary Orbits 2005*, 16th ed. (Cambridge, MA: Minor Planet Center).

Meech, K. J. and Belton,. M. J. S. 1990. "The Atmosphere of 2060 Chiron," *Astronomical Journal*, **100**, 1323–1338.

Milam, S. N., Savage, C., Brewster, M. A., Ziurys, L. M., and Wyckoff, S. 2006. "The $^{12}C/^{13}C$ Isotope Gradient Derived from the Millimeter Transitions of CN: The Case for Galactic Chemical Evolution," *Astrophysical Journal*, **634**, 1126–1132.

Milone, E. F., and Wilson, W. J. F. 2008. *Solar System Astrophysics: Background Science and the Inner Solar System.* (New York: Springer).

Moulton, F. R. 1914. *An Introduction to Celestial Mechanics.* (New York: MacMillan) (10th printing, 1951).

Mumma, M. J., DiSanti, M. A., Magee-Sauer, K., Bonev, B. P., Villanueva, G. L., Kawakita, H., Dello Russo, N., Gibb, E. L., Blake, G. A., Lyke, J. E., Campbell, R. D., Aycock, J., Conrad, A., and Hill, G. M. 2005. "Parent Volatiles in Comet 9P/Tempel 1: Before and After Impact," *Science*, **310**, 270–274.

Murad, E. and Williams, I. P., eds 2002. *Meteors in the Earth's Atmosphere. Meteoroids and Cosmic Dust and their Interactions with the Earth's Upper Atmosphere.* (Cambridge: Cambridge University Press).

Murray, C. D. and Dermott, S. F. 1999. *Solar System Dynamics.* (Cambridge: University Press) (2nd printing, 2001).

Newton, I. 1687. Philosophiae Naturalis Principia Mathematica. See Cajori, 1934.

Origen of Alexandria. 3rd c. *Contra Celsus*, tr. Crombie, in Roberts and Donaldson/Coxe, repr. 1994, pp. 422–423; excerpt appears in Kelley and Milone 2005, p. 134.

Poynting, J. H. 1903. "Radiation in the Solar System: Its Effect on Temperature and Its Pressure on Small Bodies," *Proceedings of the Royal Society of London*, **72**, 265–267.

Rietmeijer, F. J. M. 2002. "Collected Extraterrestrial Materials: Interplanetary Dust Particles, Micrometeorites, Meteorites, and Meteoric Dust," in *Meteors in the Earth's Atmosphere*, Murad, E. and Williams, I. P., (Cambridge: Cambridge University Press), pp. 215–245.

Robertson, H. P. 1937. "Dynamical Effects of Radiation in the Solar System," *Monthly Notices of the Royal Astronomical Society*, **97**, 423–438.

Ronan, C. A. and Needham, J. 1981. *The Shorter Science and Civilisation in China: An Abridgement of Joseph Needham's Original Text* (Cambridge: University Press).

Sekanina, Z. 2005. "Comet 73P/Schwassmann-Wachmann Nucleus Fragmentation, Its Light-Curve Signature, and Close Approach to Earth in 2006," *International Comet Quarterly*, **27**, 225–240.

Stagg, C. R. and Bailey, M. E. 1989. "Stochastic Capture of Short-Period Comets," *Monthly Notices of the Royal Astronomical Society*, **241**, 507–541.

Strange, H. L. and Longuski, J. M. 2002. "Graphical Method for Gravity-Assist Trajectory design," *Journal of Spacecraft and Rockets*, **39**, 9–16.

Sugita, S. et al. (22 co-authors). 2005. "Subaru Telescope Observations of Deep Impact," *Science*, **310**, 274–278.

Sunshine, M. et al. (21 co-authors). 2005. "Exposed Water Ice Deposits on the Surface of Comet 9P/Tempel 1," *Science*, **311**, 1453–1455.

Taylor, S. R. 1992. *Solar System Evolution: A New Perspective* (Cambridge: University Press).

Williams, I. P. 2002. "The Evolution of Meteoroid Streams" in *Meteors in the Earth's Atmosphere*, eds E. Murad & I. P. Williams, (Cambridge: University Press), pp. 13–32.

Wyckoff, S., Kleine, M., Peterson, B.A., Wehinger, P.A., and Ziurys, L.M. 2000. "Carbon Isotope Abundances in Comets," *Astrophysical Journal*, **535**, 991–999.

Yeomans, D. K. 1991. *Comets: A Chronological History of Observation, Science, Myth, and Folklore* (New York: Wiley).

Challenges

[14.1] Organize the historical theories of the nature of comets. Is there more evidence for smooth historical progression or for abrupt paradigm shifts?

[14.2] On the basis of information in this chapter and from the orbital and physical data of Tables 13.1 and 13.2, which, if any, of the present day moons of the planets are likely to have been comets originally?

[14.3] Discuss the limitations of Tisserand's criterion and the precision with which the left and right sides of the equality need to agree.

[14.4] (a) Compute the expected brightness of a slow (12 km/s), 10 g meteor of silicate composition; assume a vertical path. Make any other necessary assumptions, describing what they are.

(b) How many meteors of this or greater brightness are potentially visible during the year all over the entire Earth?

[14.5] Suppose the particle of Q. [14.2] were a fast (72 km/s) meteor with a mass of 10^{-2} g; (a) how bright would it get and (b) how long could it be seen?

[14.6] If the particle of Q. [14.5] avoided the Earth, what other destiny/destinies might await it, and on what time scale?

15. Meteorites, Asteroids and the Age and Origin of the Solar System

In the previous chapter, we discussed sporadic meteors and meteor showers and indicated that some of the former and most of the latter likely originate from comets. Here we discuss the source of the other meteors, especially those that survive their fiery passage through the atmosphere and impact the Earth. These *meteorites* have become a primary source of knowledge about the age and origin of the solar system. Another important source is the increasing number of objects being detected in the outer solar system, so we will also describe what has been learned about these objects in recent years. Finally we consider the birth of the solar system in the context of what we know about proto-stellar disks.

15.1 Stones from Heaven

"It is easier to believe that two Yankee professors would lie than that stones would fall from heaven." So Thomas Jefferson was reputed to have said, when he heard of a meteorite that was observed to fall on December 14, 1807 in Weston, Connecticut. Wasson (1985, p. 4), however, finds no evidence that Jefferson ever said it. Instead, in keeping with Jefferson's reputation as a careful student of the natural sciences, he cites a letter by Jefferson to Daniel Salmon, written in February, 1808 about the meteorite that he described.

"its descent from the atmosphere presents so much difficulty as to require careful examination," but, he added, "We are certainly not to deny whatever we cannot account for." ..."The actual fact however is the thing to be established." He hoped this would be done by those qualified and familiar with the circumstances.

Scientists at the end of the eighteenth century were living in an age of rationalism; they usually dismissed claims of such phenomena as incredible "folk tales". Nevertheless, Ernst Florenz Friedrich Chladni (1756–1827) argued that the evidence was very strong that meteors (bolides in particular) did produce meteorites sometimes, and argued that prehistoric meteorites might be present in the form of large iron masses. Subsequent carefully documented falls in the early nineteenth century, such as that in Wethersfield, helped turn the tide of opinion. In retrospect, we know of meteorites in the remote past that were carefully cached away because they were indeed perceived to fall from heaven. Hence, in the New Testament book of Acts (19:35), Paul's preaching of Christianity in the Asia Minor city of Ephesus infuriated its merchants and artisans, whose businesses were tied in with the worship of Artemis. Ephesus was described as the "temple Keeper," and the city of "the great stone that fell from the sky." Artemis was often depicted as a multi-breasted goddess, a possible reference to the bossed and dimpled appearance of an iron meteorite (see Wood 1968 and Wasson 1985 for similar views). Chinese records of meteorite observations date back to the seventh century BC; these and other historical data are reviewed in Kelley and Milone (2005, Section 5.6). For meteorite craters, see Hodge (1994).

15.1.1 Categories and Nomenclature of Meteorites

What were *meteoroids* in space, survived their passage through the atmosphere as *meteors* to land on Earth as *meteorites*. The term "meteor" is from the Greek word μετεορον (plural: μετεορα, "things in the air"). We discussed meteors in Chapter 14.10. Meteoroids will be discussed in Section 15.7, in the course of identifying the parent source bodies for the meteorites.

15.1.1.1 Broad Categories Older astronomy textbooks usually referred to only three types of meteorites:

- *Aerolites* (more commonly known as *stony* meteorites or *stones*)
- *Siderites* (*irons*)
- *Siderolites* (*stony-irons*)

Although this categorization is still somewhat useful in describing broadly what type of meteorite one may find, today the classes are more numerous, and many subclasses of meteorites are recognized.

Classification identifies the nature of the meteorite, but another important aspect is the circumstance under which it is found: has it been lying around on the Earth's surface for a lengthy and often unknowable amount of time where it was subjected to terrestrial contamination or was it recovered and carefully protected just after it fell? The answers to these questions may help to establish the reliability of compositional abundances in the meteorite. A "weathering grade" has been devised for Antarctic meteorites (Wlotzka 1993), for example, and, the degree to which the rock has been shocked also has been quantized, as "shock stages".

15.1.1.2 Another Distinction: *Falls* **and** *Finds* If meteorites are detected as meteors and are subsequently collected, they are called *falls*. They are overwhelmingly *stony*. *The Catalogue of Meteorites* (Grady 2000) indicates that of the 22,507 meteorites known as of December 1999, 1005 were falls. Of these, 940 are stones, 48 are irons, and 12 are stony-irons; 5 are unknown.

Meteorites which were *not* seen as meteors but which were discovered in the ground are known as *finds*. Formerly, the finds included irons in large numbers (42%), but there were still more numerous stones (52%) and some stony-irons (5%) found. The number of stony meteorites among the finds was classically underestimated because they look much like terrestrial rocks, and so may go unnoticed, especially after the thin fusion crust (acquired from atmospheric heating during entry) has worn away. These meteorites also erode more easily than other types. Discoveries of huge numbers of meteorites on the ice sheets of Antarctica (17,808 as of December, 1999: 79% of the total!) have helped to redress the balance and have produced some interesting surprises, too. With the discoveries in Antarctica, the *finds'* statistics have greatly improved: 95.59% are stony; 3.84% are irons, and 0.52% are stony-irons. Finally, meteorites have also been found in the hot deserts of Earth, on the Moon [Wood (1990, p. 244), states that up to 4% by weight of lunar soils are meteoritic], and even on Mars, thanks to the Spirit and Opportunity Rovers.

15.1.1.3 Nomenclature Historical development has dictated the familiar names of individual meteorites, which are traditionally called by the geographic region in which the discovery was made. Thus, the *Millarville* meteorite was found on April 28, 1977, on a farm near Millarville, Alberta. In the 1970s, the Meteoritical Society established rules for the naming of meteorites, and its Committee on Meteorite Nomenclature makes decisions regarding assignment of and changes to meteorite names. Its decisions are published in the *Meteoritical Bulletin*. For the numerous finds in Antarctica and in desert areas of the world, a new naming convention provides the year of recovery followed by a sequence number of discovery, hence, *ALH 84001*, the first specimen collected during the 1984–(1985) (summer) field season in the Alan Hills region of Antarctica.

15.1.2 Petrographic Categories

Basically, *petrography* is the description and classification of rocks; *petrology* is, strictly, the study of rocks, and therefore a discipline within geology. The old categories of meteorites mentioned in Section 15.1.1 have not proven adequate to reveal the full range of information that they contain. Generally, classification techniques make extensive use of chemical composition and petrographic (sometimes called petrologic) properties such as:

- Texture
- Water content
- Igneous glass state
- The presence of metallic minerals
- The structure and homogeneity of certain mineral compositions

More recent scholarship also suggests that the material of which meteorites are composed can take different forms depending on the physical conditions under which the rock was formed. Consequently, Urey and Craig (1953) began a systematic study of chondrite compositional differences; this was extended by John Wasson, who carried out similar work on the iron meteorites. On this basis, three main kinds of meteorites are recognized:

1. Non-differentiated meteorites
2. Differentiated silicate-rich meteorites (DSRs)
3. Iron meteorites

The latter two types require material to be *differentiated* or separated out, usually in a high temperature and pressure environment.

Meteorites may be classified by *group*, roughly analogous to spectral classification in stars, distinguished by mineral and chemical or isotopic composition. For example, there are 13 groups of iron meteorites.

A group may be further distinguished by *type*, or *class*, roughly analogous to the subclasses of stellar spectral types.[1]

Groups of meteorites may also be broadly linked in various ways; these linkages are termed *clans*. Table 15.1 (adapted from Wasson 1985, Appendix A), indicates the principal clans, groups, and types. A more recent summary of the classification of meteorites is given by Hutchison (2004).

15.1.3 Meteorite Groupings and Subgroupings

15.1.3.1 Undifferentiated Meteorites The principal (and under more recent classification schemes, *all*) undifferentiated meteorites are called *chondrites*, which can be defined as "microbreccias", composed of matrices of material, each with its own "genetic" background history onto which varying

---

[1] The analogy is limited because the spectral subclasses are refinements to the main classification criteria that are primarily dependent on the temperature of the photosphere. Here, the numbers represent different groups of properties, not completely dependent on the chemical composition.

Table 15.1. Classifications of meteorites[a]

Clan	Group	Other names	%Falls[b]	Examples
Chondrites				
Refractory-rich	CV	Carbonaceous, C3	1.1	Allende
Minichondrule	CO	″ C3	0.9	Ornans
″	CM	″ C2	2.0	Murchison
Volatile-rich	CI	″ C1	0.7	Orgueil
Ordinary	LL	Amphoterite	7.2	St. Mesmin
″	L	Hypersthene	39.3	Bruderheim
″	H	Bronzite	32.3	Ochansk
IAB inclusion	IAB chondrite		. . .	Copiapo
Enstatite	EL	Enstatite	0.9	Indarch
″	EH	″	0.7	Khairpur
. . .	Other chondrites	CH, CR, R	0.30 / . . .	Kakangari / Acfer 182; Renazzo; Rumuruti
Differentiated meteorites (non-irons)				
Enstatite	Aubrites	Enstatite achondrites	1.1	Norton County, Aubres
Igneous	Eucrites	Basaltic, Ca-rich, Pyroxene-plagioclase achondrite	2.8	Juvinas, Pasamonte, Piplia Kalan, Sioux County
″	Howardites	{same as Eucrites}	2.5	Kapoeta
″	Diogenites	Hypersthene achondrites	1.1	Johnstown, Roda, Shalka
″	Mesosiderites	Stony-irons	0.9	Estherville
″	Pallasites	Stony-irons	0.3	Krasnoyarsk
. . .	Urelites	Olivine-pigeonite achondrite	0.4	Novo Urei
. . .	Other differentiated, silicate-rich meteorites	Angrites; SNC; lunar; . . .	1.0	Angra dos Reis; Shergotty, Nakhla, Chassigny, ALH 84001; ALH 81005
Differentiated meteorites (irons)				
. . .	IAB irons	. . .	0.9	Canyon Diablo
. . .	IC irons	. . .	0.1	Bendegó
. . .	IIAB irons	. . .	0.5	Coahuila
. . .	IIC irons	. . .	0.1	Ballinoo
. . .	IID irons	. . .	0.1	Needles
. . .	IIE irons	. . .	0.1	Weekeroo
. . .	IIF irons	. . .	0.1	Monahans
. . .	IIIAB irons	. . .	1.5	Henbury
. . .	IIICD irons	. . .	0.1	Tazewell
. . .	IIIE irons	. . .	0.1	Rhine Villa
. . .	IIIF irons	. . .	0.1	Nelson County
. . .	IVA irons	. . .	0.4	Gibeon
. . .	IVB irons	. . .	0.1	Hoba
. . .	Other irons	. . .	0.6	Mbosi

[a] Mainly from Wasson (1985, Table A-1), among other sources.

[b] Extrapolated from a limited sample; see Wasson (1985) for details.

Table 15.2. Chondrite groups[a]

Group	Mg/Si	Ca/Si	Fe/Si	Fayalite (%)[b]	Si (%)	Fe(met/tot)[c]	$\delta^{18}O$	$\delta^{17}O$
CI	1.07	0.062	0.87	...	10.5	0.0	~16.4	~8.8
CM	1.04	0.069	0.82	...	12.9	0.0	~12.2	~4.0
CO	1.05	0.070	0.78	...	15.9	0−0.2	~−1.0	~−5.1
CV	1.07	0.085	0.76	...	15.6	0−0.3	~0	~−4.0
CK	1.13	0.080	0.79	...	15.1	0.0		
CR	1.05	0.063	0.79	...	15.3	0.4		
CH	1.06	0.060	1.52	...	13.3	0.9		
H	0.95	0.052	0.82	16−20	16.9	0.58	4.1	2.9
L	0.93	0.050	0.58	23−26	18.5	0.29	4.6	3.5
LL	0.93	0.048	0.49	27−32	18.9	0.11	4.9	3.9
EH	0.73	0.036	0.87	...	16.7	0.76	5.6	3.0
EL	0.87	0.038	0.59	...	18.6	0.83	5.3	2.7
R	0.93	0.053	0.77	...	15.8	0.0		

[a] Data taken mainly from Taylor (1992, Table 3.6.1, p. 108), Sears and Dodd (1988, Table 1.1.3, p. 15), and Hutchison (2004, Table 2.1 and 2.2, pp. 29–30).
[b] Fayalite (Fe_2SiO_4) proportion of the total olivine.
[c] Fraction of metallic iron to total iron.

amounts of molten drops (*chondrules*) of material were splashed. The detailed composition distinctions will be treated in Section 15.2.

The chondrite groups are indicated in Table 15.2, taken from Sears and Dodd (1988, p. 15), Taylor (1992, p. 108), and updated with material from Hutchison (2004, p. 30). The petrologic (or petrographic) classes of chondrites are given in number sequence in Table 15.3 (adapted from tables by Taylor 1992, Wasson 1985, Hutchison (2004, p. 41), and other sources). The petrologic class is often given with the chondrite designation, e.g., CV3 for the Allende meteorite.

15.1.3.2 Differentiated Meteorites The *differentiated silicate-rich* (DSR) and *iron* meteorites are *differentiated*: when the parent body was molten, contents were able to separate, depending on density, in a gravitational field. The silicates and metals were immiscible and thus separated quickly, with the denser molten metal ($\leq 7.9\,g/cm^3$) sinking to the centers of the parent bodies, and the silicates, much less dense, floating to the top.

Further differentiation in the silicate material, where densities range from 2.6 to $3.6\,g/cm^3$, led to crustal and mantle silicate compositions and densities. We will discuss this process further in Section 15.3.

DSRs include all those groups which are neither chondrites nor irons.

The groups of differentiated silicate-rich meteorites are given in Table 15.4, taken mainly from Wasson (1985, Table II-3, p. 33). The extent to which differentiation occured varied greatly. These meteorites are roughly of two kinds:

Table 15.3. Petrologic classes of chondrites[a]

Property	1	2	3	4	5	6
Olivine & pyroxene variation (%)		Pyroxene ≥ 5 / Olivine ≥ 3		0<Pyroxene <5		Uniform / Fe-Mg minerals
Low-Ca pyroxene, Structural state		Mostly monoclinic		>20%	Much Monoclinic / <20% / Orthorhombic	Orthorhombic
Secondary feldspar (grain size devel.)		Absent		>20% / <2 µm / Mostly micro-crystalline aggregates	<20% / <50 µm	>50 µm / Grains clear, interstitial
Igneous glass		Clear & isotropic / variable abundance		Turbid if present		Absent
Metallic minerals (max. Nickel content)		Taenite absent or minimal (<20%)		Kamacite & taenite present / (>20%)		
Sulfide minerals (<Nickel> content)	None	>0.5%		<0.5%		
Chondrules (overall texture)	None	Very well-defined	Opaque	Well-defined	Delineated	Poorly defined
Matrix texture	Fine-grained; opaque	Much opaque	Opaque	Transparent microcrystalline	Recrystallized / coarsening	
Bulk carbon content (%)	3–5	≲1–3	0.2–1	<0.2		
Bulk water content (%)	~20	3–11	0.3–3.0	<1.5		

[a] Based on Taylor (1992, Table 3.6.2, p. 108), van Schmus and Wood (1967, Table 2, p. 757), Sears and Dodd (1988, 1.1.4, pp. 23–24), Wasson (1985, Table II-2, p. 31), and Hutchison (2004, Table 2.4, p.41)

Table 15.4. Selected groups of differentiated silicate-rich meteorites[a][b]

Group	Mineral	Concentration as silicates frac.	FeO/[FeO+MgO]%	Fe–Ni %	$\langle\delta^{18}O\rangle$ $(^0/_{00})$	$\langle\delta^{17}O\rangle$ $(^0/_{00})$	$\langle\Delta^{17}O\rangle$ $(^0/_{00})$	Breccia type
Eucrites	Pigeonite	400–800	45–70	<1.0	3.61(5)	1.63(4)	−0.24(2)	Monomict
Howardites	Orthopyroxene	400–800	25–40	~1.0	3.25(11)	1.43(7)	−0.27(3)	Polymict
Diogenites	Orthopyroxene	~950	25–27	<1.0	3.41(4)	1.46(3)	−0.27(2)	Monomict
Mesosiderites	Orthopyroxene	400–800	23–27	30–55	3.41(6)	1.53(5)	−0.25(3)	Polymict
Pallasites	Olivine	~980	11–14	28–88	2.92(3)[c]	1.23(2)[c]	−0.28(1)[c]	Monomict
Aubrites	Clinopyroxene (low calcium)	~970	.01–.03	~1.0	5.26(5)	2.75(3)	+0.02(1)	Polymict
Ureilites	Olivine	~850	10–25	1.-6.	6.93(14)	2.41(15)	−1.20(8)	Polymict

[a] From Wasson (1985, Table II-3, p. 33)

[b] del values are means calculated from data of Clayton and Mayeda (1996, Table 1), who define $\Delta^{17}O = \delta^{17}O - 0.526 \times \delta^{18}O$. This quantity indicates the departure from the Terrestrial Fractionation Line in Fig.15.7, and so indicates different places of origin or formation process. Parentheses contain the standard deviations of the means, in units of the last decimal place.

[c] Main group pallasites. Clayton and Mayeda list two other pairs of Pallasites with very different del values: (−2.80, −6.15, −4.68 for one pair and +2.26, 0.41, and −0.76 for the other, respectively).

1. Those with high metal content, usually referred to as *stony-irons*

2. Those with low metal content and igneous origin, often referred to as *achondrites*

The stony-irons include the groups: *pallasites* and *mesosiderites*.

The achondrites include the groups: *ureilites, diogenites, aubrites, howardites, and eucrites.*

One small, but very important DSR achondrite group is the *SNC* (pronounced "snick") meteorites—[*S* for *shergottites* (after the prototype *Shergotty* meteorite), *N* for *nakhlites* (after the *Nahkla* meteorite), and *C* for *chassignite* (after the *Chassigny* meteorite)]—all named for the sites in which they were found. There are now several others, including the Alan Hills meteorite (Antarctica meteorite ALH84001). These meteorites have much younger radiogenic ages than most meteorites, indicating substantial remodification in a major body. In fact they appear to have Martian composition. The atmospheric samples embedded in glass globules embedded in the meteorites have isotope ratios of the Martian atmosphere. This is interesting in itself—because it indicates that over the time since the meteorites were blasted off the surface of Mars, the atmosphere has not changed greatly. Of course, if it had changed, and the composition were different from present day Mars, some other venue of origin might have been sought for them.

Another small achondrite group, discovered in Antarctica, has lunar composition.

Finally, the irons have large iron and nickel content. They consist of two main groups, distinguished by crystal structure: the *octahedrites* and the *hexahedrites*. There are, however, many possible groupings based on the relative abundances of such elements as nickel, germanium, gallium, and iridium. Table 15.5, adapted from Wasson (1985, Fig. II-4, p. 43) lists some of the properties of 13 groups of iron meteorites, and of the mesosiderite and the pallasite stony-iron groups. The irons are described further in Section 15.4.

15.2 Undifferentiated Meteorites: the Chondrites

15.2.1 Defining the Chondrites

Chondrites make up the bulk of both the finds (66.8%) and the falls (81.7%), and thus are the most common meteorites. Grady (2000) lists 15,190 chondrites, of which 821 are falls.

These meteorites have three distinctive components:

Table 15.5.[a] Iron Meteorites (& two Stony-Iron[b]) groups

Group	Bandwidth (mm)	Structure %	Ni 10^{-6}	Ga 10^{-6}	Ge 10^{-6}	Ir	Ge–Ni	Frequency correlation(%)	$\langle \Delta^{17}O \rangle$ ($^0/_{00}$)
IA	1.0–3	Om–Ogg[c]	6–9	55–100	190–520	0.6–5.5	Negative	17.0	}−0.48(2)
IB	0.01–1.0	Dom	9–25	11–55	25–190	0.3–2.0	Negative	1.7	
IC	<3	Anomalous, Og	6–7	49–55	212–247	0.07–2.1	Negative	2.1	...
IIA	>50	Hexahedrite	5–6	57–62	170–185	2–60	Positive?	8.1	...
IIB	5–15	Ogg	~6	46–59	107–183	0.01–0.9	Negative	2.7	...
IIC	0.06–0.07	O plessitic	9–12	37–39	88–114	4–11	Positive	1.4	...
IID	0.4–0.8	Of–Om	10–11	70–83	82–98	3.5–18	Positive	2.7	...
IIE	0.7–2	Anomalous	8–10	21–28	62–75	1–8	Absent	2.5	+0.59(2)
IIF	0.05–0.21	Ataxite-Of	11–14	9–12	99–193	0.8–23	Positive	1.0	...
Mes.[b]	~1	Anomalous	6–10	9–16	37–56	2.2–6.2	Absent	...	−0.25(3)
Pal.[b]	~0.9	Om	8–13	14–27	29–71	0.01–27	Negative?	...	−0.28(1)
IIIA	0.9–1.3	Om	7–9	17–23	32–47	0.15–20	Positive	24.8	}−0.21(2)
IIIB	0.6–1.3	Om	8–11	16–21	27–46	0.01–15	Negative	7.5	
IIIC	0.2–3	Off-Ogg	6–13	11–92	8–380	0.07–2.1	Negative	1.4	...
IIID	0.01–0.05	Ataxite-Off	16–23	1.5–5.2	1.4–4.0	0.02–07	Negative	1.0	}−0.43(6)
IIIE	1.3–1.6	Og	8–9	17–19	34–37	0.05–6	Absent	1.7	
IIIF	0.5–1.5	Om–Og	7–9	6.3–7.2	0.7–1.1	0.006–7.9	Negative	1.0	...
IVA	0.25–0.45	Of	7–9	1.6–2.4	0.09–14	0.4–4	Positive	8.3	+1.17(5)
IVB	0.006–0.03	Ataxite	16–18	0.17–0.27	0.03–07	13–38	Positive	2.3	...

[a] Mainly from Wasson (1985, Table II-5). Del values are averages from data of Clayton and Mayeda (1996, Table 1) for Iron Groups IAB, IIE, for Mesosiderites, Pallasites, and for iron groups IIIAB, IIICD, and IVA. Consult this source for which meteorites are included in the averages. The parentheses contain the standard deviation of the means in units of the last decimal place

[b] Mesosiderites and Pallasites

[c] O≡Octahhedrite; the designations gg, g, m, f, ff are gradations on a course-to-fine scale.

1. Chondrules (not always present)

2. A matrix of material

3. Calcium–aluminum-rich inclusions (*CAI*s)

Although it has been broadened to include all undifferentiated meteorites, the term "chondrite" was originally applied only to meteorites containing small spherical objects called *chondrules* (from the Greek word for grain) which were once molten droplets. The degree to which they are present, their appearance, and their composition are important classifying parameters. They vary in size, composition, and texture.

The variety suggests that the chondrules made their appearance under varied conditions and places in the solar system.

Chondrules range in size from 0.5 to 2.5 mm, vary in texture and composition, and are often grouped into at least two types with less (type I) and more (type II) FeO content, other minerals, and differing oxidation states. See Hewins (1997); and Hutchison (2004) for details.

The space between chondrules is filled with a matrix, thought to be fine-grained material from the original solar nebula which somehow avoided the heating undergone by the chondrules. Mixed in with this material are broken chondrule and coarse-grained, high-temperature material. The matrix is rich in FeO and is composed of fine-grained ($\sim 1\,\mu$m) dark, opaque material.

The main compositional ingredients of chondrules of carbonaceous H3 and L3 chondrites (see below for the explanation of these types of meteorites) are:

- SiO_2 (45–50% by weight)
- MgO (28–36%)
- FeO (9–16%)
- Al_2O_3 (3–5%)
- CaO (2–3%)

with oxides of Na, K, Cr, Ti, and Mn taking smaller shares.

Most chondrules are depleted in the non-exclusive categories of:

- *Siderophilic* elements (having affinity for iron and nickel) [These include: Pd, Au, Rh, Pt, Ir, Ru, Re, and Os; and to a lesser degree: Mo, Ge, As, Sb, Ni, Ag, Co, W, P, Ga, Cr, V, and Mn]; and in
- *Chalcophilic* elements (preferring the company of sulfur-bearing minerals) [These include: Ag, Bi, Cd, Cu, Hg, In, Pb, Sn, Te, Tl, and Zn)]; but they are *not* depleted in:

- *Lithophilic* elements (having affinity for minerals rich in oxygen)

  [These include: Al, Ba, Ca, Ce, Cr, Cs, K, La, Mg, Na, Rb, Se, Sr, Th, U, and Yb].

In the Earth's crust, lithophilic elements dominate, chalcophilic elements are found concentrated in ore deposits, and the siderophilic elements are largely missing. In chondritic meteorites generally, the three major groups of elements are found much more mixed than on Earth's surface, with the matrix being much richer in iron content than the chondrules. However, different kinds of chondrules may be found in one and the same meteorite, and the composition can vary among the kinds of chondrules. These results strongly suggest the presence of a very non-uniform medium in the early solar system, and the presence of short-lived radioactive nuclides such as ^{26}Al which decays to ^{26}Mg with a half-life of 0.72 My (See Gounelle & Meibom, 2008 about the origin of the ^{26}Al).

The chondrules' textures suggest rapid heating and cooling, and there is a large enough population of fused chondrules to suggest that during formation their number density was relatively high. Experiments suggest that chondrule textures can be reproduced by temperatures of ~ 1550 K for ~ 15 min followed by cooling at rates ~ 500 K/h. However, some appear to have been heated to temperatures as high as 1700 K.

For the most part, chondrites have solar composition, but some chondrites differ in having fewer volatile elements.

A *volatile* substance is relatively fast to melt and then evaporate with increasing temperature while *refractory* material is relatively slow to melt and then evaporate. Conversely, on cooling, refractory materials condense out and solidify before volatiles. Another way to put it is that the refractory elements will condense out at relatively high temperatures, while the volatiles will remain gaseous until the temperature decreases further.

The *abundance* (formally defined as the ratio of the number of atoms of an element relative to those of another element) of volatile elements is one of the chief discriminators among groups of meteorites, and this includes chondrites.

The CAIs, which contain refractory elements (such as Ca, Al, Re) are further subdivided in their rare-earth-element composition. In general, the CAIs have refractory minerals, such as:

- Perovskite $(CaTiO_3)$
- Melilite $[Ca(Al, Mg)(Si, Al)_2O_7]$
- Spinel $(MgAl_2O_4)$
- Hibonite $[Ca(Al,Mg,Ti)_{12}O_{19}]$

These minerals follow a condensation sequence, in the sense that they condense from vapor at successively lower temperatures in the order listed above, but the overall composition suggests episodes of condensation and melting in a complex early solar system environment.

15.2.2 Carbonaceous Chondrites

Grady (2000) lists 561, of which 36 are falls. Some of the groups are labeled CI, CV, CM, CO. The "C" in the groups *CI, CM, CO*, and *CV* means *carbonaceous*, i.e., containing carbon in great abundance. The second letter indicates the prototype: "I" of "CI" stands for *Ivuna*; the "V" of "CV", for *Vigrano*; the "M" of "CM" for *Migei*; and the "O" of "CO" for *Ornans*. Additional groups are CH ("H" for High-Fe), CK ("K" for *Karoonda*), and CR ("R" for *Renazzo*). Numbers are added to indicate petrologic character.

The CO and CV types contain metallic iron; the CI and CM types do not; but the CI have more iron than most other chondrites, in a form such as FeO. The CI also contain more volatiles than most others [in Wasson's (1985) classification, this is the *volatile-rich clan*]; for example, the water content may reach 10% (Taylor, 1992, p. 109). The CV group (half of the $C3$ grouping) is the most refractory-rich (Wasson's *refractory-rich clan*), followed by the CO and CM groups (Wasson's *minichondrule clan*, and the other half of the $C3$ and the $C2$ groupings, respectively) which are about equal in refractory element composition. All contain iron in relatively large amounts compared to non-carbonaceous chondrites.

One of the most well-known of the carbonaceous meteorites is the *Allende*, observed to fall in 1969 from a bright bolide depositing an estimated tonne (1000 kg) of material over a *strewn field* ellipse of $\sim 50 \times 10$ km size near the Mexican village Pueblito de Allende. The fall occurred while preparations were under way to study the rocks to be returned from the Moon, and the Allende furnished a large sample of extraterrestrial test material, essentially uncontaminated by terrestrial weathering and chemical processes. Petrologically, Allende belongs to the most primitive class[2] (3), and thus can, in principle, give the best evidence for conditions in the early solar system.

Calcium–aluminum-rich inclusions (*CAIs*) in a matrix interspersed with chondrules provide a rich mixture of information about the nature and origin of the source of the Allende material.

The abundance ratio $^{87}Sr/^{86}Sr$ is significantly lower than the value for achondrites and because this ratio increases with time, the carbonaceous chondrites may represent the oldest datable material in the solar system.

---

[2] Classes (1) and (2) seem to have been subject to aqueous alteration, and (4) and (6) to alteration by heating.

Fig. 15.1. A small slice of the Allende meteorite, viewed from both sides, with each side photographed under slightly different lighting conditions. Note the fusion crust, the round chondrules, the surrounding matrix, and the white calcium–aluminum-rich inclusions (CAI)

Figure 15.1 shows the two faces of a slice of the Allende meteorite as photographed under slightly different lighting conditions. Notice the CAI and chondrules as well as the appearance of the fusion crust.

15.2.3 Ordinary Chondrites

The *ordinary chondrites* are the most common among both the chondrites and all meteorites. Grady (2000) lists 14,265 ordinary chondrites, of which 739 are falls. They are further divided according to iron content, (parentheses gives the % of this type among the ordinary chondrites):

- "H" for high (48.8%)
- "L" for low (43.6%)
- "LL" for low iron and low metal (7.3%)

The chondrule compositions in these meteorites are more or less uniform across subgroups. This means that the matrix and inclusions are the main sources of the differences. Ordinary chondrites are further separable from the carbonaceous chondrites and other meteorites on the bases of oxygen isotope and carbon content.

15.2.4 Enstatites

The *enstatite chondrites* contain an abundance of enstatite, $MgSiO_3$. "EH" and "EL" indicate high and low iron content, respectively. These two subgroups have similar oxygen isotope ratios and solar volatile composition, but have different bulk properties, suggesting different parent bodies, though probably from the same general region of the solar nebula. Grady (2000) lists 201 enstatite meteorites of which 15 are falls.

15.2.5 The R Group

Finally, the R or *Rumurutiite* group is named after the *Rumuruti*, the only fall in the group of 19, listed by Grady (2000). They are similar in bulk chemical composition to H chondrites, but with twice as much sulfur and roughly half the carbon content. They are also distinguished by isotope composition.

15.2.6 Former Members, from the IAB Clan

Finally, among Wasson's clans was one that included a slightly differentiated group of chondrites (currently classified, according to Hutchison (2004, p. 25) as a group of "primitive achondrites") and silicate inclusions of similar isotopic composition in a group of iron meteorites (IAB/IIICD).

15.2.7 Origins of the Chondrites

The state of the iron is a further clue to the origins of these meteorites. The ratio of reduced iron (in the form of metallic iron or bound in FeS) to oxidized forms (in silicates and Fe_3O_4), and the abundance ratio $FeO/(FeO + MgO)$ indicate the relative abundance of iron oxides.

Within each group there are six or sometimes seven stepped petrologic classes of each of 10 bulk properties (see Table 15.3). On this basis, the Allende meteorite is classified as *CV3* (or, more precisely, *CV3.2*), the Bruderheim meteorite as *L3*, and the Abee chondrite as *EH4*.

The degree of *equilibration* (equilibration means that grains of the same material, such as olivine or orthopyroxene, have the same composition) may indicate (in an inverse manner) how closely a particular chondrite resembles the original solar nebula. Wasson suggests that the least equilibrated, Type 3 chondrites of the *CO, CV, H, L, LL,* and *EH* groups, most closely approximate those conditions, and therefore may represent the earliest primordial solar system material.

What distinguishes these types of meteorites is that they do not seem to have formed in a body larger than ~ 100 km in diameter, because their parent body did not melt, permitting the materials in it to *fractionate* or *differentiate* (separate out).

It has been estimated that the heat of accretion and radioactive isotopes combined could produce the fully melted condition—if the accretion body were large enough. Recall, however, from Chapters 8 and 13 that the Earth's moon and several other moons, both large and small have provided many surprises; it is quite possible, therefore, that there is a piece of the chondrite puzzle that had not yet been found. Nevertheless, as we have noted, the radioisotope studies suggest the chondrites to be truly ancient, and this is reinforced by the ratio of the number of atoms of deuterium to those of hydrogen (D/H). In chondrites, D/H varies, from $\lesssim 10^{-5}$ (for some CM2 chondrites) to above terrestrial (for, e.g., some CV3 and LL chondrites). Compare these to the ratio for the interstellar medium (ISM), $\sim 1.5 \times 10^{-5}$, or that of terrestrial hydrogen (1.6×10^{-4}). Although an ISM composition is expected of the early solar nebula that eventually became the solar system, fractionation clearly occurred. The separation of isotopes depends on the reaction rates, and these depend on the rms speeds and thus on temperature and the atomic weights of the atoms within the compounds. Thus, D/H was higher in the inner solar nebula and decreased with distance from the Sun as the solar wind blew away more H than D atoms. But D/H of water that condensed during accretion beyond ~ 5 au (the "snow line") increased with increasing distance. See Bertotti et al. (2003, pp. 229ff) for a fuller discussion.

The chondrules, and the CAIs, however, are relatively depleted in volatiles, therefore they must have been formed when the sun had already begun to disperse the volatile gases of the early solar system.

A number of theories for the origin of chondrules have been put forward:

1. Impacts on surfaces of planets

2. Impacts on chondrite parent bodies

3. Collisions between molten proto-planets

4. Volcanism on planetary surfaces

5. Solar nebula origin

The fifth is the least objectionable on the basis of available evidence and isotopic data. In this model, grains in the disk melted through one or more mechanisms to create the chondrules. Such mechanisms could be flares in the T Tauri stage of the Sun's development, electrical discharges, and radioactive heating due to short-lived radioisotopes. Subsequent accretion with grains to form a matrix, the CAIs, along with metals, sulfides, minerals, and perhaps ices, formed the chondritic parent bodies. The accretion process is the primary process. Subsequent pressure and temperature variations produced different degrees of metamorphosis, the secondary processes. Finally, shock events and weathering constituted tertiary processes. The earlier classification methods, emphasizing physical color and texture, tended to result in classification based on the shock and weathering events more than the origins and metamorphoses.

Now we begin an examination of the differentiated meteorites.

15.3 DSR Meteorites

The hypothesis mentioned in Section 15.1 that the suggested sources of DSR meteorites are the crusts and mantles of molten parent bodies has not been proven. It is, however, a logical and widely accepted hypothesis. In addition to accounting for the existence of irons, stony-irons, and that group of stony meteorites known as *achondrites*, there is still more evidence, which we present below. We discuss the achondrites and the stony-irons first, and then the irons.

15.3.1 The Igneous Clan

Grady (2000) lists 610 achondrites (including the *aubrites*, which we describe separately, later), of which 78 are falls.

Wasson links the five types: *howardites*, *eucrites*, *diogenites*, *mesosiderites*, and *pallasites* in the *igneous* clan. Their oxygen isotope ratios are similar. The compositions are consistent with origin in a common parent body. Hutchison, however, combines the first three into a petrogenetic association called *HED*. His other associations include the *main group pallasites*, the *mesosiderites*, the *IIIAB iron association*, and the Martian *SNC* meteorites. Other achondrite groups are the *acapulcoites*, *angrites*, *brachinites*, *lodranites*, and *winonaites*.

On Earth, the melting of the mantle produces basalts. The structure of uncrushed *eucrites* resembles vesicular terrestrial basaltic lava.

The mantle itself is thought to be composed of "ultramafic" material—mostly olivine with some pyroxene. *Diogenites* have just this composition.

These groups of DSR meteorites are usually found brecciated, i.e., they are from rock that became fragmented or crushed. If the breccias are of the same rock type, they are said to be *monomict*; if from different rock types, *polymict*. Additionally Hutchison (2004, p. 48) describes two other types of brecciated rock, *genomict* and *regolith* breccias. The former contains more than one petrologic type but only one chemical group. The latter are polymict breccias that formed on the surfaces of small bodies (no atmosphere), irradiated by solar wind, and consolidated by micrometeorites and impact products.

Howardites are polymict breccias that have the same dominant rock type as the eucrites and diogenites.

Among the "stony-iron" meteorites, the major types are the *mesosiderites* and the *pallasites*. The number of stony-irons is 116 (at present writing), with only 12 falls. Of these, mesosiderites number 66, of which seven are falls, and pallasites, 50, of which five are falls.

However they are classified, the *mesosiderites* and the *howardites* appear to be regolith breccias—and the *mesosiderites* may have originated on the impacted stony surfaces of bodies which were impacted by an iron meteorite (or vice versa).

The *pallasites* have equal parts of metal and olivine (the predominant mineral in the Earth's mantle) and thus may represent material at the interface of a mantle and core. A few pallasites of unusual composition may have resulted from large impact events on the surfaces of previously undifferentiated bodies. See Table 15.4.

15.3.2 Other DSR Meteorites

The *aubrites* consist almost entirely of the pyroxene mineral *enstatite* ($MgSiO_3$). In *aubrites*, the oxygen isotope ratios, among other properties, are similar to those in *EH* and *EL* chondrites, suggesting a link among these groups. Grady (2000) gives 46 of these, of which nine are falls.

Ureilites are mixtures of ultramafic silicates and carbon-rich material. The carbon material contains diamonds, created through very high shockwave-driven pressure. The *ureilites* have oxygen isotope ratios similar to those of *CM* chondrites. There are 92 of these known, five of which are falls.

Another group of DSRs includes the *Shergotty–Nakhla–Chassigny* (SNC) clan. Grady (2000) lists 15 SNC meteorites, of which four were falls:

1. Shergottites have the minerals *pigeonite* [$Ca_x(Mg, Fe)_{(1-x)}SiO_3$, with $x \leq 0.1$], *augite* (pyroxenes with some alumina), *maskelynite* (a type of plagioclase glass), and *magnetite*. The four earliest known specimens are:

   The *Shergotty* meteorite, itself
   Zagami (the largest, at 23 kg) with very similar composition (see Figure 15.2 and color Plate 21)
   ALH A77005 (collected by expedition "A" in the Alan Hills, Antarctica)
   EET A79001 (collected by expedition "A" in the Elephant Moraine region, Antarctica)

Fig. 15.2. Fragments of the Zagami DSR achondrite Shergottite fall near Zagami, Nigeria, October 3, 1962. Noble gas analysis suggests a Martian source. See Plate 21.

2. Nakhlites are composed of *augite* ($\sim$79% by weight!), *olivine, feldspars,* and *magnetite*. There are three known specimens:

Nakhla (the most massive: 40 kg)
Lafayette (Indiana)
Governador Valadares

3. One known chassignite, *Chassigny*, a fall, of 4 kg weight, is primarily *olivine* (88.5%), with small amounts of *augite, plagioclase* [$(NaSi)_x(CaAl)_{(1-x)}AlSi_2O_6$, with $0.1 \leq x \leq 0.9$], and *orthopyroxene* ([$Ca_x(Mg, Fe)_{(1-x)}SiO_3$, with $x \leq 0.5$]

4. One of the SNC meteorites is *ALH 84001*. Grady (2000) lists it as a separate subgroup and summarizes the work done on this meteorite in this excerpt (we indicate omissions with ellipses):

"...collected ... 1984-85 ...originally classified as a diogenite. It was not recognised as an unusual additional member of the SNC group until 1993. It differs from the other martian meteorites in age, composition, texture and shock history, but has an oxygen isotope composition characteristic of the SNCs. ...Almost every aspect of the mineralogy, petrography, mineral chemistry, chronology and origin of ALH 84001 has been the subject of heated debate, fuelled in 1996 by an announcement that structures identified within the carbonate assemblages had been identified as a possible fossilised martian biota... As of December 1999, the biological, or otherwise, origin of the features had not been conclusively established to the satisfaction of the entire scientific community, and the arguments continue. Completely aside from its potential role in any Life on Mars debate, ALH 84001 is a fascinating meteorite in other ways. It has a much older crystallisation age [4.50–4.56 Ga]... than the other SNC meteorites, and has a complex chronology and shock history... multiple shock events."

ALH 84001 is classified now as an *SNC orthopyroxenite* (the bulk is orthopyroxene, $(Mg, Fe)SiO_3$). Elemental and isotopic composition of volatiles in its embedded glasses indicates a Martian origin. It is among the oldest known rocks on Earth. After its ejection from Mars in an impact event, it orbited the Sun for $\sim$16 My before falling as a meteorite,

$\sim$13,000 y BP. It has 3.6-4.0 Gy old carbonate globules and associated magnetite crystals in fissures, some of which are purported to be biogenic in origin (by magnetotactic bacteria), and it has PAHs (multiple aromatic carbon rings) also of possible Martian biogenic origin.

As noted, SNC meteorites include volatile abundances and, for most, isotope ratios consistent with an igneous origin on Mars. Figure 15.2 shows tiny fragments (the scale is labeled in cm) of the Zagami fall, as authenticated by the Bethany Sciences Company. An ejection age of 2.81 million years was obtained by Eugster et al. (1997). Drake et al. (1994) and Bogard and Garrison (1998) found evidence of Martian composition. The degree of oxidation suggests possible oxidation of upwelling (mantle) magma through water-bearing material in the crust; exposure to an aqueous environment has also been suggested for ALH 84001.

Taylor (1992, Table 3.12.3, p. 132) provides a list of 11 lunar meteorites. Seven are classified as anorthositic breccia, two as basaltic breccia, and two as Mare gabbro. Grady (2000) list 18 lunar meteorites, all of them finds. MET 01210, a lunar meteorite discovered in 2001 in the Meteorite Hills region of Antarctica, is made up primarily of very-low-titanium (VLT) basalt material (Zeigler et al. 2005; Joy et al. (2006)). It is shown in Figure 15.3.

Lastly, we mention the *tektites*; these have terrestrial composition (except for a few that are "spiced" by meteoritic composition) and are almost certainly the crystallized droplets of impact melts which were strewn away from a terrestrial crater site. In at least one case (the *moldavite tektites* found in Bohemia), the associated crater is known (the Ries Crater in SE Germany).

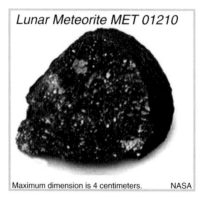

Lunar Meteorite MET 01210

Maximum dimension is 4 centimeters.   NASA

Fig. 15.3. Lunar meteorite MET 01210. A basaltic regolith breccia, it was collected in 2001 in the Meteorite Hills, Antarctica, and described in the Antarctic Meteorite Newletter, 27, No. 1, Feb. 2004. It is composed mainly of very-low-titanium crust, which can be seen at the right. It is basically a collection of stone and mineral clasts or fragments in a glassy matrix. NASA photo, curator.jsc.nasa.gov/antmet/amn/amn.cfm. See color Plate 22

15.4 Iron Meteorites

Irons are distinguished basically by their metallic crystalline structure:

- *Hexahedrites* consist entirely of single crystals of *kamacite* [Fe_x-$Ni_{(1-x)}$, with $0.96 \leq x \leq 0.93$], in cubic structures. The nickel content of hexahedrites is always in the narrow range 53–58 mg/g.

- Most irons are *octahedrites*. *Kamacite* crystals grow in lamellae along four sets of crystalline planes of *taenite* [Fe_x-$Ni_{(1-x)}$, with $0.5 \leq x \leq 0.8$]. Exposed on a planar surface, the sections of kamacite lamellae appear as "kamacite bands." The widths of these bands ("bandwidths" in Table 15.5) form a sequence from <0.1 to >3.3 mm. Etching with a dilute solution of nitric acid in ethyl alcohol of a polished plane section reveals the discontinuities in the octahedral structure; the structure is called the *Widmanstätten figures* or *patterns*. The longer the cooling time, and the lower the nickel content, the thicker the lamellae.

- FeS and carbon inclusions are sometimes found in some groups (for example, IAB and IIICD irons).

- *Plessites* are mixtures of fine kamacite and taenite which have nickel concentrations exceeding ~12%, and the iron meteorites which have plessite as well as a few fine kamacite spindles are called *plessitic octahedrites*. If they consist entirely of plessite, they are called *ataxites*, meaning "structureless." This is a misnomer, based on poor-resolution microscopy; viewed with adequate resolution, small "sparklets" of kamacite can be seen in such rocks.

A principal classification tool for the irons is the relative abundance of certain trace elements, especially gallium, germanium, and iridium. The presence of iridium (Ir) in the Cretaceous–Tertiary interface (the geologic periods terminating and beginning the Mesozoic and Cenozoic Eras, respectively) suggested to Luis Alvarez (1980) that a large meteoritic impact was responsible for the demise of the dominant planetary life forms of the era.

Whereas the Ga and Ge concentrations usually occupy a narrow range for a particular find, the Ir concentration varies widely, but the correlation with Ni shows a slope which can help in further identification.

The results of the different iron groups suggest that different groups originated in different parent bodies, which were located at different distances from the Sun. Typical cooling rates for the iron meteorites are ~1 K/My, which suggests parent bodies of the order of 300 km radii. Reheating may have occurred in some cases; and the cooling age is relative only to the most recent time of heating, when presumably all other clocks were set/reset.

Fig. 15.4. A fragment of the Henbury iron meteorite find from the Northern Territory, Australia, 1931, according to the Bethany Sciences Company. It is classified as a medium octahedrite (IIIAB), with bandwidth 0.9 mm. See Plate 23

A fragment of the Henbury iron (IIIAB) medium octahedrite, a 1931 find from the Northern Territory, Australia, as authenticated by Bethany Sciences, can be seen in Figure 15.4 and color Plate 24.

Figure 15.5 displays two faces of a fragment of the Sikhote-Alin (eastern Siberia) fall of February 12, 1947, authenticated by the Bethany Sciences

Fig. 15.5. A fragment of the Sikhote-Alin meteorite fall in eastern Siberia, February 12, 1947. It is classified a Type IIAB iron, coarsest octahedrite, with bandwidth of 9 mm. The circumstances of this spectacular meteoritic fall are described by Fessenkov (1955). See also color Plate 24

Company. This iron meteorite is classified as Type IIAB, coarsest octahedrite, with bandwidth 9 mm. Fessenkov (1955) describes the "meteorite shower" of this impressive fireball fall.

15.5 Ages and Origins of Meteorites

The age of a meteorite may be reckoned in a number of ways. The meaning of "age" depends, in a sense, on the technique used to determine it. Fundamentally the age refers to the time elapsed since the material came together in a common, closed system. However, time elapsed since a certain major alteration took place can also qualify as an "age," hence, there are radiogenic, gas-retention, cosmic-ray exposure, and terrestrial "ages."

15.5.1 Radiogenic Ages

The most important ages are *radiometric*, i.e., determined from the decay of radioactive isotopes. The idea is this:

If N_D is the number of daughter atoms at time t, N is the number of radioactive nuclei producing the daughter atoms at time t, and if N_{D0} is the number of daughter atoms at time t_0, then

$$N_D = N_{D0} + \alpha N(e^{\lambda(t-t_0)} - 1) \tag{15.1}$$

where α represents the fraction of parent radioisotope atoms that will decay into this particular daughter and λ is the decay constant. In the case of $^{87}\text{Rb} \rightarrow {}^{87}\text{Sr}$, $\alpha = 1$; using the element name to represent numbers of atoms, and normalizing to the numbers of atoms of the stable isotope ^{86}Sr, we have:

$$^{87}\text{Sr}/^{86}\text{Sr} = ({}^{87}\text{Sr}/^{86}\text{Sr})_0 + ({}^{87}\text{Rb}/^{86}\text{Sr})(e^{\lambda(t-t_0)} - 1) \tag{15.2}$$

If the parent melt is uniformly-mixed, then at the time of solidification the initial ratio $({}^{87}\text{Sr}/^{86}\text{Sr})_0$ will be the same for all minerals condensing from the melt. However, different minerals may contain different initial amounts of ^{87}Rb, so over time the $^{87}\text{Sr}/^{86}\text{Sr}$ ratio increases at different rates for different minerals. At any given instant of time, (15.2) is a linear equation in the isotope abudances, so a plot of the abundance ratio $({}^{87}\text{Sr}/^{86}\text{Sr})$ vs. $({}^{87}\text{Rb}/^{86}\text{Sr})$ for different minerals in the meteorite then yields two important results:

Table 15.6. Radionuclides for age determinations[a]

Parent	Daughter	Decay constant	Half-life
^{40}K	^{40}Ar	$5.55 \times 10^{-10} \, \text{y}^{-1}$	$1.25 \, \text{Gy}$
^{40}K	^{40}Ca	5.55×10^{-10}	1.25
^{87}Rb	^{87}Sr	1.42×10^{-11}	48.8
^{147}Sm	^{143}Nd	6.54×10^{-12}	106.
^{176}Lu	^{176}Hf	1.94×10^{-11}	35.7
^{187}Re	^{187}Os	1.52×10^{-11}	45.6
^{232}Th	^{208}Pb	4.95×10^{-11}	14.0
^{232}Th	^{4}He	4.95×10^{-11}	14.0
^{235}U	^{207}Pb	9.85×10^{-10}	0.704
^{235}U	^{4}He	9.85×10^{-10}	0.704
^{238}U	^{206}Pb	1.55×10^{-10}	4.47
^{238}U	^{4}He	1.55×10^{-10}	4.47

[a] Data from Wasson (1985, Table E-1), Kirsten (1978), and Hutchison (2004, Table 6.1, p. 157)

1. The zero point, $(^{87}\text{Rb}/^{86}\text{Sr}) = 0$, gives the initial ratio
2. The slope yields a constant dependent on the time and the decay constant (or *half-life*, to which it is related[3])

The decay constant of $^{87}\text{Rb} = 1.42 \times 10^{-11} \, \text{y}^{-1}$, corresponds to a half-life, $\tau_{\frac{1}{2}} = 48.8 \, \text{Gy}$. Other radionuclides and their daughters are given in Table 15.6.

The mean age of the samples of different falls yields $t \approx 4.5 \times 10^9$ years. The radiometric ages of some $^{87}\text{Rb} \rightarrow \, ^{87}\text{Sr}$ determinations are given in Table 15.7, from Wasson (1985).

The effect of age on the ratios of (15.2) is illustrated in Figure 15.6, with data drawn from a variety of meteorites of different origins in the solar system; these include an iron meteorite, several chondrites, and two achondrites (a mesosiderite and a nakhlite). Note that in a very few cases the ages are below

---

[3] The half-life is the time interval $(t - t_0)$ at the end of which $N/N_0 = \frac{1}{2}$, where N and N_0 are the numbers of radioactive nuclides at instants t and t_0, respectively. It is related to the decay constant as follows: the decay equation is

$$\text{d}N/N = -\lambda \, \text{d}t \tag{15.3}$$

the integration of which yields $N = N_0 \, e^{-\lambda(t-t_0)}$ so that $\ln(N/N_0) = -\lambda(t - t_0)$ and so when $N/N_0 = 1/2$, $\lambda(t - t_0)_{\frac{1}{2}} = \ln 2$; from $(t - t_0)_{\frac{1}{2}} = 0.693/\lambda$, we may write,

$$N = N_0 \exp[-1.443(t - t_0)/(t - t_0)_{\frac{1}{2}}] \tag{15.4}$$

Table 15.7. Selected meteoritic age determinations[a]

Method	Meteorite	Class	Age (gy)	Ratio $= (^{87}Sr/^{86}Sr)_0$
^{87}Rb-^{87}Sr	Indarch	EH4	4.39 ± 0.04	0.7005 ± 0.0009
"	Tieschitz	H3/L3.6	4.53 ± 0.06	0.69880 ± 0.00010
"	Soko-Banja	LL4	4.45 ± 0.02	0.69959 ± 0.00024
"	Allende[b]	CV3	4.5	0.69877 ± 0.00002
"	Norton County	Aub	4.39 ± 0.04	0.7005 ± 0.0004
"	Juvinas	Euc	4.50 ± 0.07	0.69898 ± 0.00005
"	Kapoeta[c]	How	4.44 ± 0.12	0.69885
"	Colomera	IIE	4.51 ± 0.04	0.69940 ± 0.00004
"	Kodaikanal	IIE	3.7 ± 0.1	0.713 ± 0.020
"	Nakhla	SNC	1.34 ± 0.02	0.70232 ± 0.00006

[a] Mainly from Wasson (1985, Table III-1, pp. 52–53) and Hutchison (2004, p. 73)
[b] Isochrone is not well defined in Rb-Sr analysis; age from Pb-Pb analysis is 4.553 ± 0.004 Gy (Tatsumoto et al., 1976).
[c] Different parts of the breccia yield different values, presumably metamorphic ages, in the range 3.5–3.8 Gy.

the ages of the solar system by significant amounts. It is not known exactly how the cosmic "clocks" were restarted in these cases. In the case of Nakhla and the shergottites (where a "whole-rock," Sm–Nd isochron indicates $\Delta t = 1.3$ Gy, while Rb–Sr, referred to (Taylor 1992, p. 132) as "internal isochrons," gives ~ 180 My, the very low age attributed to shocks) it is reasonable to suppose that these have something to do with the impact event that caused its ejection from Mars.[4]

---

[4] The question about the origin of the SNC meteorites aroused considerable controversy in past decades. Wasson (1985, p. 79) states that the escape velocity (in m/s) from planetary bodies with the density of ordinary chondrites, ~ 3500 kg/m^3, is:

$$v_{esc} \approx 1.4R \qquad (15.5)$$

where R is the planetary radius in meters. In general, this quantity is:

$$v_{esc} = v_{parabolic} = [2GM/R]^{\frac{1}{2}} = [(8\pi G/3)\rho R^2]^{\frac{1}{2}} \qquad (15.6)$$

which indeed yields ~ 1.40 for the constant in Wasson's equation, with his assumptions. Actually, the density of Mars is ~ 3900 kg/m^3, so the appropriate coefficient for Mars is ~ 1.48. He asserts that if an object impacts a planetary body at several times the escape velocity, the mass of the ejecta will exceed the mass of the object. Mars' escape velocity is 5.01 km/s and it has a mean orbital velocity of about 24 km/s (26.5 at perihelion), so that an object approaching at a high enough velocity to result in ejection of Martian material would need a velocity within ~ 15 km/s of this orbital speed. Such a difference is certainly plausible. In any case, the chemical evidence for Martian origin is fairly persuasive, and taken together the arguments support a Martian origin for the SNC meteorites.

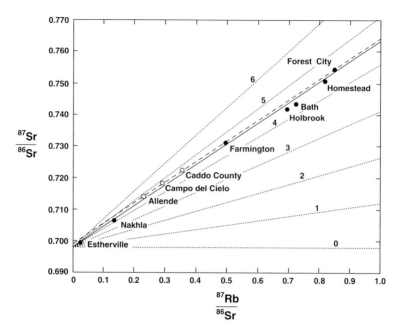

Fig. 15.6. Measured values of $^{87}\text{Sr}/^{86}\text{Sr}$ vs. $^{87}\text{Rb}/^{86}\text{Sr}$ for seven stony meteorites (solid dots, Pinson et al. 1965, Tables 1 and 2), a whole-rock sample from the Allende CV3 carbonaceous chondrite (open square, Shimoda et al 2005), and a mineral sample from each of two iron meteorites (open circles, Liu et al 2003). The dotted lines are isochrons labeled with ages in units of 10^9 years measured from when the material in the meteorites solidified from the solar nebula and was fully-mixed with $(^{87}\text{Sr}/^{86}\text{Sr})_0 = 0.698$, the value derived from the stony meteorites (Pinson et al. 1965). The solid dots lie close to a mean isochron of age $\sim 4.5 \times 10^9$ yr (solid line). The dashed line shows the 4.5×10^9 yr isochron for $(^{87}\text{Sr}/^{86}\text{Sr})_0 = 0.699$, the value derived from Allende and the iron meteorites (Shimoda et al 2005, Liu et al 2003), and provides a better fit to these meteorites. Figure inspired by Wood 1968, Figure 4-1, p. 58

15.5.2 Gas Retention Ages

When a gas is the daughter atom, the analyses yield *gas retention ages*.

In Table 15.6, the processes are:

- $^{40}\text{K} \rightarrow \ ^{40}\text{Ar}$
- $^{232}\text{Th} \rightarrow \ ^4\text{He}$
- $^{235,238}\text{U} \rightarrow \ ^4\text{He}$

Sometimes the ages are the same as for the less volatile daughters, often they are younger, but none have been found to be older.

According to Wood (1968, p. 61), at temperatures above about $250\,\mathrm{K}$, $^{40}\mathrm{Ar}$ is able to diffuse out of mineral lattices and escape, but at lower temperatures it remains trapped.

For helium, the temperature of escape must be much lower, perhaps $100\,\mathrm{K}$ or less.

Thus gas retention ages indicate the interval since the last instant that the meteoroid was subjected to temperatures above these limits. Passage through the atmosphere is not such a time, because the penetration depth for the ablation temperatures is a few centimeters at most for iron, and millimeters for chondritic meteorites.

15.5.3 Cosmic Ray Exposure Ages

Another type of dating is revealed by *cosmic ray exposure* ages. Cosmic rays (CR) bombard all exposed surfaces (and the Earth, where, however, they usually collide with atmospheric particles creating cascading cosmic ray showers of decreasing energies as they approach the surface of Earth).

On fully exposed surfaces, high energy particles may collide with nuclei of atoms, creating spallation products. For example, a high energy cosmic ray proton may collide with an iron nucleus to produce:

$$^1\mathrm{H} + \ ^{56}\mathrm{Fe} \rightarrow \ ^{36}\mathrm{Cl} + \ ^3\mathrm{H} + 2\,^4\mathrm{He} + \ ^3\mathrm{He} + 3\,^1\mathrm{H} + 4\mathrm{n}$$

where n refers to neutrons. The resulting nuclides are called *cosmogenic nuclei.*

Cosmic rays may penetrate to a meter or so of rocky material, creating *spallation* products which are radioactive. These decay over time, but more are created through subsequent CR collisions. Wood (1968, p. 65) suggests that the best dates are obtained from comparison between two cosmogenic nuclei, one of which is radioactive, the other stable. The abundance of the stable isotope will increase with time, the other will approach a steady-state after a few half-lives, where the decrease due to decay is matched by the increase due to continuing exposure. Such a pair is $^{38}\mathrm{Ar}$, which is stable, and $^{39}\mathrm{Ar}$, which is radioactive with a half-life of only 325 years. The CR exposure ages tend to be much shorter than $^{87}\mathrm{Rb} \rightarrow \ ^{87}\mathrm{Sr}$ and even gas-retention ages. This supports the idea that meteoroids are the results of repeated collisions and subsequent fragmentations, so that material once protected from cosmic rays by overburden subsequently becomes exposed to cosmic radiation.

15.5.4 Case Study: The Zagami SNC Basaltic Shergottite

This meteorite was a fall (October 3, 1962), so its *terrestrial age* (present year – fall year) is known exactly. The recovered mass of this meteorite is

18.1 kg, of which 13.74% is iron (in all forms). A microscopic cross-section shows dark gray pyroxene grains and light plagioclase glass in the form of maskelynite (Hutchison 2004, Fig. 9.8, where cross-sections of other SNC meteorites are also shown). Zagami contains amphibole (a group of hydrated silicate minerals), indicating formation in an aqueous environment, possibly a Martian "hydrosphere" (Karlsson et al. 1992), but an environment with 10% of the water content of Earth. Studies of shock-melted pockets and veins of impact glass in Zagami indicate that the noble gas, and nitrogen and argon, isotope ratios resemble those of the Martian atmosphere; Zagami is enriched in ^{15}N and ^{40}Ar relative to ^{14}N and ^{36}Ar (Marti et al. 1995). The glasses were formed under a pressure of $\sim 31 \pm 2$ GPa and a temperature of $\sim 220°C$. The isotope ^{17}O is enhanced relative to Earth rocks, and this, too, is the case for Mars. The oxides have yielded the amount of free oxygen (the *fugacity*) in the environment in which the basalt originated, presumably magma from the Martian mantle. From the noble gases, an ejection age (from Mars) of 2.81 My has been derived (Eugster et al. 1997). Similar ejection ages have been found for two other basaltic shergottites: Shergotty and QUE 94201, suggesting a common ejection event for all three. Bogard (1999) obtained an Ar–Ar age of 242 My. Finally studies of the U–Pb, Rb–Sr, and Sm–Nd isotopes (Borg et al. 2005) have yielded two ages: a formation age of 4550 ± 10 My and a *differentiation age* (the interval between Mars' formation and the differentiation in a molten environment of the meteorite source material) of 163 ± 4 My, consistent with a *crystallization age* (the interval between source formation and the partial melting event that produced the rock in the crust of Mars).

15.6 Other Sources of Evidence for Meteoritic Origins

The evidence briefly summarized here strongly implies that meteorites of different composition originated in different places in the solar system, and that the differentiated meteorites have undergone further modification.

Isotope evidence further supports their disparate origins. The quantity $\delta^{17}O$ $(^0/_{00})$ is defined as

$$\delta^{17}O \, (^0/_{00}) \equiv 1000[(^{17}O/^{16}O)_{sample} - (^{17}O/^{16}O)_{standard}]/[(^{17}O/^{16}O)_{standard}]$$
$$(15.7)$$

and similarly for the $\delta^{18}O(^0/_{00})$ quantity (called "del" values). These quantities are usually normalized to SMOW (*standard mean ocean water*). The $\delta^{17}O \, (^0/_{00})$ vs. $\delta^{18}O \, (^0/_{00})$ plot provides a discriminant for various groups of carbonaceous chondrites, and for DSRs. This suggests that different conditions prevailed at different places in the original solar nebula to allow different degrees of fractionation. The del values that are given for the principal groups of DSRs in Table 15.4, for some irons in Table 15.5, and plotted in Figure 15.7 [based on data from Clayton & Mayeda (1984; 1996) and Clayton et al. (1991)], illustrates this well.

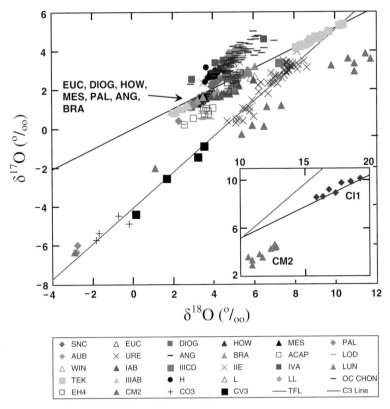

Fig. 15.7. Fractional oxygen isotopes of various meteorites, the Earth–Moon, and two lines: one approximating the theoretical mixing line for anhydrous minerals and the other the terrestrial fractionation line TFL. This sort of evidence points to heterogeneous origins in the early solar nebula and proto-planetary disk. The ratios of oxygen isotopes plotted are from analyses by Clayton & Mayeda (1984; 1996), and Clayton et al. (1991). The terrestrial fractionation line has a slope 0.52; the "C3 line" is a linear fit to the combined CO3 and CV3 data with slope 0.918 and zero point −4.036; similarly, the carbonaceous chondrite chondrules lay along a line with a slope ∼1. These lines typify the distributions of the meteorites, and suggest that exchanges between at least two reservoirs (Clayton, 1993; Wasson (1985, esp. pp. 68–74, 215–217), involving gas and dust components of the solar nebula, had occurred. The heating source is not yet agreed on. Achondrites: ACAP = Acapulcoites; ANG = Angrites; AUB = Aubrites; BRA = Brachinites; DIOG = Diogenites; EUC = Eucrites; HOW = Howardites; LOD = Lodranites; LUN = Lunar; SNC = Martian (Shergottites, Nakhlites Chassigny, Orthopyroxene); URE = Ureilites; WIN = Winonaites. Stony-irons: MES = Mesosiderites; PAL = Palasites Chondrites: OC CHON = Ordinary chondrite chondrules. TEK = tektites (Earth). For more chondrite group identifications see Table 15.2 (recall that numbers indicate petrologic classes), and for the irons', see Table 15.5. See color Plate 25.

When the values for the Allende CAIs are examined in this way, they fall along a line, and this line has a different slope from the fractionation line of terrestrial materials. Some of the Allende chondrules fall along the same line, others do not. Relations that involve del values of isotopes of other elements, for example $\delta^{48}Ca$ vs. $\delta^{50}Ti$ or $\delta^{30}Si$ vs. $\delta^{18}O$, similarly provide discriminants of this sort. For further details, see Taylor (1992, p. 101) and Wasson (1985).

Deuterium/hydrogen abundance is another indicator of origin, but its main value is in demonstrating the fractionation of hydrogen in the early solar nebula. Jupiter and Saturn have D/H ratios that most resemble the interstellar medium ($\sim10^{-5}$). Those of Titan, Comet Halley, and Earth are all $\sim2 \times 10^{-4}$; those of Uranus and Neptune lie somewhere in between.

15.7 Parent Bodies and the Asteroids

The pre-terrestrial encounter orbits of meteoroids that were sufficiently well observed as meteors (including some that were found as meteorites) demonstrate that these objects originate in the asteroid belt. They can be assumed to be, therefore, products of the "asteroid mill" which fragments larger bodies into smaller ones through multiple collisions over long periods of time.

15.7.1 The Discovery of Ceres

The initial detection of the asteroid (1) Ceres was made by Giuseppe Piazzi (1746–1826) on the first night of the nineteenth century (January 1 1801). The discovery was later realized to fit perfectly in the scheme of Johann Daniel Titius von Wittenberg (1729–1796) and published by him in his 1766 translation of a book on astronomy by Charles Bonnet. The relation went unnoticed, however, until it was mentioned in a book by Johann Elert Bode (1747–1826) in 1772. Up to the discovery of Uranus, little was made of the progression, at least outside of Germany, as Bode himself complained. In retrospect, astronomers realized that Uranus fit the law well and, on this basis, Baron Francis X. von Zach (1754–1832), court astronomer at Gotha, began a campaign to look for the missing planet at 2.8 au, as predicted by the Titius–Bode law (see Milone & Wilson 2008, Chapter 1). In 1800 he had organized the sky into 24 zones, had organized a network of astronomers to search each one, and had just sent out the maps to the searchers, when announcement came of Piazzi's discovery.

Thus, the "law" appeared to be verified, and indeed it was used in the theoretical calculations of the location of the trans-Uranian planet later to be called Neptune.

Titius himself had suggested that there might be undiscovered *satellites* of Mars in the gap between Mars and Jupiter ("satellites" presumably on the grounds that a major planet would have been known from antiquity). Kepler had expressed a similar idea with respect to the location of undiscovered planets in this and other spaces. The idea that these would be too far from Mars to have stable orbits around that planet apparently did not occur to them.

The orbit was not determined when first observed because Piazzi was not able to observe it over a sufficiently large arc, as required with the methods of the day, before it came into conjunction with the Sun.

However, Karl Friedrich Gauss (1777–1855) invented a method by which an orbit could be computed from just three observations of position and velocity. Moreover, he invented a new computational scheme—the method of least squares—with which calculations could be expedited and parameters determined optimally.

With this method Gauss produced an ephemeris for Ceres; armed with this, von Zach and Heinrich Wilhelm Matthäus (or Matthias) Olbers (1758–1840) relocated Ceres, and thus confirmed the discovery. Ceres was seen to have a semi-major axis of 2.767 au.

The mystery of the fourth "planet" was solved, but another mystery was encountered! Other new "planets" in this region began to be discovered: Pallas was discovered by Olbers on March 28, 1802; Juno in 1807, by Karl Ludwig Harding (1765–1834); and Vesta by Olbers in 1807. The law had predicted a planet, but due to some cataclysm, there were only fragments of the original planet left—according to a theory put forward by Olbers.

It was 40 years, however, before more of the planetary "fragments" would be discovered—a very puzzling time in astronomy. More were indeed found, though, and the number of discoveries has been increasing ever since.

15.7.2 Nomenclature

Before a name is assigned (by the IAU), a provisional designation is given: year followed by a half-month code, and by another letter indicating the order of discovery in that half-month, somewhat similar to the (current) naming of comets. When the discovery is confirmed, a sequence number and a name are assigned. Thus 1 Ceres, 2 Pallas, etc., for established asteroids but 1999 CB (second to be discovered in the interval February 1–15 of the year 1999) for newly discovered objects.

The main asteroid belt is a region between the orbits of Mars and Jupiter, roughly centered at 2.8 au from the Sun. There are many groupings of asteroids in this region; we describe the principal families in Section 15.7.3. Several gaps in the distribution of asteroids with semi-major axis (the

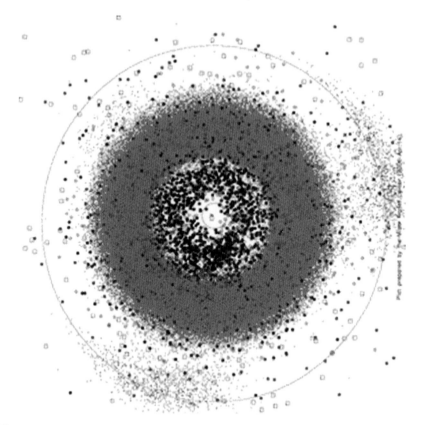

Fig. 15.8. The small bodies of the inner solar system (color view: Plate 26). *Black*: asteroids that cross the orbits of the Earth and of Mars; *green*: the "main belt" asteroids between the orbits of Mars and Jupiter. The Trojan asteroids cluster along Jupiter's orbit (*the outer circle*), centered on stations preceding and following Jupiter by 60°. The Hilda group of asteroids (Section 15.7.3 and Figure 15.10) lies between the Trojans and the main belt asteroids, and also at the upper left, opposite Jupiter, which is indicated by a *filled circle* at the lower right. Courtesy Minor Planet Center, SAO, Cambridge, MA

Kirkwood gaps), as well as peaks, demonstrate the existence of commensurabilities with the planets, especially Jupiter. The population of small bodies of the solar system can be seen as a snapshot in late April, 2006, for the inner and outer parts of the solar system (Figures 15.8 & Plate 26 and 15.9 & Plate 27, respectively). Asteroids are continually being discovered, and from ever more distant parts of the solar system.

Large numbers of objects have been detected in the outer solar system over the past 15 years and the realization that the Pluto–Charon system has more in common with the large icy moons of some of the giant planets has thrown

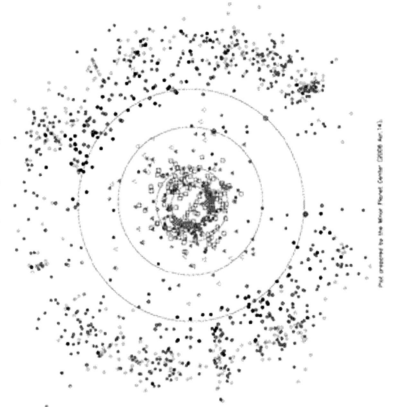

Fig. 15.9. The small bodies of the outer solar system, from the Jovian orbit to beyond Neptune. A color view is shown in Plate 27. *Filled* and *unfilled squares* indicate numbered periodic comets and all other comets, respectively. *Triangles* represent Centaurs. *Black circles* near the orbit of Neptune are Plutinos (the largest of which is Pluto, near 3 o'clock, and just outside Neptune's orbit, the *outer blue circle*). Neptune is seen at ∼1 o'clock. The outermost *filled* and *unfilled red circles* are main Edgeworth–Kuiper Belt and scatter objects, respectively. Courtesy Minor Planet Center, SAO, Cambridge, MA

the nomenclature into some confusion. The meaning of the term "planet," in particular, has required considerable discussion and although the IAU has resolved the issue for the moment, increasing numbers of substellar objects are being discovered both within and outside the solar system. As they become better studied, these objects may well inspire further refinement of what we mean by the word "planet."

Between Jupiter and Neptune, we find *"Centaurs,"* thought to have escaped from the *Kuiper Belt* [or the *Edgeworth–Kuiper Belt*, to honor not only Gerard Peter Kuiper (1950–1973), but also the Scottish astronomer Kenneth Edgeworth

(1880–1972) who independently suggested versions of it], located between $\sim$30 and $\sim$50 au from the Sun. Bodies more distant than Neptune are called, unsurprisingly, "*Trans-Neptunian Objects.*" The best known object of this group is the dwarf planet Pluto. Pluto is also trapped in a 3:2 resonance[5] with Neptune; any object sharing this trait is a "*Plutino.*" The term "*Cubewano*" has been applied to an object in the main Edgeworth-Kuiper Belt, not trapped in a planetary resonance and with a typical distance of $\sim$41 au.

15.7.3 Families of Orbits

Although there are many exceptions, for the most part, asteroids are found in the region between Mars and Jupiter. These are the *main belt asteroids*, which can be seen as a thick torus in Figure 15.8.

Asteroids have direct orbits, but exhibit a wide variety of inclinations and eccentricities. There are orbital groups of asteroids, somewhat arbitrarily separated primarily on the basis of semi-major axis and/or range of distance from the Sun; Figures 15.10 (the *Hildas*), 15.11 (the *Koronis* family), and 15.8 (the *Trojan* asteroids, associated with Jupiter) illustrate this. The inner solar system asteroids have been grouped according to their orbital dispositions with respect to the Earth's orbit. The groups are:

- *Amors*, "Mars-crossers" (est. $\sim$1500 asteroids) which approach the Earth's orbit but do not cross it, i.e., they have perihelia outside the Earth's aphelion distance. Examples include the asteroids 1221 Amor, 1943 Anteros, 2061 Anza, 433 Eros (see Figure 15.12), 1915 Quetzalcoatl, 2608 Seneca, and 2004 VB61

- *Apollo* group (est. $\sim$1000 asteroids) which are "Earth-crossers," and which have $a > 1.00$ au. Examples are: 2101 Adonis, 1862 Apollo, 1865 Cerberus, 1864 Daedalus, 2212 Hephaistos, 3200 Phaethon, 1566 Icarus, and 2004 WS2

- *Atens* (est. $\sim$100 asteroids), which have $a < 1.00$ au, and most, though not all, have their aphelia greater than 1 au. Examples: 2062 Aten, 2340 Hathor, 3554 Amun, 5381 Sekhmet, and 2004 WC1.

Elsewhere, asteroids are grouped into "families" by virtue of similar orbital parameters, following work initially carried out by Kiyotsugu Hirayama (1918a,b). The families include the following (with defining orbital elements; the semi-major axes are given in au):

---

[5] This is the ratio of the sidereal orbital periods: $3P_{\text{Neptune}} = 2P_{\text{Plutinos}}$.

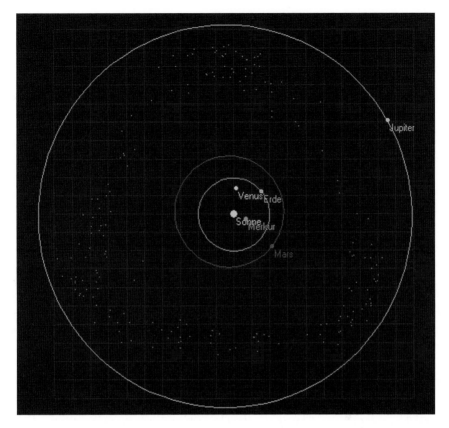

Fig. 15.10. The Hilda family, in 3:2 resonance with Jupiter, cropped from an *Easy-Sky* demo image, based on data from the MPC; courtesy Matthias Busch

- Hungaria $(1.2 < a < 2.00)$
- Flora $(a \approx 2.2)$
- Main Belt I $(2.06 < a < 2.50)$
- Main Belt II $(2.50 < a < 2.82)$
- Main Belt III $(2.82 < a < 3.27)$
- Main Belt IV $(3.27 < a < 3.65)$

and other families which overlap with these but have similarities in other elements:

- Phocaea $(a \approx 2.4,\ e \approx 0.25,\ i \approx 23°)$
- Nysa $(a \approx 2.43,\ e \approx 0.17,\ i \approx 3°)$

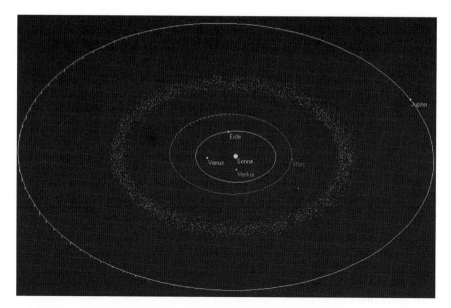

Fig. 15.11. The distribution of the Koronis family of asteroids, from an *Easy-Sky* demo image, based on data from the MPC; courtesy Matthias Busch

- Koronis $(a \approx 2.85,\ e \approx 0.05,\ i \approx 2°)$
- Eos $(a \approx 3.01,\ e \approx 0.17,\ i \approx 10°)$
- Themis $(a \approx 3.13,\ e \approx 0.15,\ i \approx 1°)$
- Hilda $(a \approx 4,\ e \approx 0.15,\ i \approx 8°)$
- Trojans $(a \approx 5.2,\ e \lesssim 0.1,\ i \lesssim 20°)$

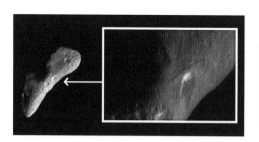 

Fig. 15.12. Asteroid 433 Eros, as viewed from the NEAR Shoemaker spacecraft. **a** A view from February 12, 2000 at a distance of 1748 km and higher resolution view (*inset*) from February 29, 2000 at a distance of 283 km. NASA/JPL image PAI02492. **b** A swath of Eros, only 1.4 km across as seen from an altitude of 35 km. NASA/JPL image PAI03132. Both figures were produced by the Applied Physics Laboratory of John Hopkins University

Other families that have been suggested include: Budosa, Eugenie, Leto, Lydia, and Maria. However, not all authorities are in agreement with what constitutes a real family. Classification by reflectance spectroscopy does not necessarily solve the disagreements because fragments of the same parent body or bodies may present considerably different properties (if, for example, they were differentiated through some fractionation process). The existence of families suggests a limited number of parent bodies. The Hilda family is a dynamical family only, however, characterized by a 3:2 resonance with Jupiter (see Figure 15.10). In this sense they are related to the Trojan asteroids, trapped at the stable Lagrangian points, L_4 and L_5, $\sim60°$ from Jupiter's location, and in a 1:1 resonance with Jupiter.

In the realm of the gas giants of the outer solar system there are the *Centaurs* (Section 14.7; Figure 15.9), believed to be objects perturbed from the Kuiper Belt.

The number of known outer solar system objects is increasing rapidly. The asteroidal objects referred to as 2060 *Chiron* and 944 *Hidalgo* lay well beyond the main belt. Chiron has been identified as a Centaur, from its location, and as a cometary body on the basis of its coma observed during perihelion passage. In addition, recent observations of the Centaur 5145 *Pholus* indicate that it has a cometary surface; it is not impossible that all Centaurs are in fact cometary bodies and that they are related to the short-period comets. What is clear is that they are icy bodies.

We have already mentioned in Section 13.4.2 the "trans-Neptunian objects" (TNOs), a group of which (the *Plutinos*) are locked into a 3:2 orbital resonance with Neptune. As of 2002, there were 400 TNOs known and by 2007, 1000; a discovery rate of ~10/month ensures many more discoveries. Jewitt and Luu (2000) estimated that the number of such objects with diameters $\gtrsim100$ km is of order $\sim10^4$; they are seen in Figure 15.9. Figure 15.13 where the TNO eccentricities are plotted against their semi-major axes, identifies the Plutinos at ~39 au. The Kuiper Belt objects (KBOs) have been further identified as "classical" or "cold" and "scattered" or "excited" or "hot" KBOs. The classical variety is characterized by the upper limits, $i < 32°$ inclination and $e < 0.2$, whereas the "excited" objects may have these values pumped up through interactions with the outer gas giants. See Morbidelli and Brown (2004) for further discussion of current theories of the KBOs origin and evolution.

15.7.4 Dimensions and Masses of Asteroids

15.7.4.1 Asteroid Dimensions and Albedo The diameters of asteroids can be determined in any of four ways, depending on distances from the Earth and from the Sun. All of them require knowledge of the distance. These methods are:

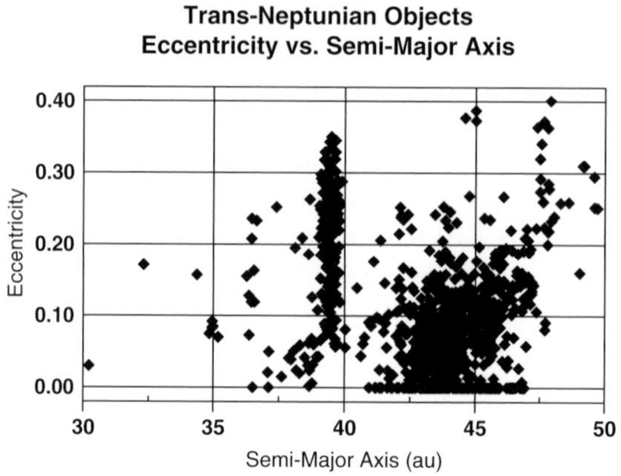

Trans-Neptunian Objects
Eccentricity vs. Semi-Major Axis

Fig. 15.13. The plot of eccentricities vs. semi-major axes for the Trans-Neptunian Objects of the solar system according to the list on the Minor Planet Center's website in September, 2007. Note the large range of eccentricities for the objects near 39 au, characteristic of the "Plutinos." The higher eccentricity objects beyond this distance are members of the "scattered" population

- Micrometer measurement of the angular diameter
- Stellar occultation
- Radiometric measurement
- Doppler radar observations

Because asteroids are neither particularly large nor close to us, there are few reliable micrometer measures of them. The relevant relation, is:

$$\theta = D/r \qquad (15.8)$$

where θ is the angular diameter, D is the asteroid diameter, and r its distance.

The largest asteroid is Ceres, with a diameter of ~950 km and a mean opposition distance from Earth of ~1.8 au; therefore, the angular diameter of Ceres is only $206265 \times 950/(1.8 \times 1.5 \times 10^8) = 0.73$ arc-sec. The angular resolution of any telescope is $1.22\ \lambda/D$, so in order to obtain a resolving power of 0.1 of this value or ~0.07 arc-sec, a telescope of only ~1.9 m is needed, but this is a theoretical value. Atmospheric seeing at even the best sites is limited to ~0.2 arc-sec; consequently, the precision of even this relatively easy asteroid observation is less than ~25% for a single ground-based observation. To be sure, N observations can beat down the noise by the factor $\sqrt{N}$, and the use of adaptive and, even more usefully, active optics can greatly improve the precision of such measurements. But the usefulness of this technique is clearly limited. From space, however, the theoretical limit is approached more easily.

Stellar occultations are a more promising technique, but here the observer cannot choose the targets and is at the mercy of the motions of the asteroid and the direction of the asteroid's shadow. Nevertheless, David W. and Joan Bixby Dunham and colleagues have encouraged worldwide participation in occultation observations, with improved astrometry of asteroid and star positions as an important byproduct, since these must be improved for each event so that observers in various parts of the world can be alerted. Up to 1998, 170 ground-based occultations had been studied. By 2001, more than 280 had been observed thanks in part to the HIPPARCOS astrometric satellite mission. At present writing, this number is now ∼1000 (Dunham, private commun., 2007). A spectacular example of the success of the technique was the high precision determination of the size and shape of (2) Pallas by Dunham et al. (1990).

The positions and thus the orbits of asteroids are becoming better and better known thanks to the operations of the Minor Planet Center, in Cambridge, MA. The MPC encourages worldwide participation of asteroid observations and position reduction of CCD observations. The immense distance of a star means that the light rays are essentially parallel, and the length of the shadow of the asteroid is, very closely, the size of the asteroid; so a measurement of the duration of the event and the orbital speed of the asteroid at the time of the occultation and its distances to the Sun and Earth suffice.

The radiometric method requires optical and infrared photometry so that both the radius and the bolometric albedo can be determined. The method is approximate, but one can pick one's targets—sufficient light gathering power, and thus size, for the available telescope being the only requirement. The relevant relations are as follows. The light reflected by the asteroid at all wavelengths is:

$$\ell_{\text{refl(bol)}} = [\mathcal{L}_\odot / (4\pi r_\odot{}^2)] \, a \, A \tag{15.9}$$

where $\mathcal{L}_\odot$ is the solar luminosity, $r_\odot$ is the asteroid's distance from the Sun, a is the cross-sectional area, and A is the bolometric albedo. Ignoring phase effects, observations made from the Earth are diluted by the inverse square law, and they are made in discrete passbands of effective wavelength λ. Setting $a = \pi R^2$, where R is an effective radius, the reflected flux observed at the Earth is:

$$\mathfrak{F}_{\text{refl}(\lambda)} = \ell_{\text{refl}(\lambda)} / (2\pi r_\oplus{}^2) = [\mathcal{L}_{\odot,\lambda} / (8\pi r_\oplus^2 \, r_\odot{}^2)] \, R^2 A_\lambda \tag{15.10}$$

where $\mathcal{L}_{\odot,\lambda}$, and $\ell_{\text{refl}(\lambda)}$ are the solar luminosity and the reflected power in the passband,[6] and $r_\oplus$ is the asteroid's distance from the Earth. Note

---

[6] These quantities made may be thought of as incremental amounts per wavelength interval, thus: $l_\lambda = \Delta l / \Delta \lambda$.

that the factor 2π in the denominator of the left-hand side of (15.10) is not strictly correct because the phase may not be exactly 0; however this factor is more accurate than 4π because the reflection is not isotropic. Now the solar luminosity at each wavelength is known, $\mathfrak{F}_{\text{refl}(\lambda)}$ is observed, and the distances known, so the quantity $R^2 A_\lambda$ is found. An integration of the flux of (15.10) over all wavelengths gives:

$$\mathfrak{F}_{\text{refl(bol)}} = \int \ell_{\text{refl}(\lambda)}/(2\pi r_\oplus{}^2)\mathrm{d}\lambda = \int [\mathcal{L}_{\odot,\lambda}/(8\pi r_\oplus{}^2 r_\odot{}^2)]\, R^2 A_\lambda\, \mathrm{d}\lambda$$

$$= \text{const } AR^2 \tag{15.11}$$

If the integration can be carried out adequately, the product of A and R^2 can be found.

The factors can be separated with the help of infrared observations. The radiation which is not reflected is absorbed. The absorbed power must be reradiated, as noted earlier. The emitted light at the asteroid is:

$$\ell_{\text{em, bol}} = f\,\pi R^2 \sigma T^4 = [\mathcal{L}_\odot/(4\pi r_\odot{}^2)]\, a(1-A) \tag{15.12}$$

where f is a factor that takes the rotation of the object into account (for example, the factors 4 or 2 in the rapid vs. slow rotator models and somewhere in between for intermediate cases.), and T is the effective equilibrium temperature of the asteroid.

We observe passband fluxes in the infrared also (despite the use of the term "bolometer" for detectors in the thermal infrared), because careful corrections for terrestrial atmospheric absorption and emission must be made, and these are best done when the wavelength range over which the observations are made is stipulated. Then (for rapidly spinning asteroids) the emitted flux observed at the Earth is:

$$\mathfrak{F}_{\text{em}(\lambda)} = \ell_{\text{em}(\lambda)}/(4\pi r_\oplus{}^2) = [\mathcal{L}_{\odot,\lambda}/(16\pi r_\oplus{}^2 r_\odot{}^2)]\, R^2(1-A_\lambda) \tag{15.13}$$

where we have used $a = \pi R^2$ Integrating,

$$\mathfrak{F}_{\text{em(bol)}} = \ell_{\text{em(bol)}}/(4\pi r_\oplus{}^2) = \int [\mathcal{L}_{\odot,\lambda}/(16\pi r_\oplus^2 r_\odot^2)]\, R^2\,(1-A_\lambda)\mathrm{d}\lambda$$

$$= \text{const}'(1-A)R^2 \tag{15.14}$$

Setting

$$AR^2 = \chi_1 \tag{15.15}$$

and

$$(1-A)R^2 = \chi_2 \tag{15.16}$$

$$(1-A)/A = \chi_2/\chi_1 \tag{15.17}$$

whence:

$$A = \chi_1/(\chi_1 + \chi_2) \tag{15.18}$$

and R is found from either (15.15) or (15.16).

Thus, in principle, the radii and the albedos are determinable.

Radar observations have been successfully applied to asteroids that approach Earth closely enough to be able to reflect sufficient power back to Earth for Doppler ranging analyses. The small asteroid 4769 Castalia was found to be double in this way.

15.7.4.2 Asteroid Masses and Densities Masses are another story entirely.

Until recently, no binary asteroids were known, so masses have been essentially based on determined diameters and on estimated densities. This situation is changing and the number is growing rapidly:

- The Galileo spacecraft observed the binary asteroid 243 Ida (a member of the Koronis family of Figure 15.11) and its companion, Dactyl (see Figure 15.14)

- The asteroid 624 Hektor (Hector) seems to be a contact binary asteroid (in analogy with a class of eclipsing binary stars where two stars are in varying degrees of contact, and are known as "overcontact" binaries), in which two lobes of material appear joined by a narrow neck

- Direct imaging of Edgeworth-Kuiper or "scattered" objects have shown several with companions (Brown et al. 2006).

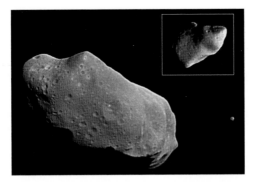

Fig. 15.14. *Main panel*: 243 Ida, an S-class asteroid and member of the Koronis family, and its small companion, 1993 (243)1 Dactyl, as seen on August 28, 1993 by the Galileo spacecraft. Ida: $58 \times 23 \times 15$ km; Dactyl: $1.2 \times 1.4 \times 1.6$ km. *Inset*: 951 Gaspra, $18 \times 10 \times 9$ km, visited by Galileo on Oct. 29, 1991. NASA/JPL image PIA00333

- Finally ground-based radar and Very Large Telescope observations have revealed an increasing number of asteroidal companions.

In addition, masses of some planetary moons thought to be captured asteroids, such as Phobos and Deimos, have been determined by mutual interaction and through spacecraft observations. The asteroids 1 Ceres and 2 Pallas have close orbits and undergo mutual interactions; consequently, their masses have been determined. Moreover, the asteroid 4 Vesta has also undergone interaction with 197 Arete, which permitted Vesta's mass to be determined. The Amor asteroid Eros was visited by the NEAR Shoemaker spacecraft (which landed on it!). However, the number of such determinations is small. The masses and densities of four asteroids are:

- 1 Ceres : $1.2 \pm 0.6 \times 10^{24}$ kg; 2100 ± 800 kg/m^3
- 2 Pallas : $2.2 \pm 0.4 \times 10^{23}$ kg; 2600 ± 900 kg/m^3
- 4 Vesta : $2.8 \pm 0.2 \times 10^{23}$ kg; 3200 ± 1400 kg/m^3
- 433 Eros : 6.687×10^{15} kg; 2700 kg/m^3

Until a large number of independently determined masses are made, densities are in the same situation as the masses. So, at present we do not have really good estimates of the true densities of most asteroids.

15.7.5 Asteroids and Meteorites

The meteoroid orbits determined from meteors strongly suggest origins in the asteroid belt. Reflectance spectra support the association as well. On the basis of the reflectance spectra, which reveal band structures attributed to various minerals, and photometric albedos, asteroids have been assigned classes. These are shown in Table 15.8, based, in part, on Chapman (1990, Table 1, p. 238) and Wetherill and Chapman (1988, esp., Table 2.1.3, p. 54).

Work carried out by Binzel, Bus, and Burbine (1998) indicates that ordinary chondrites are derived from the S-class main belt asteroids, and that previous disagreement between asteroid and meteorite spectral features is likely due to size effects. The lack of D and P analogs is very likely due to the fragility of those icy bodies. The density of Halley's comet is determined to be ~ 200–500 kg/m^3. Such material is easily destroyed by the pressures encountered in entry into a planetary atmosphere (of course a large amount of it can still trigger a great catastrophe!).

Asteroids have clearly undergone substantial modification. Evidence now suggests that many asteroids have some degree of aqueous modification as

Table 15.8. Asteroids and meteorites[a]

Asteroid Type (class)	Location (au) Peak	Range	Spectral data	A	Associated meteorites group/clan/type	Examples
Q (primitive)	1	?	Similar to S; deeper absorption	mod.	Ordinary chondrites	1862 Apollo
E (igneous)	1.9	1.9–3.1	Flat; slightly red	0.3–0.6	Aubrites (enstatite chondrites)	44 Nyssa 434 Hungaria
S (igneous)	2.4	1.9–3.5	Red in visible; 9 subtypes; abs. near 1 μm	0.1–0.2	Ordinary chondrites; may be pallasites	433 Eros 5 Astraea
V (igneous)	2.55	…	~S with pyroxene abs.	High	Basaltic achondrites	4 Vesta
T (metam.)	2.7	2.5–3	Red in visible	0.04–0.1	Altered carbon. chondrites?	
A (igneous)	2.8	2.7–3.2	UV absorption; olivine absorption.	0.1–0.4	Pallasites? Chassignites Olivine-rich achondrites	
R (igneous)	2.9	…	Pyroxene, olivine similar to V type	~0.4	Olivine-rich achondrites	349 Dembowska
C (primitive)	3.0	2.3–4.0	Red in visible; flat, dark; 3.07 μm absorption band (ices)	0.03–0.07	Carbonaceous chondrites (CM/CI)	176 Iduna
K (primitive)	3.1	…		Low	CV/CO Chondrites	221 Eos
M (igneous)	3.0	2.3–3.5	Somewhat red	0.1–0.2	Irons	16 Psyche; 21 Lutetia

Asteroid Type (class)	Location (au) Peak	Range	Spectral data	A	Associated meteorites group/clan/type	Examples
B (metamorphic)	~3	2.3–4	Like C but slightly brighter	0.04–0.08	Modified carbonaceous chondrites	
G (metamorphic)	~3	2.3–4	Like C but strong UV absorption	0.05–0.09	"	1 Ceres
F (metamorphic)	~3	2.3–4	Like C, flat; no UV	0.03–0.06	"	
P (primitive)	4.0	3.0–5.2	Similar to M but darker	0.02–0.06	None known	
D (primitive)	>5	3–>5	Redder than P	0.02–0.05	None known	

[a] based in part on Chapman (1990, Table 1, p. 236), and on Wetherill and Chapman (1988, esp. Table 2.1.3 on p. 54).
General information on locations of super classes:
Igneous (<2.7 a.u.); Metamorphic (~3.2 a.u.); Primitive (>3.4 a.u.)

well. This could have taken place early in the solar system during the T Tauri stage of the Sun's evolution when, it is thought, the hydrated carbonaceous chondrites may have been exposed to short-lived sporadic heating and were subject to a variety of temperatures, thus resulting in the variety of petrologic types of Table 15.3.

Asteroids undergo dynamical or orbital evolution also, especially those in near resonances with Jupiter, i.e., near the Kirkwood gaps, which are subject to chaotic variation. Recent work has shown that objects moving in these orbits may do so for 10^5 years or more, but may suddenly be perturbed into very different orbits. It is conjectured that just such processes have led to the creation of the Earth- and Mars-crossing asteroids.

In addition to the perturbations exerted by other objects, the radiation pressure of the Sun and the impact of the solar wind have non-gravitational effects which could be observable (such effects have certainly been seen in comets).

Finally, the absorption of sunlight and decreasing radiation by an asteroid with distance from the evening terminator on the night side may cause small but systematic dynamical effects on the motion of the asteroid, through the *Yarkovsky effect*. The heating of a surface by sunlight and the cooling by reradiation is a process which is not isotropic. If the object being heated is a rotating spherical object on the order of a meter in diameter, its orbit can be affected by the anisotropy. The object is coolest on the night side of the sunrise terminator by an amount ΔT, and so will have a temperature $T - \Delta T$; on the sunlit side of the sunset terminator, on the other hand, the temperature will be $T + \Delta T$. The resulting force on the object from this effect will be

$$F_Y = (8/3)(\pi R^2 \sigma T^4/c)(\Delta T/T) \cos \psi \qquad (15.19)$$

sometimes called the *Yarkovski force*, where R is the particle's radius and ψ is its "obliquity," the angle between its orbital plane and its equator. When $0° \leq \psi < 90°$, the rotation is prograde (CCW as seen from the north ecliptic pole of the solar system), and when $90° < \psi \leq 180°$ it is retrograde. When prograde, both eccentricity and semi-major axis will be increased. When retrograde, the net result will be a decrease of the semi-major axis, and the object will spiral into the Sun. If $\Delta T << T$, the effect will be small, and the time scale will be very long. It is problematic if this effect will dominate over other perturbative effects, but in low-density regions of the solar system, it may be an important factor, and may even be important for somewhat larger objects.

The D/H ratios have shown that different meteorites originated in different places in the solar system, and indicate that the Moon was born very close to the Earth. The picture confirms the view that volatiles were driven out of the inner solar system at an early date, resulting in the availability of ices to form a profusion of bodies in the outer solar system.

The short-period comets, with direct orbits and moderate eccentricities, may be derived from icy bodies that may have been perturbed through collisional encounters in the Kuiper Belt. This belt may, in turn, be the evolved remnant of the planetary disk of the early solar system. The existing TNO population show both a "classical" and a dynamically perturbed component, in which eccentricities and inclinations are more varied.

Dynamical studies have shown that long-period comets originate in very far orbits in the solar system (the *Oort Cloud*, after Jan Oort) and are perturbed, perhaps by a star traveling in the same direction but providing a slight negative acceleration resulting in an inward falling direction for the comet [see Chapter 14.7, and for a discussion of current views of the origin and changes to the Oort Cloud, see Dones et al. (2004)].

Consequently, we have no evidence at present that anything seen on Earth has come from outside the material present in the original solar nebula except for photons and cosmic rays.

15.8 Implications for the Origin of the Solar System

The current picture of the origin of the solar system is of a rotating solar nebula, which developed a disk in conservation of angular momentum as it gravitationally collapsed. Bipolar outflows may have accompanied the development of a thick disk. High temperatures would have characterized the inner nebula, at the core of which a star would be born, and the temperature gradient through the disk would have insured different rates of condensation with distance from the core. In the middle part of the disk, strong magnetic and electrostatic fields would have produced electric currents and electrical discharges that would have melted the aggregating dust particles and caused them to rain down on other components, giving rise to the undifferentiated bodies that became the chondrites. Fractionation processes would cause distributions of isotopes of various chemical species.

As aggregation proceeded and as the solar wind and radiation pressure of the young star drove away the dust and gas of the inner solar system, aggregates massive enough to differentiate did so. Subsequent major collisions disrupted many of these, giving rise to the iron and silicate-rich differentiated bodies, which were further impacted in most cases. The major planets themselves had disks and developed satellite systems and rings, some of which may be still present today.

In an initially hot environment, there is an expected sequence of condensations as the more refractory materials condense out of vapor first, and the more volatile materials later, when the temperature has decreased further. In this way, minerals appeared in a *condensation sequence*, the order in which the minerals condensed out of the solar nebula.

The outer sub-giant planets migrated into their present positions, possibly by absorbing orbital energy from encountered bodies. But most of the icy bodies were ejected or wound up in the Kuiper Belt. The remnants of the solar nebula are to be found still today in the Oort Cloud, some of the denizens of which were flung out there by interactions with the major planets early in the history of the solar system.

15.9 The Solar Nebula

Stars typically have their origins in the midst of a molecular cloud, which collapses to an association of stars. The association is usually born with an excess of dynamical energy, so the stellar system eventually dissipates. The Sun may have had its beginning in just such an environment.

Here we concentrate on the collapse leading to a single star, of solar mass, starting with a condensation from a molecular cloud, in order to see how the size and mass of this condensing cloud depend on its temperature and the density. For a cloud of material that is gravitationally bound, it may be shown[7] that

$$2 <K> = - <U> \tag{15.20}$$

where $<K>$ is the mean kinetic energy and $<U>$ is the mean potential energy. Equation (15.20) is a form of the *Virial Theorem*, an expression widely used in astrophysics, especially to determine the mass of an ensemble of objects, ranging from particles to galaxies. Because the total energy, E, is merely the sum of the kinetic and potential energies, we then have:

$$<E> = <K> + <U> = - <K> < 0 \tag{15.21}$$

Thus, the total energy of a bound ensemble is negative. In an ensemble of gaseous molecules, the kinetic energy may be expressed as

$$(3/2)\, kT = 1/2\, m <v^2> \tag{15.22}$$

where $k = 1.38 \times 10^{-23}$ is the Boltzmann constant and T is the temperature. The kinetic energy of the total ensemble of N molecules is then

$$K_{\text{ens}} = \Sigma K_i = 1/2\, N\, m <v^2> = (3/2)\, NkT \tag{15.23}$$

---

[7] For example, by Karttunen et al. (1994, pp. 146–148) or Zeilik and Gregory (1998, p. P12).

If we use the approximation that the cloud is pure molecular hydrogen, so that

$$N = M/(2m_H), \tag{15.24}$$

where M is the total mass and m_H is the mass of the hydrogen atom, (15.23) becomes

$$K_{ens} = (3/2)\,kT\,M/(2m_H) \tag{15.25}$$

Applying the Virial Theorem to (15.25), and writing out the potential energy explicitly for the ensemble of mass M = $N(2m_H)$ (Milone & Wilson 2008, Chapter 5.3), we obtain from (15.20)

$$2(3/2)\,kT\,M/(2m_H) = G\,M^2/\Re \tag{15.26}$$

where $\Re$ is the radius of the cloud. This simplifies to

$$(3/2)\,kT/m_H = GM/\Re \tag{15.27}$$

and, substituting for the mass,

$$M = (4/3)\pi\,\Re^3\rho \tag{15.28}$$

we obtain

$$kT/m_H = (8/9)\pi\,GR^2\rho \tag{15.29}$$

so that the radius, expressed in terms of the temperature and density of the cloud, becomes

$$\Re = \{kT/[(8/9)\pi\,G\rho\,m_H]\}^{\frac{1}{2}} \tag{15.30}$$

In SI units,

$$\Re = 6.65 \times 10^6\,(T/\rho)^{1/2} \tag{15.31}$$

With typical values $T = 10\,K$, and $\rho = 10^{-15}\,kg/m^3$, the radius is about $\Re \approx 7 \times 10^{14}\,m = 4700\,au = 0.023\,pc$. Finally, from (15.28), the mass of such a cloud would be $M \approx 1.4 \times 10^{30}\,kg$, or of the order of a solar mass.

The collapse of the cloud may not have been purely gravitational, however. In the presence of magnetic fields, expected because they are in fact present in the interstellar medium, ambipolar diffusion (cf. Chapter 11.1.1.5) can levitate ions while neutral atoms, molecules, and grains more readily fall into a gravitational well. A shock-wave triggered collapse from a supernova explosion was suggested by Cameron and Truran (1977) to account for the ^{26}Al, so useful for the melt required for chondrules, and primordial planetary heat sources. Another source, however, could be a red giant wind. In a recent study, Gounelle and Meibom, 2008, argue that the radionuclides arose from the proto-sun environment itself. According to Boss (1995), a speed for the shock of $\sim$25 km/s would suffice to trigger a collapse. The presence of even a slight amount of rotation in the initial cloud would result in sufficient rotation in the collapsed nebula to flatten it into a disk.

A strong bipolar outflow is observed in some young pre-stellar objects, presumably emanating from the directions of the rotational poles. The mechanism for the outflow may be a strong wind emanating from the proto-Sun, the equatorial wind component of which is obscured by the disk. A discussion of other mechanisms, and a thorough review of the subject, can be found in Boss (2004).

As the disk rotates, shearing must take place, as suggested originally by Laplace, leading to Keplerian orbits for the condensation products.

We next consider the disks themselves.

15.10 The Proto-Planetary Disk

The planets of the solar system and planetary systems of other stars had their beginnings in a thick disk around the proto-star. The discovery of a disk around the bright star β Pictoris (Smith and Terrile 1984) called attention to this phenomenon as a general stage of all stars. *Remnant* or *debris disks* were subsequently found around Vega (viewed pole-on), Fomalhaut, and ε Eridani. Since then the number of stars found to have disks has burgeoned.

In young associations, such as in Orion, in a rich star-forming region of the Milky Way, *proplyds* (for *proto-planetary disks*) may appear as elongated nebulae (Figures 15.15 and 15.16) often with tail-like structures, subject to erosion by the intense UV and x-ray radiation and stellar winds from hot, new stars in the vicinity. That the objects contain stars, and are young, is generally accepted. O'Dell and Wen (1994) argue that proplyds are flattened envelopes or circumstellar disks, supporting an argument made earlier by Meaburn (1988), and demonstrate this with a rectangular profile for a dark, silhouetted object designated (on the basis of position) 183-405. It is clearly optically thick and was measured to have dimensions 1.2×0.9 arc-secs, corresponding to 550×410 au if at a distance of 460 pc. A rough calculation can provide a lower limit on the mass of this object. Following, O'Dell and Wen, the optical depth along a line of sight is given by

$$\tau = N_{\rm d}\, \pi R^2\, Q(2\pi R/\lambda) \qquad (15.32)$$

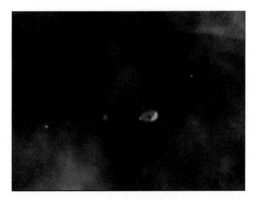

Fig. 15.15. Possible proplyds in Orion, as imaged with the HST Wide Field and Planetary Camera 2 on December 29, 1993 (O'Dell and Wen 1994, Figure 3c on plate 7). 183-405 is the dark object on the right, seen in silhouette. Courtesy, C. R. O'Dell, Vanderbilt University, NASA, and ESA. Reproduced by permission of the AAS. See also color Plate 28

where N_d is the number of particles in a column of radius R, and $Q(2\pi R/\lambda)$ is the extinction efficiency. The mass of material in this column is,

$$m = N_d\, 4\pi R^3 \rho/3 \qquad (15.33)$$

where ρ is the density of the material, in $kg \cdot m^{-3}$. Substitution then yields,

$$m = 4R\rho\tau/[3Q(2\pi R/\lambda)] \qquad (15.34)$$

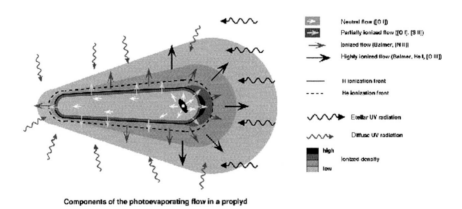

Components of the photoevaporating flow in a proplyd

Fig. 15.16. A model for a proplyd, from Henney and O'Dell (1999, Figure 5, p. 2359), showing the effect of ionizing radiation from nearby hot stars. Note the circumstellar disk, a possible example of which O'Dell and Wen suggested was seen in silhouette in Figure 15.15. Courtesy, C. R. O'Dell, Vanderbilt University, NASA, and ESA. Reproduced by permission of the AAS

With values of $\rho/[Q(2\pi R/\lambda)] = 1000\,\text{kg/m}^3$ and $R = 200\,\text{nm}$, summing over all columns in the line of sight to the object and using an average value of τ calculated over the entire rectangle, O'Dell and Wen obtained a minimum mass of about $4 \times 10^{23}\,\text{kg}$, compared to $\sim 6 \times 10^{24}\,\text{kg}$ for the Earth. Even if an order of magnitude too small, the computed mass indicates that this object would not be an equivalent of the solar system.

True, the Sun is relatively massive and luminous in its neighborhood, but as we have seen in the previous section, solar mass objects can be expected to form from molecular clouds.

The current view of planetary formation is that material will accrete through numerous collisions, and that objects in dense zones will grow to form proto-planets. This will create gaps in the disks; gaps indeed have been observed in stellar remnant disks. The presence of gas, dust, and orbiting debris of various sorts will create a drag, however, and planets will tend to migrate inward at rates that depend on the viscosity of the disk as well as the mass of the planet, and, eventually, to fall into the star. Matsumura et al. (2007) have determined that the presence of a "dead zone," a region of low viscosity, between the star and the proto-planet, can slow and even halt the migration of the planet. They have simulated the fates of both terrestrial and Jovian mass planets, showing that the presence of dead zones interior to their orbits allows planets time to create a gap within the disk, as accretion proceeds. This will be discussed further in Chapter 16, in the discussion of the origin of "hot Jupiters."

We now proceed to discuss the known extra-solar planetary systems, many of which appear to involve planets that are even more massive than our own. In that context, we will reconsider the origin of planetary systems.

References

Alvarez, L. 1980. "Extraterrestrial Cause for the Cretaceous–Tertiary Extinction," *Science*, **208**, 1095–1108.

Binzel, R. P., Bus, S. J., and Burbine, P. H. 1998. "Relating S-Asteroids and Ordinary Chondrite Meteorites: The New Big Picture," *BAAS*, **30**, 1041.

Bertotti, B., Farinella, P., and Vokrouhlicky, D. 2003. *Physics of the Solar System.* (Dordrecht: Kluwer).

Bogard, D. D. and Garrison, D. H. 1998. "Relative Abundances of Ar, Kr, and Xe in the Martian Atmosphere as Measured by Martian Meteorites," *Geochimica et Cosmochimica Acta*, **62**, 1829–1835.

Bogard, D. D. and Garrison, D. H. 1999. "Argon 39–Argon 40 'ages' and Trapped Argon in Martian Shergottites, Chassigny, and Allen Hills 84001," *Meteoritics and Planetary Science*, **34**, 451–473.

Borg, L. E., Edmunds, J. E., and Asmerom, Y. 2005. "Constraints on the U–Pb
Isotopic Systematics of Mars Inferred from a Combined U–Pb, Rb–Sr, and Sm–Nd
Isotopic Study of the Martian Meteorite Zagami," *Geochimica et Cosmochimica
Acta*, **69**, 5819–5830.

Boss, A. P. 1995. "Collapse and Fragmentation of Molecular Cloud. II. Collapse
Induced by Stellar Shock Waves," *Astrophysical Journal*, **439**, 224–236.

Boss, A. P. 2004. "From Molecular Clouds to Circumstellar Disks" in *Comets II*,
eds M. C. Festou, H. U. Keller, and H. A. Weaver (Tucson, AZ: University of
Arizona Press), pp. 67–80.

Brown, M. E., van Dam, M. A., Bouchez, A. H., Le Mignant, D., Campbell, R. D.,
Chin, J. C. Y., Conrad, A., Hartman, S. K., Johansson, E. M, Lafon, R. E.,
Rabinowitz, D. L., Stomski, Jr., P. J., Summer, D. M., Trujillo, C. A., and
Wizinowich, P. L. 2006. "Satellites of the Largest Kuiper Belt Objects," *Astro-
physical Journal*, **639**, L43–L46.

Cameron, A. G. W. and Truran, J. W. 1977. "The Supernova Trigger for Formation
of the Solar System," *Icarus*, **30**, 447–461.

Chapman, C. R. 1990. "Asteroids," in *The New Solar System*, eds. J. K. Beatty
and A. Chaikin. (Cambridge, MA: Sky Publishing Corp.; Cambridge, UK: Press
Syndicate of University of Cambridge), 3rd ed., pp. 231–240.

Clayton, R. N. 1993, "Oxygen Isotopes in Meteorites," *Annual Review of Earth and
Planetary Sciences*, **21**, 115–149.

Clayton, R. N. and Mayeda, T. K. 1984, "The Oxygen Isotope Recod in Murchison
and Other Carbonaceous Chondrites," *Earth & Planetary Science Letters*, **67**,
151–161.

Clayton, R. N. and Mayeda T. K. 1996, "Oxygen Isotope Studies of Achondrites,"
Geochimica et Cosmochimica Acta, **60**, No. 11, 1999–2017.

Clayton, R. N., Mayeda T. K., Goswami, J. N., and Olsen, E. J. 1991, "Oxygen
Isotope Studies of Ordinary Chondrites," *Geochimica et Cosmochimica Acta*, **55**,
2317–2337.

Dones, L., Weissman, P. R., Levison, H. F., and Duncan, M. J. 2004. "Oort Cloud
Formation and Dynamics" in *Comets II*, eds M. C. Festou, H. U. Keller, and H.
A. Weaver (Tucson, AZ: University of Arizona Press), pp. 153–174.

Dunham, D. W., Bixby Dunham, J., et al. (44 other co-authors) 1990. "The Size
and Shape of (2) Pallas from the 1983 Occultation of 1 Vulpeculae," *Astronomical
Journal*, **99**, 1636–1662.

Eugster, O., Weigel, A., and Polnau, E. 1997. "Ejection Times of Martian
Meteorites," *Geochimica et Cosmochimica Acta*, **61**, 2749–2757.

Fessenkov, V. G. 1955. "Sikhoté-Aline Meteorite" in *Meteors*, ed. T. R. Kaiser (New
York: Pergamon Press), pp. 179–183.

Gounelle, M. and Meibom, A. 2008. "The Origin of Short-Lived Radionuclides
and the Astrophysical Environment of Solar System Formation," *Astrophysical
Journal*, **680**, 781–792.

Grady, M. M. 2000. *Catalogue of Meteorites*, 5th Ed. (London: The Natural History
Museum).

Henney, W. J., and O'Dell, C. R. 1999. "A Keck High-Resolution Spectroscopic
Study of the Orion Nebula Proplyds," *Astronomical Journal*, **118**, 2350–2368.

Hewins, R. H. 1997. "Chondrules," *Annual Review of Earth and Planetary Sciences*,
25, 61–83.

Hirayama, K. 1918a. "Groups of Asteroids Probably of Common Origin," *Proc. Phys. Math. Soc. Japan, Ser.* 2, No. 9, 354–361.

Hirayama, K. 1918b. "Groups of Asteroids Probably of Common Origin," *Astronomical Journal*, **31**, 185–188.

Hodge, P. 1994. *Meteorite Craters and Impact Structures of the Earth.* (Cambridge: The University Press).

Hutchison, R. 2004. *Meteorites* (Cambridge, UK: University Press).

Jewitt, D. and Luu, J. 2000. "Physical Nature of the Kuiper Belt" in *Protostars and Planets. IV*, eds V. Mannings, A. P. Boss, and S. S. Russell (Tucson, AZ: University of Arizona Press), pp. 1201–1229.

Joy, K. H., Crawford, I. A., Russell, S. S., Swinyard, B., Kellett, B., and Grande, M. 2006. "Lunar Regolith Breccias MET01210, PCA02007 and DAG400: Their Importance in Understanding the Lunar Surface and Implications for the Scientific Analysis of D-CIXS Data." *Lunar and Planetary Sciences*, **37**, No. 1274, Suppl. p. 5221.

Karlsson, H. R., Clayton, R. N., Gibson, Jr., E. K., and Mayeda, T. K. 1992. "Water in SNC Meteorites: Evidence for a Martian Hydrosphere," *Science*, **255**, 1409–1411.

Karttunen, H., Kroger, P., Oja, H., Poutanen, M., and Donner, K. J., eds. 2003. *Fundamental Astronomy* (Berlin: Springer-Verlag).

Kelley, D. H. and Milone, E.F. 2005. *Exploring Ancient Skies: An Encyclopedic Survey of Ancient Astronomy* (New York: Springer-Verlag).

Kirsten, T. 1978. "Time and the Solar System," in *Origin of the Solar System*, ed. S. F. Dermott, (New York: John Wiley), pp. 267–346.

Liu, Y., Nyquist, L., Wiesmann, H., Shih, C., Schwandt, C., and Takeda, H. 2003. "Internal Rb-Sr Age and Initial $^{87}Sr/^{86}Sr$ of a Silicate Inclusion from the Campo Del Cielo Iron Meteorite," *Lunar and Planetary Science* XXXIV (2003).

Marti, K., Kim, J. S., Thakur, A. N., McCoy, T. J., and Keil, K. 1995. "Signatures of the Martian Atmosphere in Glass of the Zagami Meteorite," *Science*, **267**, 1981–1984.

Matsumura, S., Pudritz, R. E., and Thommes, E. W. 2007. "Saving Planetary Systems: Dead Zones and Planetary Migration," *Astrophysical Journal*, **660**, 1609–1623.

Meaburn, J. 1988. "An Extended High-speed Flow from a Compact, Ionized Knot in the Orion Nebula (M42)," *Monthly Notices, Royal Astronomical Society*, **233**, 791–800.

Morbidelli, A. and Brown, M. E. 2004. "The Kuiper Belt and the Primordial Evolution of the Solar System" in *Comets II*, eds M. C. Festou, H. U. Keller, and H. A. Weaver (Tucson, AZ: University of Arizona Press), pp. 175–191.

O'Dell, C. R. and Wen, Z. 1994. "Postrefurbishment Mission Hubble Space Telescope Images of the Orion Nebula: Proplyds, Herbig-Haro Objects, and Measurement of a Circumstellar Disk," *Astrophysical Journal*, **436**, 194–202.

Pinson, W. H., Jr., Schnetzler, C. C., Beiser, E., Fairbairn, H. W., and Hurley, P. M. 1965. "Rb-Sr Age of Stony Meteorites," *Geochimica et Cosmochimica Acta*, **29**, 455–466.

Righter, K. 2007. *The Lunar Meteorite Compendium.* http://www-curator.jsc.nasa.gov/antmet/lmc/index.cfm

Sears, D.W.G. and Dodd, R. T. 1988. *Meteorites and the Early Solar System* (Tucson, AZ: University of Arizona Press).

Shimoda, G., Nakamura, N., Kimura, M., Kani, T., Nohda, S., and Yamamoto, K. 2005. "Evidence from the Rb-Sr System for 4.4 Ga Alteration of Chondrules in the Allende (CV3) Parent Body," *Meteoritics and Planetary Science*, **40**, Nr 7, 1059–1072.

Smith, B. A. and Terrile, R. J. 1984. "A Circumstellar Disk around β Pictoris," *Science*, **226**, 1421–1424.

Tatsumoto, M., Unruh, D. M., and Desborough, G. A. 1976. "U-Th-Pb and Rb-Sr Systematics of Allende and U-Th-Pb Systematics of Orgueil," *Geochimica et Cosmochimica Acta*, **40**, 617–634.

Taylor, S. R. 1992. *Solar System Evolution: A New Perspective* (Cambridge: University Press).

Urey, H. C. and Craig, H. 1953. "The Composition of Stone Meteorites and the Origin of the Meteorites," *Geochimica et Cosmchimica Acta*, **4**, 36–82.

Van Schmus, W. R. and Wood, J. A. 1967. "A Chemical-Petrologic classification for the Chondritic meteorites." *Geochimica et Cosmochimica Acta*, **31**, 747–765.

Wasson, J. T. 1985. *Meteorites: Their Record of Early Solar System History* (New York: Freeman).

Wetherill, G. W., and Chapman, C. R. 1988. "Asteroids and Meteorites," in *Meteorites and the Early Solar System*, eds. Kerridge, J. F., and Matthews, M. S. (Tucson: The University of Arizona Press), pp. 35–67.

Wlotzka, F. 1993. "A Weathering Scale for the Ordinary Chondrites," *Meteoritics*, **28**, 460.

Wood, J. A. 1968. *Meteorites and the Origin of Planets* (New York: McGraw-Hill).

Wood, J. A. 1990. "Meteorites" in *The New Solar System*, eds J. K. Beatty and A. Chaikin, (Cambridge, MA: Sky Publishing Corp.; Cambridge, UK: Press Syndicate of University of Cambridge) 3rd Ed., pp. 241–250.

Zeigler, R. A., Korotev, R. L., Jolliff, B. L., and Haskin, L. A. 2005. "Petrography of Lunar Meteorite MET 01210," *Lunar and Planetary Science*, **XXXVI**, Abstract No. 2385.

Zeilik, M. and Gregory, S. A. 1998. *Introductory Astronomy and Astrophysics* (Fort Worth, TX: Saunders College Publishing).

Challenges

[15.1] Discuss how points along the plot of the ratios $^{87}Sr/^{86}Sr$ vs. $^{87}Rb/^{86}Sr$ change with time. What do we mean by "time" here anyway? (See Figure 15.6)

[15.2] Examine the validity of equations (15.5) and (15.6), defining all quantities, and evaluate the dynamic evidence that SNC meteorites come from Mars. Assume the correctness of Wasson's assertion that an impactor with substantially greater speed than v_∞ may result in planetary mass loss.

[15.3] Summarize the different types of ages that a meteorite can have and associate each with a stage of a meteorite's history.

[15.4] Examine the list of the principal types of asteroids and their associated meteorites. What can you conclude about the origin of those meteorites.

[15.5] Discuss what meteorites could be expected from cometary impact on Earth and on meteoroids or their parent bodies. Do we have any evidence that such impacts occur in any known meteorite specimens?

[15.6] Describe the time-line of the development stages of the material in the Zagami meteorite from the original aggregation of elements to the fall and recovery.

16. Extra-Solar Planetary Systems

16.1 Historical Perspective

Our solar system can no longer be considered unique! Planets are common among stars like the Sun, especially those richer in metals[1] and they are not entirely absent near other types of stars, such as sub-giants, giants, a certain type of pulsar, and even around very cool, low-mass M dwarfs.

Toroidal circumstellar disks have been discovered around main sequence stars through detection of extended infrared regions around young main sequence stars and through direct visual imaging with an occulting disk blocking the light of the star. The first star about which a disk was detected with the coronographic technique, as the latter is called, was β Pictoris. Beginning in 1995, planetary discoveries have increased monthly. Virtually every issue of the major journals now contains papers on circumstellar disks, disk remnants, or extra-solar planets. The history of pursuit of such planets did not start in 1995, however.

In the 1960s Peter Van de Kamp of the Sproul Observatory claimed to discover one or two companions (depending on the assumed orbital eccentricity) for Barnard's Star on the basis of proper motions. This study has not been replicated and the planets remain in doubt. Other searches over the years included that by Campbell and Walker (1979), and Campbell et al. (1988) using a spectrograph at the Canada–France–Hawaii Telescope. At those times the precision achievable with the technique that they employed was almost sensitive enough to reveal the radial velocity variations of the star in its absolute orbit with planetary objects.

In 1992, planets were detected around a pulsar. Although some claimed detections of planetary-sized masses around such an object have been found to be due to artifacts of the observing and data reduction processes, in the case of PSR 1257+12, the claim has thus far survived all scrutiny. In 1995, the discovery of a planet around a main sequence star, 51 Pegasi, was announced by Michel Mayor at Geneva (Mayor and Queloz 1995). Confirmation of the

---

[1] The term "metals" in astrophysics frequently, as here, refers to all elements heavier than hydrogen and helium.

planetary nature of this object, as well as the discovery of about a dozen other candidates, were announced in short order by a team headed by Geoffrey Marcy (see Marcy and Butler 1998 for a review). In these cases, radial velocity variations of the stars indicated the presence of massive Jupiter-like planets.

The kinds of methods that can be used to find planets around other stars are described in Section 16.2; each is limited in particular ways, but all are promising, and observers have claimed detections with each one of them. At this writing, one method (spectroscopy) has thus far been the most successful (Section 16.2.1). The vast majority of suspected planets have been detected by this method. Increasingly, however, other methods are yielding detections. The first planet to be observed by photometric transits (Section 16.2.2) of the surface of a star (HD 209458b, discussed in detail in Section 16.4.1) had been discovered previously through radial velocity variations of the star. Increasing numbers of other candidate planets are being discovered by photometric as well as radial velocity and astrometric techniques, and most of these have been or are in the process of being confirmed with radial velocity techniques.

In this chapter, no systematic attempt has been made to identify the original discoverers of all planets or planetary systems, if more recent studies are available, but such identifications can be found in the online list, http://obswww.unige.ch/~naef/who_discovered_that_planet.html.

Table 16.1 lists the known planetary systems and Table 16.2 is the list of planets themselves, as of mid-2006, updated to include some of the more recent transit and occultation discoveries, and the fifth planet of the 55 Cnc system. Notably not included are the transit candidates found by the HST SWEEP project and the COROT space mission. The comprehensive website maintained by Jean Schneider (CNRS-LUTH, Paris Observatory), http://vo.obspm.fr/exoplanetes/encyclo/catalog.php and that maintained by the extra-solar planet finding group, http://exoplanets.org, have proven invaluable for compiling these tables, and should be consulted for updates. As we were completing the compilation, the definitive catalog of Butler et al. (2006) appeared, and we revised the table extensively to include data from there. Table 16.1 also includes data from the Bright Star Catalogue and from the SIMBAD database accessed through: http://simbad.u-strasbg.fr/Simbad and from other sources in the literature, some of which are identified in the comments column of Table 16.2. The data are arranged according to right ascension and are thus suitable for observational use. The observable quantities are emphasized, but some absolute parameters, such as the true size of the orbit, the planetary radius, and the mass, when known, are also given. The columns of Table 16.1 give: the most familiar designations (1, 2); the position in 2000.0 coordinates (3–8); the spectral type (9); the V (or other, as indicated) magnitude (10); the (B-V) color index (11); the star's parallax (12) and uncertainty in it (13) in units of milliarc-sec; the corresponding distance (14), and lower and upper bounds

(15, 16), respectively; the radial velocity (17); and the proper motion in RA and Dec, rounded to the nearest milliarc-sec per year (18, 19); finally, the number of planets discovered in that system to date (20). We have not entered several types of data which would be useful to know, namely the mass, radius, effective temperature, rotational velocity indicator (v sin i, where i is the inclination of the star's rotation pole to the line of sight), and an indication of chemical composition ([Fe/H]). These quantities tend to be model specific, so we refer the reader to, for example, Santos et al. (2003, 2004a), for sets of these data. In some specific discussions of planetary systems, below, we cite relevant additional data from the authors who carried out the analyses.

The columns of Table 16.2 display: the name of the system (1–3); the component (4; the star itself is component "a", so companions are designated "b" for the first discovered, "c" for the second, and so on); the period of the system and the uncertainty in it (5, 6); the semi-major axis, and its uncertainty, in au (7, 8); the eccentricity and its uncertainty (9, 10); the projected mass (or the actual mass if the inclination is derived through a non-spectroscopic method) and its uncertainty, given in units of Jupiter's mass (11, 12); the inclination and its uncertainty in degrees (13, 14); the radius and its uncertainty in terms of Jupiter's radius (15, 16); an epoch, T_0, usually a moment of periastron passage, and its uncertainty (17, 18), in modified Julian Day numbers (JDN-24 400 000); a predicted moment of transit, T_t, and its uncertainty (19, 20); and two columns of comments (21, 22). Some of the planets in this list may actually be brown dwarfs or, perhaps, members of some intermediate but not yet recognized class. This is the case because the inclinations, i, of their orbits are not usually known and spectroscopic analysis yields only the projected masses, $M_{1,2}$ sin i. Where the companion may be a brown dwarf, we designate the system with a "BD?" in column 21, and provide references for these cases in column 22. Column 21 also indicates if the planet has been detected photometrically through a transit eclipse of its star ("T"), through gravitational lensing ("G"), or by direct imaging ("I"). In those three cases, the masses given are not projected, whereas those determined through the radial velocity method, namely M sin i, are lower limits to M. When known, the full masses are indicated in column 11.

An interesting fact to emerge from Table 16.2 is the number of cases where the eccentricity is large. If all planets formed within a disk and with coplanar, concentric orbits, how could such eccentricities arise? One suggestion is that some may arise through the *Kozai mechanism*, where angular momentum exchanges with a distant companion star in a high-inclination orbit give rise to long-period oscillations in the eccentricity and inclination of the planet. A recent discussion can be found in Takeda and Rasio (2005). Most stars with planetary systems are not known to be binary stars (although some certainly are). Interactions among planets, both detected and undetected, and with objects in the the equivalent of an Edgeworth-Kuiper Belt or within a disk, may be responsible in other cases.

Table 16.1. Stars with planets

Henry Draper Catalog Number	Names Bayer/Flamsteed/HR/ Other Designation	Position (2000.0)						Spectral Type
		RA			Dec			
		h	m	s	Deg	′	″	
HD 142	HR 6, HIP 522	00	06	19.2	−49	04	31	G1 IV
HD 1237	GJ 3021, HIP 1292	00	16	12.7	−79	51	04	G6 V
−	WASP-1	00	20	40.	+31	59	24	F7 V
HD 2039	HIP 1931	00	24	20.3	−56	39	00	G2/G3 IV-V
HD 2638	HIP 2350	00	29	59.9	−05	45	50	G5
HD 3651	54 Psc, HR 166, HIP 3093	00	39	21.9	+21	15	02	K0 V
HD 4208	HIP 3479	00	44	26.7	−26	30	56	G0 V/G5 V
HD 4308	HIP 3497	00	44	39.3	−65	38	58	G5 V
HD 4203	HIP 3502	00	44	41.2	+20	26	56	G5
HD 6434	HIP 5054	01	04	40.2	−39	29	18	G3 IV/G2/G3 V
HD 8574	HIP 6643	01	25	12.5	+28	34	00	F8
HD 9826	50 upsilon And, HR 458, HIP 7513	01	36	47.8	+41	24	20	F8 V
HD 10647	HR 506, HIP 7978	01	42	29.3	−53	44	27	F8 V
HD 10697	109 Psc, HR 508, HIP 8159	01	44	55.8	+20	04	59	G5 IV
HD 11977	eta^2 Hyi, HR 570, HIP 8928	01	54	56.1	−67	38	50	G8III-IV/G8.5 III/G5 II
HD 11964	HIP 9094	01	57	09.6	−10	14	33	G5
HD 12661	HIP 9683	02	04	34.3	+25	24	52	G6 V/K0 III/K0 V
HD 13189	HIP 10085	02	09	40.2	+32	18	59	K2 III
HD 13445	HR 687, GJ 86, 10138	02	10	25.9	−50	49	25	K1 V
HD 16141	79 Cet, HIP 12048	02	35	19.9	−03	33	38	G5 IV
HD 17051	iota Hor, HR 810, HIP 12653	02	42	33.4	−50	48	01	G0 V/G3 IV
−	HIP 14810, BD +20 518	03	11	14.2	+21	05	50	G5
HD 19994	94 Cet, HR 962, HIP 14954	03	12	46.4	−01	11	46	F8 V
HD 20367	HIP 15323	03	17	40.0	+31	07	37	G0 V
HD 20782	HIP 15527	03	20	03.6	−28	51	15	K0/G3 V
HD 22049	18 epsilon Eri, HR 1084, HIP 16537	03	32	55.8	−09	27	30	K2 V
HD 23079	HIP 17096	03	39	43.1	−52	54	57	F8/G0 V
HD 23596	HIP 17747	03	48	00.4	+40	31	50	F8 V
HD 27442	epsilon Ret, HR 1355, HIP 19921	04	16	29.0	−59	18	08	K2 IVa/gK5
HD 27894	HIP 20277	04	20	47.0	−59	24	39	K2 V
HD 28185	HIP 20723	04	26	26.3	−10	33	03	G5 IV
HD 30177	HIP 21850	04	41	54.4	−58	01	15	G8 V
HD 33283	HIP 23889	05	08	01.0	−26	47	51	G5 IV/G3/G5 V
HD 33636	HIP 24205	05	11	46.4	+04	24	13	G0 V
HD 33564	HIP 25110	05	22	33.5	+79	13	52	F6 V
HD 37124	HIP 26381	05	37	02.5	+20	43	51	G4 V/IV-V
HD 39091	pi Men, HR 2022, HIP 26394	05	37	09.9	−80	28	09	GI IV/G3 IV
HD 37605	HIP 26664	05	40	01.7	+06	03	38	K0 V
HD 38529	HIP 27253	05	46	34.9	+01	10	05	G4 IV/G4 V
HD 41004A	HIP 28393	05	59	49.7	−48	14	23	KI V
HD 41004B	HDS 814B, HIP 28393	05	59	49.7	−48	14	23	M2
HD 40979	HIP 28767	06	04	29.9	+44	15	38	F8 V
HD 44627	AB Pic	06	19	12.9	−58	03	16	K2 V
HD 45350	HIP 30860	06	28	45.7	+38	57	47	G5 IV
HD 46375	HIP 31246	06	33	12.6	+05	27	47	KI IV
HD 47536	HR 2447, HIP 31688	06	37	47.6	−32	20	23	K0 III/KI III/K2 III
HD 49674	HIP 32916	06	51	30.5	+40	52	04	G5 V/G0 V
HD 50499	HIP 32970	06	52	02.0	−33	54	56	GI IV/G0 V
HD 50554	HIP 33212	06	54	42.8	+24	14	44	F8 V
HD 52265	HIP 33719	07	00	18.0	−05	22	02	G0 V/F8/C0 III-IV
HD 59686	HR 2877, HIP 36616	07	31	48.4	+17	05	10	K2 III/K0 III
HD 63454	HIP 37284	07	39	21.9	−78	16	44	K4 V
HD 62509	78 beta Gem, HR 2990, NSV3712	07	45	19.0	+28	01	34	K0 IIIb
−	XO-2, GSC -3413-00005	07	48	06.5	+50	13	33	K0 V
HD 65216	HIP 38558	07	53	41.3	−63	38	50	G5 V

					d error			PM		
Stellar Properties										
V	B-V	p mas	p error mas	d pcs	−	+	V_r km/s	RA	Dec mas	No of Planets
5.70	0.52	39.00	0.64	25.64	0.41	0.43	+2.60	575	−40	1
6.59	0.75	56.76	0.53	17.62	0.16	0.16	−6.10	434	−58	1
11.8	−	−	−	−	−	−	−13.46	−	−	1
9.01	0.61	11.13	1.13	89.85	8.28	10.15	+8.40	79	15	1
9.44:	0.90	18.62	1.35	53.71	3.63	4.20	+9.55	−107	−224	1
5.87	0.85	90.03	0.72	11.11	0.09	0.09	−34.20	−461	−371	1
7.79	0.67	30.58	1.08	32.70	1.12	1.20	+55.40	314	150	1
6.54	0.65	45.76	0.56	21.85	0.26	0.27	+97.70	158	−742	1
8.68	0.73	12.85	1.27	77.82	7.00	8.53	−	125	−124	1
7.72	0.60	24.80	0.89	40.32	1.40	1.50	+22.40	−169	−528	1
7.80	−0.20	22.50	0.82	44.44	1.56	1.68	+18.60	253	−159	1
4.09	0.54	74.25	0.72	13.47	0.13	0.13	−28.30	−173	−381	3
5.52	0.53	57.63	0.64	17.35	0.19	0.19	+2.90	167	−107	1
6.29	0.67	30.71	0.81	32.56	0.84	0.88	−43.50	−45	−105	1
4.70	0.92	15.04	0.47	66.49	2.01	2.14	−16.20	76	73	1
6.42	0.83	29.43	0.91	33.98	1.02	1.08	−6.90	−368	−243	1
7.44	0.72	26.91	0.83	37.16	1.11	1.18	−	−108	75	2
7.57	1.49	0.54	0.93	1851.5	1171.	712.	−	1	6	1
6.17	0.77	91.63	0.61	10.91	0.07	0.07	+53.10	2093	654	1
6.78	0.71	27.85	1.39	35.91	1.71	1.89	−53.00	−157	−437	1
5.40	0.57	58.00	0.55	17.24	0.16	0.17	+15.50	334	219	1
8.51	0.75	18.91	1.45	52.88	3.77	4.39	−	−3	−54	2
5.06	0.57	44.69	0.75	22.38	0.37	0.38	+18.30	193	−69	1
6.41	0.52	36.86	1.08	27.13	0.77	0.82	+5.30	−103	−57	1
7.38	0.65	27.76	0.88	36.02	1.11	1.18	+39.50	35	−65	1
3.73	0.88	310.74	0.85	3.22	0.01	0.01	+15.50	−976	18	1
7.10	0.50	28.90	0.56	34.60	0.66	0.68	+0.50	−194	−92	1
7.24	0.59	19.24	0.85	51.98	2.20	2.40	−10.20	54	21	1
4.44	1.08	54.84	0.50	18.23	0.16	0.17	+29.30	−48	−168	1
9.42	1.00	23.60	0.91	42.37	1.57	1.70	−	182	273	1
7.81	0.71	25.28	1.08	39.56	1.62	1.77	+49.60	206	−37	1
8.41	0.75	18.28	0.77	54.70	2.21	2.41		66	−12	1
8.05	0.61	11.51	0.90	86.88	6.30	7.37	+4.10	56	−46	1
7.06	0.58	34.85	1.33	28.69	1.05	1.14	+5.30	181	137	1
5.10	0.45	47.66	0.52	20.98	0.23	0.23	−9.90	−79	161	1
7.68	0.67	30.08	1.15	33.24	1.22	1.32	−12.00	−80	−420	3
5.67	0.58	54.92	0.45	18.21	0.15	0.15	+9.40	293	−30	1
8.69	0.83	23.32	1.31	42.88	2.28	2.55	−22.05	199	−13	1
5.94	0.75	23.57	0.92	42.43	1.59	1.72	+28.90	−80	−142	2
8.65	0.84	23.24	1.02	43.03	1.81	1.98	+42.20	−42	65	1
12.33	1.52	23.24	1.02	43.03	1.81	1.98	−	−	−	1
6.75	0.52	30.00	0.82	33.33	0.89	0.94	+32.80	95	−152	1
9.19	0.84	21.97	0.82	45.52	1.64	1.76	+22.20	14	45	1
7.88	0.74	20.43	0.98	48.95	2.24	2.47	−21.30	−44	−53	1
7.84	0.86	29.93	1.07	33.41	1.15	1.24	+1.00	114	−97	1
5.26	1.19	8.24	0.56	121.36	7.72	8.85	+78.80	109	64	1
8.10	0.71	24.55	1.14	40.73	1.81	1.98	+11.80	35	−123	1
7.22	0.58	21.16	0.68	47.26	1.47	1.57	+36.50	−69	69	1
6.86	0.53	32.23	1.01	31.03	0.94	1.00	−4.20	−37	−96	1
6.30	0.54	35.63	0.84	28.07	0.65	0.68	+53.25	−116	80	1
5.45	1.14	10.81	0.75	92.51	6.00	6.90	−40.20	43	−75	1
9.40	1.00	27.93	0.86	35.80	1.07	1.14	−	−21	−40	1
1.15	1.00	96.74	0.87	10.34	0.09	0.09	+3.30	−626	−46	1
11.18	0.82	−	−	150	2.	4.	+47.4	−35	−154	1
7.98	0.64	28.10	0.69	35.59	0.85	0.90	+42.30	−122	146	1

(Continued)

Table 16.1. (Continued)

Henry Draper Catalog Number	Names Bayer/Flamsteed/HR/ Other Designation	Position (2000.0) RA			Dec			Spectral Type
		h	m	s	Deg	′	″	
HD 66428	HIP 39417	08	03	28.7	−01	09	46	G5
HD 68988	HIP 40687	08	18	22.2	+61	27	39	G0
HD 69830	HIP 36616	08	18	23.9	−12	37	56	G8 V/K0 V
HD 70642	HIP 40952	08	21	28.1	−39	42	19	G5 IV-V/G6 IV-V
HD 72659	HIP 42030	08	34	03.2	−01	34	06	G0 V
HD 73256	HIP 42214	08	36	23.0	−30	02	15	G8/K0 V
HD 73526	HIP 42282	08	37	16.5	−41	19	09	G6 V
HD 74156	HIP 42723	08	42	25.1	+04	34	41	G0 V
HD 75289	HIP 43177	08	47	40.4	−41	44	12	G0 V/G0 Ia
HD 75732	55 rho[1] Cnc, HR 3522, HIP 43587	08	52	35.8	+28	19	51	G8 V
HD 76700	HIP 43686	08	53	55.5	−66	48	04	G6 V
HD 80606	NSV 4463, IDS 09158+5102B, HIP 45982	09	22	37.6	+50	36	13	G5
HD 81040	HIP 46076	09	23	47.1	+20	21	52	G0 V
HD 82943	HIP 47007	09	34	50.7	−12	07	46	G0 V
HD 83443	HIP 47202	09	37	11.8	−43	16	20	K0 V
HD 86081	HIP 48711	09	56	05.9	−03	48	30	F8 V
HD 88133	HIP 49813	10	10	07.7	+18	11	13	G5 IV
HD 89307	HIP 50473	10	18	21.3	+12	37	16	G0 V
HD 89744	HIP 50786	10	22	10.6	+41	13	46	F7 V/F5 V
HD 92788	HIP 52409	10	42	48.5	−02	11	02	G5 IV
HD 93083	HIP 52521	10	44	20.9	−33	34	37	K2 V
–	OGLE-TR 132	10	50	34.7	−61	57	26	–
–	OGLE-TR 113	10	52	24.4	−61	26	49	–
–	OGLE-TR 111	10	53	17.9	−61	24	20	–
–	OGLE-TR 109	10	53	40.7	−61	25	15	∼F0
–	BD-10 3166*	10	58	28.8	−10	46	13	K0 V
HD 95128	47 UMa, HR 4277, HIP 53721	10	59	28.0	+40	25	49	G0 V
HD 99109	HIP 55664	11	24	17.4	−01	31	45	K0
HD 99492	83 Leo B, Wolf 394, HIP 55848	11	26	46.3	+03	00	23	K2 V
–	GJ 436, HIP 57087	11	42	11.1	+26	42	24	M2.5
HD 101930	HIP 57172	11	43	30.1	−58	00	25	K1 V
HD 102117	HIP 57291	11	44	50.5	−58	42	13	G6 V/G5 IV
HD 102195	HIP 57370	11	45	42.3	+02	49	17	K0
HD 104985	HIP 58952	12	05	15.1	+76	54	21	G9 III/K0 III-IV
–	2MASSW J1207334-393254	12	07	33.5	−39	32	54	M8
HD 106252	HIP 59610	12	13	29.5	+10	02	30	G0 V
HD 107148	HIP 60081	12	19	13.5	−03	19	11	G5
HD 108147	HIP 60644	12	25	46.3	−64	01	20	F8/G0 V
HD 108874	HIP 61028	12	30	26.9	+22	52	47	G5
HD 109749	HIP 61595	12	37	16.4	−40	48	44	G3 V
HD 111232	HIP 62534	12	48	51.8	−68	25	31	G5 V/G8 V
HD 114386	HIP 64295	13	10	39.8	−35	03	17	K3 V
HD 114762	HIP 64426	13	12	19.7	+17	31	02	F9 V
HD 114783	HIP 64457	13	12	43.8	−02	15	54	K0 III
HD 114729	HIP 64459	13	12	44.3	−31	52	24	G3 V
HD 117176	70 Vir, HR 5072, HIP 65721	13	28	25.8	+13	46	44	G2.5 Va/G5 V
HD 117207	HIP 65808	13	29	21.1	−35	34	16	G5 IV/G8 IV-V
HD 117618	HIP 66047	13	32	25.6	−47	16	17	G0 V/G2 V
HD 118203	HIP 66192	13	34	02.5	+53	43	43	K0
HD 120136	4 tau Boo, HR 5185, HIP 67275	13	47	15.7	+17	27	25	F6 IV/F7 V
HD 121504	HIP 68162	13	57	17.2	−56	02	24	G2 V/G5 IV
HD 122430	HIP 68581	14	02	22.8	−27	25	45	K0 III/K3 III
HD 128311	HIP 71395	14	36	00.6	+09	44	47	K0 III
HD 130322	HIP 72339	14	47	32.7	−00	16	53	K0 III
HD 134987	23 Lib, HR 5657, HIP 74500	15	13	28.7	−25	18	34	dG4/G5 V
HD 136118	HIP 74948	15	18	55.5	−01	35	33	F8 V

Stellar Properties								PM		
					d error			RA	Dec	No of
V	B-V	p	p error	d			V_r			Planets
		mas	mas	pcs	−	+	km/s		mas	
8.25	0.71	18.17	1.24	55.04	3.52	4.03	+44.30	−69	−208	1
8.21	0.62	17.00	0.96	58.82	3.14	3.52	−69.70	128	32	1
5.95	0.79	79.48	0.77	12.58	0.12	0.12	+304.00	279	−989	3
7.18	0.71	34.77	0.60	28.76	0.49	0.51	+48.10	−202	226	1
7.48	0.57	19.47	1.03	51.36	2.58	2.87	−18.40	−114	−98	1
8.07	0.76	27.38	0.77	36.52	1.00	1.06	+29.50	−181	66	1
9.00	0.69	10.57	1.01	94.61	8.25	10.00	+26.10	−60	162	2
7.62	0.54	15.49	1.10	64.56	4.28	4.93	+3.70	25	−200	2
6.36	0.58	34.55	0.56	28.94	0.46	0.48	+14.00	−21	−228	1
5.95	0.87	79.80	0.84	12.53	0.13	0.13	+26.60	−484	−234	5
8.13	0.75	16.75	0.66	59.70	2.26	2.45	+36.60	−283	121	1
8.93	0.72	17.13	5.79	58.38	14.75	29.81	+3.30	47	7	1
7.74	0.63	30.71	1.24	32.56	1.26	1.37	+48.90	209	42	1
6.54	0.59	36.42	0.84	27.46	0.62	0.65	+8.10	2	−174	2
8.24	0.79	22.97	0.90	43.54	1.64	1.78	+27.61	22	−121	1
8.74	0.61	10.97	1.21	91.16	9.06	11.30	−	−67	17	1
8.06	0.82	13.43	1.16	74.46	5.92	7.04	−3.53	−13	−264	1
7.06	0.64	32.38	0.99	30.88	0.92	0.97	+22.60	−273	−39	1
5.74	0.49	25.65	0.70	38.99	1.04	1.09	−6.50	−120	−139	1
7.10	0.50	30.94	0.99	32.32	1.00	1.07	−5.00	−13	−223	1
8.33	0.94	34.60	1.00	28.90	0.81	0.86	−	−93	−151	1
I=15.7	−	−	−	1500	−	−	−	−	−	1
I=14.1	V-I=1.7	−	−	−	−	−	−	−	−	1
I=15.5	V-I=1.4	−	−	850	40	40	−	−	−	1
15.8	V-I=0.8	−	−	2590	250	250	−	−	−	1
10.00	0.90	−	−	80	10	10	−	−183	−5	1
5.10	0.56	71.04	0.66	14.08	0.13	0.13	+12.60	−316	55	2
8.80	0.70	16.54	1.31	60.46	4.44	5.20	−	−180	−160	1
7.57	1.01	55.59	3.31	17.99	1.01	1.14	+1.70	−730	191	1
10.68	1.52	97.73	2.27	10.23	0.23	0.24	+10.00	896	−814	1
8.21	0.91	32.79	0.96	30.50	0.87	0.92	−	15	347	1
7.47	0.70	23.81	0.83	42.00	1.41	1.52	+48.90	−63	−70	1
8.07	0.83	34.51	1.16	28.98	0.94	1.01	+2.10	−190	−111	1
5.80	1.02	9.80	5.20	102.04	35.37	115.35	−19.80	142	−92	1
J=13.0	−	−	−	53	6	6	−	−	−	1
7.36	0.64	26.71	0.94	37.44	1.27	1.37	+15.40	24	−279	1
8.02	0.66	19.51	1.00	51.26	2.50	2.77	−	−56	−48	1
6.99	0.50	25.93	0.69	38.57	1.00	1.05	−5.10	300	−1	1
8.76	0.71	14.59	1.24	68.54	5.37	6.37	−30.70	129	−89	2
8.20	0.70	16.94	1.91	59.03	5.98	7.50	−13.70	−158	−5	1
7.61	0.68	34.63	0.80	28.88	0.65	0.68	+102.20	28	113	1
8.80	0.90	35.66	1.32	28.04	1.00	1.08	−	−138	−325	1
7.30	0.55	24.65	1.44	40.57	2.24	2.52	+49.90	−583	−2	1
7.57	0.91	48.95	1.06	20.43	0.43	0.45	−12.80	−138	10	1
6.69	0.62	28.57	0.97	35.00	1.15	1.23	+64.70	−202	−308	1
5.00	0.69	55.22	7.30	18.11	2.11	2.76	+4.90	−235	−576	1
7.30	0.60	30.29	0.92	33.01	0.97	1.03	−17.90	−205	−72	1
7.18	0.56	26.30	0.93	38.02	1.30	1.39	+0.90	25	−125	1
8.07	0.65	11.29	0.82	88.57	6.00	6.94	−	−88	−78	1
4.50	0.48	64.12	0.70	15.60	0.17	0.17	−15.60	−480	54	1
7.54	0.59	22.54	0.91	44.37	1.72	1.87	+19.30	−251	−84	1
5.48	1.35	7.51	0.73	133.16	11.80	14.34	+0.00	−32	−4	1
7.51	0.99	60.35	0.99	16.57	0.27	0.28	−9.60	205	−250	2
8.05	0.75	33.60	1.51	29.76	1.28	1.40	−12.50	−130	−141	1
6.45	0.70	38.98	0.98	25.65	0.63	0.66	+3.40	−399	−75	1
6.94	0.48	19.13	0.85	52.27	2.22	2.43	−3.60	−124	24	1

(Continued)

Table 16.1. (Continued)

Henry Draper Catalog Number	Names Bayer/Flamsteed/HR/ Other Designation	RA h	m	s	Dec Deg	′	″	Spectral Type
–	GJ 581, HO Lib, Wolf 562, HIP 74995	15	19	26.8	−07	43	20	M3
HD 137759	12 i Dra, HR 5744, HIP 75458	15	24	55.8	+58	57	58	K2 III
HD 137510	HR 5740, HIP 75535	15	25	53.3	+19	28	51	G0 IV-V
	GQ Lup	15	49	12.1	−35	39	04	K7e V
HD 330075	HIP 77517	15	49	37.7	−49	57	49	G5
HD 141937	HIP 77740	15	52	17.5	−18	26	10	G0 V/G2/G3 V
HD 142415	HIP 78169	15	57	40.8	−60	12	01	G0/G1 V
HD 143761	15 rho Crb, HR 5968, , HIP 78459	16	01	02.7	+33	18	13	G2 V/G0 Va
–	XO-1, GSC 02041−01657	16	02	11.8	+28	10	10	–
HD 142022A	HIP 79242	16	10	15.0	−84	13	54	K0 V/G8
HD 145675	14 Her, HIP 79248	16	10	24.3	+43	49	04	K0 V/dK1
HD 147506	HAT-P-2,HIP 80076	16	20	36.4	+41	02	53	F8
HD 147513	HR 6094, NSV 7680, HIP 80337	16	24	01.3	−39	11	35	dG5
HD 149026	HIP 80838	16	30	29.6	+38	20	50	G0
HD 150706	HIP 80902	16	31	17.6	+79	47	23	dG3/G0
HD 149143	HIP 81022	16	32	51.1	+02	05	05	G0
HD 154345	HIP 83389	17	02	36.4	+47	04	55	G8 V
HD 154857	HIP 84069	17	11	15.7	−56	40	51	F2 V/G5 V
HD 160691	mu Arae, HR 6585, HIP 86796	17	44	08.7	−51	50	03	G3 IV-V/G5 V
–	OGLE 2005-BLG 071	17	50	09.8	−34	53	42	–
HD 162020	HIP 87330	17	50	38.4	−40	19	06	K2 V:
–	OGLE-Tr 10	17	51	28.3	−29	52	35	G2 V
–	TrES-3, GSC 03089−00929	17	52	07.0	+37	32	46	G
–	TrES-4, GSC 02620−00648	17	53	13.1	+37	12	42	F/G
–	OGLE 2005-BLG-390L	17	51	28.3	−29	52	35	G2 V
–	OGLE-Tr 56	17	56	35.5	−29	32	21	–
HD 164922	ADS 11003A, HIP 88348	18	02	30.9	−26	18	47	K0 V
–	MOA 2003-BLG-53, EWS 2003-BLG-235	18	05	16.4	−28	53	42	G/K V
–	EWS 2005-BLG-169	18	06	05.3	−30	43	58	–
HD 168443	HIP 89844	18	20	03.9	−09	35	45	G5 IV
HD 168746	HIP 90004	18	21	49.8	−11	55	22	G5 IV
HD 169830	HR 6907, HIP 90485	18	27	49.5	−29	49	01	G0 V/F9 V
–	Denis J 184504.9 -635747, SCR J 1845-6357	18	45	05.1	−63	57	47	M8.5 V
–	TrES 1, GSC 02625−01324	19	04	09.8	+36	37	58	K0 V
HD 177830	HIP 93746	19	05	20.8	+25	55	14	dK2/K0
–	TrES-2, GSC 03549−02811	19	07	14.0	+49	18	59	G0 V
HD 178911B	HIP 94075	19	09	03.1	+34	35	59	dG1/G5
HD 179949	HIP 94645	19	15	33.2	−24	10	46	F8 V
HD 183263	HIP 95740	19	28	24.6	+08	21	29	G2 IV
HD 185269	HIP 96507	19	37	11.7	+28	30	00	G0 IV
HD 186427	16 Cyg, HR 7503, ADS 12815 A, HIP 96901	19	41	52.0	+50	31	03	GI.5 V
HD 187123	HIP 97336	19	46	58.1	+34	25	10	G5 IV
HD 187085	HIP 97546	19	49	33.4	−37	46	50	F8 V/G0 V
HD 188015	HIP 97769	19	52	04.5	+28	06	10	G5 IV
HD 188753A	HIP 98001	19	54	58.4	+41	52	17	dK0
HD 189733	HIP 98505	20	00	43.7	+22	42	39	G5 IV
HD 190228	HIP 98714	20	03	00.8	+28	18	25	G5 IV
HD 190360	GJ777A, HR 7670, HIP 98767	20	03	37.4	+29	53	49	G6 IV
HD 192263	ADS 13547 A, HIP 99711	20	13	59.8	−00	52	01	K2 V
HD 195019	ADS 13886 AB, HIP 100970	20	28	18.6	+18	46	10	G3 IV-V
–	WASP-2	20	30	54	+06	25	46	K1 V
HD 196050	HIP 101806	20	37	51.7	−60	38	04	G3 V
HD 196885	HR 7907, IDS 20351+054A	20	39	51.9	+11	14	59	F8 V/F* IV
HD 202206	HIP 104903	21	14	57.8	−20	47	21	G5 IV/G6 V
HD 208487	HIP 108375	21	57	19.8	−37	45	49	F8 V/G2 V:

Stellar Properties								PM		
					d error					
V	B-V	p mas	p error mass	d pcs	−	+	V_r km/s	RA mas	Dec	No of Planets
10.56	1.61	159.52	2.27	6.27	0.09	0.09	−9.50	−1225	−100	1
3.31	1.18	31.92	0.51	31.33	0.49	0.51	−10.70	−8	17	1
6.26	0.60	23.95	0.94	41.75	1.58	1.71	−3.30	−53	−7	1
11.40	0.96	−	−	140:	50:	50:	−	−27	−14	1
9.36	0.94	19.92	1.49	50.20	3.49	4.06	−	−236	−94	1
7.25	0.60	29.89	1.08	33.46	1.17	1.25	−3.30	97	24	1
7.34	0.60	28.93		34.57	0.00	0.00	−12.00	−114	−102	1
5.40	0.61	57.38	0.71	17.43	0.21	0.22	+18.40	−197	−773	1
11.19	0.66	−	−	200	20	20	−	−20	15	1
7.69	0.78	27.88	0.68	35.87	0.85	0.90	−10.50	−338	−31	1
6.67	0.90	55.11	0.59	18.15	0.19	0.20	−5.50	133	−298	1
8.71	0.41	−	−	−	−	−	−19.9	−	−	1
5.38	0.60	77.69	0.86	12.87	0.14	0.14	+10.10	73	3	1
8.16	0.56	12.68	0.79	78.86	4.63	5.24	−18.10	−77	53	1
7.03	0.57	36.73	0.56	27.23	0.41	0.42	−14.00	96	−88	1
7.90	0.63	15.75	1.07	63.49	4.04	4.63	−	−10	−85	1
6.74	0.76	55.37	0.55	18.06	0.18	0.18	−46.20	124	855	1
7.25	0.68	14.59	0.91	68.54	4.02	4.56	+27.90	87	−55	1
5.15	0.70	65.46	0.80	15.28	0.18	0.19	−9.00	−15	−191	3
I=19.5	−	−	−	2900	−	−	−	−	−	1
9.18	0.96	31.99	1.48	31.26	1.38	1.52	−27.60	21	−25	1
14.9	V-I=0.85	−	−	1500	−	−	−	−	−	1
12.40	0.71	−	−	−	−	−	+9.58	−22.5	32	1
11.59	0.52	−	−	440	60	60	−	−6.5	−24	1
I=15.7	−	−	−	6500	1000	1000	−	−	−	1
15.3	V-I=1.26	−	−	1500	−	−	−	−	−	1
6.99	0.80	45.61	0.71	21.93	0.34	0.34	+22.8	390	−602	1
I=21.4	−	−	−	5200	2900	200	−	−	−	1
I=20.4	−	−	−	2700	−	−	−	−	−	1
6.92	0.70	26.40	0.85	37.88	1.18	1.26	−48.85	−92	−224	2
7.95	0.69	23.19	0.96	43.12	1.71	1.86	−25.60	−22	−69	1
5.91	0.48	27.53	0.91	36.32	1.16	1.24	−17.40	−84	15	2
17.40	1.65	282	23	3.55	0.27	0.31	−	2444	696	1
11.30	0.90	−	−	156	6	6	−	−40	−22	1
7.18	1.09	16.94	0.76	59.03	2.53	2.77	−74.00	−41	−52	1
11.41	0.62	−	−	220	10	10	−	4	−3	1
7.98	0.73	21.40	4.95	46.73	8.78	14.06	−40.37	65	192	1
6.25	0.51	36.97	0.80	27.05	0.57	0.60	−25.50	115	−102	1
7.86	0.63	18.93	1.06	52.83	2.80	3.13	−50.70	−18	−33	1
6.683	0.56	21.11	0.74	47.37	1.60	1.72	+0.60	−32	−81	1
5.96	0.64	46.25	0.50	21.62	0.23	0.24	−25.60	−148	−159	1
7.86	0.61	20.87	0.71	47.92	1.58	1.69	−17.60	143	−123	1
7.22	0.53	22.23	1.14	44.98	2.19	2.43	+14.90	8	−104	1
8.22	0.70	19.00	0.95	52.63	2.51	2.77	+2.60	54	−91	1
7.43	0.79	22.31	0.78	44.82	1.51	1.62	−23.50	−53	284	1
7.67	0.93	51.94	0.87	19.25	0.32	0.33	−2.70	−2	−251	1
7.31	0.76	16.10	0.81	62.11	2.98	3.29	−48.70	105	−70	1
5.71	0.73	62.92	0.62	15.89	0.16	0.16	−43.30	683	−524	2
7.79	0.94	50.27	1.13	19.89	0.44	0.46	−11.30	−63	262	1
11.98	−	−	−	−	−	−	−	−	−	1
6.91	0.64	26.77	0.89	37.36	1.20	1.28	−92.70	349	−57	1
7.60	0.50	21.31	0.91	46.93	1.92	2.09	+60.90	−191	−64	1
6.40	0.51	30.31	0.81	32.99	0.86	0.91	−28.30	47	83	1
8.08	0.69	21.58	1.14	46.34	2.33	2.58	−	−38	−120	2
7.48	0.52	22.73	1.01	43.99	1.87	2.05	+5.30	101	−118	1

(Continued)

Table 16.1. (Continued)

Henry Draper Catalog Number	Names Bayer/Flamsteed/HR/ Other Designation	Position (2000.0)						Spectral Type
		RA			Dec			
		h	m	s	Deg	′	″	
HD 209458	V376 Peg, HIP 108859	22	03	10.8	+18	53	04	F8 V
HD 210277	HIP 109378	22	09	29.9	−07	32	55	G0 V
HD 212301	HIP 110852	22	27	30.9	−77	43	05	F8 V
HD 213240	HIP 111143	22	31	00.4	−49	26	00	G4 IV/G0/G1 IV
−	GJ 876, IL Aqr, HIP 113020	22	53	16.7	−14	15	49	M4 V
HD 216435	tau[1] Gru, HR 8700, HIP 113044	22	53	37.9	−48	35	54	G3 IV/G0 V
HD 216437	rho Ind, HR 8701, HIP 113137	22	54	39.5	−70	04	25	G1 V/G2.5 IV
HD 216770	HIP 113238	22	55	53.7	−26	39	32	K1 V
HD 217014	51 Peg, HR 8729, NSV 14374, HIP 113357	22	57	28.0	+20	46	08	G2.5 IVa
−	HAT-P-1, ADS 16402B	22	57	46.8	+38	40	30	F8
HD 217107	HR 8734, HIP 113421	22	58	15.5	−02	23	43	G5/G8 IV
HD 219449	91 psi[1] Aqr, HR 8841, ADS1663A,	23	15	53.5	−09	05	16	K0 III
HD 222404	5 gamma Cep, HR 8974, NSV 14656, HIP 116727	23	39	20.8	+77	37	56	K III-IV/K1 IV
HD 222582	HIP 116906	23	41	51.5	−05	59	09	G5 IV
HD 224693	HIP 118319	23	59	53.8	−22	25	41	G2 V

Stellar Properties								PM			
					d error						
V	B-V	p	p error	d			V_r	RA		Dec	No of
		mas	mas	pcs	−	+	km/s		mas		Planets
7.65	0.53	21.24	1.00	47.08	2.12	2.33	−14.80	29		−18	1
6.63	0.71	46.97	0.79	21.29	0.35	0.36	−24.10	85		−450	1
7.77	0.51	18.97	0.73	52.71	1.95	2.11	+4.80	76		−92	1
6.80	0.61	24.54	0.81	40.75	1.30	1.39	−0.50	−135		−194	1
10.17	1.60	212.69	2.10	4.70	0.05	0.05	+8.70	960		−676	3
6.03	0.63	30.04	0.73	33.29	0.79	0.83	−1.10	217		−81	1
6.06	0.63	37.71	0.58	26.52	0.40	0.41	−3.00	−43		73	1
8.10	0.82	26.39	1.06	37.89	1.46	1.59	+30.70	229		−178	1
5.49	0.67	65.10	0.76	15.36	0.18	0.18	−31.20	208		61	1
10.6	0.6	−	−	139	19	22	−	29.3		−51.0	1
6.18	0.72	50.71	0.75	19.72	0.29	0.30	−14.00	70		−53	2
4.21	1.11	21.97	0.89	45.52	1.77	1.92	−26.40	369		−17	1
3.23	1.03	72.50	0.52	13.79	0.10	0.10	−42.40	−49		127	1
7.70	0.60	23.84	1.11	41.95	1.87	2.05	+11.50	−145		−111	1
8.23	0.62	10.63	1.17	94.07	9.33	11.63	+1.50	149		28	1

Total no. of suspected planets   209
No. of stars with suspected planets   181

Table 16.2. Extrasolar planets

System Designation	Comp	Period (d)	P error (d)	a au	a error (au)	e	e error	M/M $(\sin i)^*$ (M_J)	M^* err (M_J)	i deg	i err deg
HR 6, HD 142	b	350.3	3.6	1.045	0.061	0.26	0.18	1.31	0.18	–	–
HD 1237, GJ 3021	b	133.71	0.20	0.495	0.029	0.511	0.017	3.37	0.49	–	–
WASP-1	b	2.51995	0.00001	0.0379	0.0042	0	–	0.79	0.10	>86.1	–
HD 2039	b	1120	23	2.23	0.13	0.715	0.046	6.11	0.82	–	–
HD 2638	b	3.44420	0.00020	0.0436	0.0025	0	0	0.477	0.068	–	–
54 Psc, HD 3651	b	62.206	0.021	0.296	0.017	0.618	0.051	0.227	0.023	–	–
HD 4208	b	828.0	8.1	1.650	0.096	0.052	0.040	0.804	0.073	–	–
HD 4308	b	15.560	0.020	0.1179	0.0068	0.000	0.010	0.047	0.007	–	–
HD 4203	b	431.88	0.85	1.164	0.067	0.519	0.027	2.07	0.18	–	–
HD 6434	b	21.9980	0.0090	0.1421	0.0082	0.170	0.030	0.397	0.059	–	–
HD 8574	b	225.0	1.1	0.759	0.044	0.370	0.082	1.96	0.22	–	–
ups And, HD 9826	b	4.617113	0.000082	0.0595	0.0034	0.023	0.018	0.687	0.058	–	–
ups And, HD 9826	c	241.23	0.30	0.832	0.048	0.262	0.021	1.98	0.17	–	–
ups And, HD 9826	d	1290.1	8.4	2.54	0.15	0.258	0.032	3.95	0.33	–	–
HR 506, HD 10647	b	1003	56	2.03	0.15	0.16	0.22	0.93	0.18	–	–
109 Psc, HD 10697	b	1076.4	2.4	2.16	0.12	0.1023	0.0096	6.38	0.53	–	–
eta^2 Hyi, HD 11977	b	711.0	8.0	1.94	0.11	0.40	0.07	6.5	1.2	–	–
HD 11964	b	2110	270	3.34	0.40	0.06	0.17	0.61	0.10	–	–
HD 12661	b	262.53	0.27	0.831	0.048	0.361	0.011	2.34	0.19	–	–
HD 12661	c	1679	29	2.86	0.17	0.017	0.029	1.83	0.16	–	–
HD 13189	b	471.6	6.0	1.85	0.35	0.27	0.06	14	6	–	–
GJ 86, A, HD 13445	b	15.76491	0.00039	0.1130	0.0065	0.0416	0.0072	3.91	0.32	–	–
79 Cet, HD 16141	b	75.523	0.055	0.363	0.021	0.252	0.052	0.260	0.028	–	–
iota Hor, HD 17051	b	302.8	2.3	0.930	0.054	0.14	0.13	2.08	0.26	–	–
BD +20 518, HIP 14810	b	6.6742	0.0020	0.0692	0.0040	0.1480	0.0060	3.91	0.55	–	–
BD +20 518, HIP 14810	c	95.2847	0.0020	0.407	0.023	0.409	0.006	0.76	0.12	–	–
94 Cet, HD 19994	b	535.7	3.1	1.428	0.083	0.300	0.040	1.69	0.26	–	–
HD 20367	b	469.5	9.3	1.246	0.075	0.320	0.090	1.17	0.23	–	–
HD 20782	b	585.860	0.030	1.364	0.079	0.925	0.030	1.78	0.34	–	–
eps Eri, HD 22049	b	2500	350	3.38	0.43	0.25	0.23	1.06	0.16	–	–
HD 23079	b	730.6	5.7	1.596	0.093	0.102	0.031	2.45	0.21	–	–
HD 23596	b	1565	21	2.83	0.17	0.292	0.023	7.80	1.10	–	–
eps Ret, HD 27442	b	428.1	1.1	1.271	0.073	0.060	0.043	1.56	0.14	–	–
HD 27894	b	17.9910	0.0070	0.1221	0.0071	0.0490	0.0080	0.618	0.088	–	–
HD 28185	b	383.0	2.0	1.031	0.060	0.070	0.040	5.72	0.93	–	–
HD 30177	b	2770	100	3.95	0.26	0.193	0.025	10.45	0.88	–	–
HD 33283	b	18.179	0.007	0.145	–	0.48	0.05	0.33	–	–	–
HD 33636	b	2127.7	8.2	3.27	0.19	0.4805	0.0060	9.28	0.77	–	–
HD 33564	b	388.0	3.0	1.124	0.065	0.340	0.020	9.1	1.3	–	–
HIP HD 37124	b	154.46	–	0.529	0.031	0.055	–	0.64	0.11	–	–
HIP HD 37124	c	2295.00	–	3.19	0.18	0.200	–	0.683	0.088	–	–
HIP HD 37124	d	843.60	–	1.639	0.095	0.140	–	0.624	0.063	–	–
pi Men, HD 39091	b	2151	85	3.38	0.22	0.6405	0.0072	10.27	0.84	–	–
HD 37605	b	54.23	0.23	0.261	0.015	0.737	0.010	2.86	0.41	–	–
HD 38529	b	14.3093	0.0013	0.1313	0.0076	0.248	0.023	0.85	0.07	–	–
HD 38529	c	2165	14	3.74	0.22	0.3506	0.0085	13.2	1.1	–	–
HD 41004A	b	963	38	1.70	0.11	0.74	0.20	2.6	1.8	–	–
HD 41004B	b	1.328300	0.000012	0.0177	0.0010	0.081	0.012	18.4	2.6	–	–
HD 40979	b	263.84	0.71	0.855	0.049	0.269	0.034	3.83	0.36	–	–
AB Pic	b	–	–	2.75	0.05	–	–	13.50	–	–	–

R (R_J)	R err (R_J)	T_0 (mod. JDN)	T_0 err (d)	T_t (mod. JDN)	T_t err (d)	Comment/ method*	Comments/references
–	–	11963	43	11737	25	–	Butler et al. (2006)
–	–	11545.86	0.64	–	–	–	Naef (2001a)
1.44	0.04	–	–	14005.75196	0.00045	T	Collier Cameron et al. (2007), Stempels et al. (2007)
–	–	12041	13	10992	26	–	Butler et al. (2006)
–	–	13323.206	0.002	–	–	–	Moutou et al. (2005)
–	–	12189.83	0.68	12176.3	1.9	–	Butler et al. (2006)
–	–	11040	120	10440	16	–	Butler et al. (2006); Vogt et al. (2002)
–	–	13311.7	2.0	–	–	–	Udry et al. (2006)
–	–	11918.9	2.7	11558.7	7.2	–	Butler et al. (2006); Vogt et al. (2002)
–	–	11490.8	0.6	–	–	–	Mayor et al. (2004)
–	–	11475.6	5.5	11504.8	7.3	–	Butler et al. (2006), Perrier et al. (2003)
–	–	11802.64	0.71	11802.966	0.033	–	Butler et al. (2006), Naef et al. (2004)
–	–	10158.1	4.5	10063.9	3.8	–	Butler et al. (2006), Naef et al. (2004)
–	–	8827	30	8127	39	–	Butler et al. (2006), Naef et al. (2004)
–	–	10960	160	10221	83	–	Butler et al. (2006), Mayor & Santos (2003)
–	–	10396	29	10350.4	5.6	–	Butler et al. (2006)
–	–	11420	–	–	–	–	Setiawan et al. (2005); $M_{star} = 1.91$ M_{sun} assumed
–	–	12290	420	11870	120	–	Butler et al. 2006
–	–	10214.1	2.9	10046.1	2.5	–	Butler et al. 2006
–	–	12130	330	12368	22	–	Butler et al. 2006
–	–	52327.9	20.2	–	–	BD?	Hatzes et al. 2005
–	–	11903.36	0.59	11895.551	0.076	–	Butler et al. 2006, Queloz et al. 2000
–	–	10338.0	3.0	10344.1	1.6	–	Butler et al. 2006
–	–	11227	46	10998	19	–	Butler et al. 2006, Naef 2001b
–	–	13694.588	0.040	11737	25	–	Wright et al. 2007
–	–	13679.585	0.040	11737	25	–	Wright et al. 2007
–	–	10944	12	–	–	–	Mayor et al. 2004
–	–	11860	18	–	–	–	Udry et al. 2003a
–	–	11687.1	2.5	–	–	–	Jones et al. 2006
–	–	8940	520	9330	200	–	Butler et al. 2006, Hatzes et al. 2000
–	–	10492	37	10551	14	–	Butler et al. 2006
–	–	11604	15	–	–	–	Perrier et al. 2003
–	–	10836	55	10692.2	8.6	–	Butler et al. 2006
–	–	13275.46	0.48	–	–	–	Moutou et al. 2005
–	–	11863	26	–	–	–	Santos et al. 2001
–	–	11437	72	11738	16	–	Butler et al. 2006
–	–	13017.6	0.3	–	–	–	Johnson et al. (2006a)
–	–	11205.8	6.4	9396	12	BD?	Butler et al. 2006; Perrier et al. 2003, Vogt et al. 2002
–	–	12603.0	8.0	–	–	–	Galland et al. 2005
–	–	10000.11	–	–	–	–	Vogt et al. 2005, Udry et al. 2003a
–	–	9606.0	–	–	–	–	Vogt et al. 2005, Udry et al. 2003a
–	–	9409.4	–	–	–	–	Vogt et al. 2005, Udry et al. 2003a
–	–	7820	170	5920	260	BD?	Butler et al. 2006; Jones et al. 2002
–	–	12994.27	0.45	–	–	–	Cochran et al. 2004
–	–	9991.59	0.23	9991.56	0.17	–	Butler et al. 2006
–	–	10085	15	10319	13	–	Butler et al. 2006
–	–	12425	37	–	–	–	Zucker et al. 2004
–	–	12434.88	0.03	–	–	BD?	Zucker et al. 2004
–	–	10748.1	8.6	10561.2	6.4	–	Butler et al. 2006
–	–	–	–	–	–	I	Chauvin et al. 2005

(Continued)

Table 16.2. (Continued)

System Designation	Comp	Period (d)	P error (d)	a au	a error (au)	e	e error	M/M $(\sin i)^*$ (M_J)	M^* err (M_J)	i deg	i err deg
HD 45350	b	967.0	6.2	1.96	0.11	0.798	0.053	1.96	0.17	–	–
HD 46375	b	3.023573	0.000065	0.0398	0.0023	0.063	0.026	0.226	0.019	–	–
HD 47536	b	712.13	0.31	1.613	0.093	0.200	0.080	5.20	0.99	–	–
HD 49674	b	4.94737	0.00098	0.0580	0.0034	0.087	0.095	0.105	0.011	–	–
HD 50499	b	2480	110	3.87	0.26	0.14	0.20	1.75	0.53	–	–
HD 50554	b	1224	12	2.28	0.13	0.444	0.038	4.46	0.48	–	–
HD 52265	b	119.290	0.086	0.504	0.029	0.325	0.065	1.09	0.11	–	–
HD 59686	b	303	–	0.911	–	0.00	–	5.25	–	–	–
HD 63454	b	2.817822	0.000095	0.0363	0.0021	0	–	0.385	0.055	–	–
78 beta Gem, HD 62509	b	589.64	–	1.64	–	0.02	–	2.30	–	–	–
XO-2	b	2.615857	0.000005	0.0369	0.0002	0	–	0.57	0.06	88.9	0.7
HD 65216	b	613	11	1.374	0.082	0.410	0.060	1.22	0.19	–	–
HD 66428	b	1973	31	3.18	0.19	0.465	0.030	2.82	0.27	–	–
HD 68988	b	6.27711	0.00021	0.0704	0.0041	0.1249	0.0087	1.86	0.16	–	–
HD 69830	b	8.667	0.030	0.0785	–	0.10	0.04	0.0322	–	–	–
HD 69830	c	31.56	0.040	0.186	–	0.13	0.06	0.0374	–	–	–
HD 69830	d	197	3	0.633	–	0.07	0.07	0.0573	–	–	–
HD 70642	b	2068	39	3.23	0.19	0.034	0.043	1.97	0.18	–	–
HD 72659	b	3630	230	4.77	0.37	0.269	0.038	3.30	0.29	–	–
HD 73256	b	2.54858	0.00016	0.0371	0.0021	0.029	0.020	1.87	0.27	–	–
HD 73526	b	187.499	0.030	0.651	0.038	0.390	0.054	2.04	0.29	–	–
HD 73526	c	376.879	0.090	1.037	0.060	0.400	0.054	2.26	0.27	–	–
HD 74156	b	51.643	0.011	0.290	0.017	0.6360	0.0091	1.80	0.26	–	–
HD 74156	c	2025	11	3.35	0.19	0.583	0.039	6.00	0.95	–	–
HD 75289	b	3.509267	0.000064	0.0482	0.0028	0.034	0.029	0.467	0.041	–	–
55 rho[1] Cnc	b	14.6516	0.0007	0.115	0.0000	0.014	0.008	0.824	0.007	–	–
55 rho[1] Cnc	c	44.345	0.007	0.240	0.0000	0.086	0.052	0.169	0.008	–	–
55 rho[1] Cnc	d	5218	230	5.77	0.11	0.03	0.03	3.84	0.08	–	–
55 rho[1] Cnc	e	2.8171	0.0001	0.038	0.000	0.07	0.06	0.034	0.004	–	–
55 rho[1] Cnc	f	260.0	1.1	0.781	0.007	<0.2	0.2	0.14	0.04	–	–
HD 76700	b	3.97097	0.00023	0.0511	0.0030	0.095	0.075	0.233	0.024	–	–
HD 80606	b	111.4487	0.0032	0.468	0.027	0.9349	0.0023	4.31	0.35	–	–
HD 81040	b	1001.7	7.0	1.94	0.11	0.526	0.042	6.9	1.1	–	–
HD 82943	b	219.50	0.13	0.752	0.043	0.39	0.26	1.81	0.21	–	–
HD 82943	c	439.2	1.8	1.194	0.069	0.020	0.098	1.74	0.19	–	–
HD 83443	b	2.985698	0.000057	0.0406	0.0023	0.012	0.023	0.398	0.035	–	–
HD 86081	b	2.13750	0.00020	0.0346	–	0.008	0.004	1.5	–	–	–
HD 88133	b	3.41587	0.00059	0.0472	0.0027	0.133	0.072	0.299	0.033	–	–
HD 89307	b	2900	1100	3.9	1.3	0.01	0.16	2.61	0.37	–	–
HD 89744	b	256.80	0.13	0.934	0.054	0.6770	0.0072	8.58	0.71	–	–
HD 92788	b	325.81	0.26	0.965	0.056	0.334	0.011	3.67	0.30	–	–
HD 93083	b	143.58	0.60	0.477	0.028	0.140	0.030	0.368	0.054	–	–
OGLE-TR-132	b	1.689868	0.000003	0.0306	–	0	–	1.14	0.12	85	–
OGLE-TR-113	b	1.4324752	0.0000015	0.0232	0.0038	0	–	1.32	0.19	>86.7	–
OGLE-TR-111	b	4.014442	–	0.047	0.001	0	–	0.52	0.13	>~86.5	–
OGLE-TR-109	b	0.589128	–	–	–	0	–	14	8	77	5
BD-10 3166	b	3.48777	0.00011	0.0452	0.0026	0.019	0.023	0.458	0.039	–	–
47 UMa, HD 95128	b	1089.0	2.9	2.13	0.12	0.061	0.014	2.63	0.23	–	–

R (R_J)	R err (R_J)	T_0 (mod. JDN)	T_0 err (d)	T_t (mod. JDN)	T_t err (d)	Comment/ method*	Comments/references
–	–	11822	13	10894	16	–	Butler et al. 2006, Endl et al. 2006
–	–	11071.53	0.19	11071.359	0.037	–	Butler et al. 2006
–	–	11599	22	–	–	–	Setiawan et al. 2003
–	–	11882.38	0.88	11880.00	0.18	–	Butler et al. 2006
–	–	11230	230	–	–	–	Vogt et al. 2005
–	–	10646	16	10767	18	–	Butler et al. 2006; Perrier et al. 2003, Fischer et al. 2002b
–	–	10833.7	4.2	–	–	–	Butler et al. 2006, Naef 2001b
–	–	–	–	–	–	–	Mitchell et al. (2003)
–	–	13111.129	0.005	–	–	–	Moutou et al. 2005
–	–	–	–	–	–	–	Hatzes et al. 2006
0.98	0.02	–	–	14147.7490	0.0002	T	Burke, et al. 2007
–	–	10762	25	–	–	–	Mayor et al. 2004
–	–	12139	16	12012.1	7.1	–	Butler et al. 2006
–	–	11548.84	0.16	11549.663	0.040	–	Butler et al. 2006; Vogt et al. 2002
–	–	13496.8	0.6	–	–	–	Lovis et al. 2006
–	–	13469.6	2.8	–	–	–	Lovis et al. 2006
–	–	13358.0	34.0	–	–	–	Lovis et al. 2006
–	–	11350	380	10707	48	–	Butler et al. 2006
–	–	11673	89	10060	240	–	Butler et al. 2006
–	–	12500.18	0.28	–	–	–	Udry et al. 2003b
–	–	10038	15	–	–	–	Tinney et al. 2006
–	–	10184.5	8.6	–	–	–	Tinney et al. 2006
–	–	11981.321	0.091	–	–	–	Naef et al. 2004
–	–	10901	10	–	–	–	Naef et al. 2004
–	–	10830.34	0.48	10829.872	0.038	–	Butler et al. 2006, Udry et al. 2000
–	–	10002.9	1.2	–	–	–	Fischer et al. 2007
–	–	9989.3	3.3	–	–	–	Fischer et al. 2007
–	–	12500.6	230	–	–	–	Fischer et al. 2007
–	–	9999.8364	0.0001	–	–	–	Fischer et al. 2007
–	–	10080.9	1.1	–	–	–	Fischer et al. 2007
–	–	11213.32	0.67	11213.89	0.12	–	Butler et al. 2006, Fischer et al. (2002a)
–	–	13199.052	0.006	13093.11	0.09	–	Butler et al. have 4.31; Naef et al. 2001a
–	–	12504	12	–	–	–	Sozzetti et al. 2006b
–	–	–	–	–	–	–	Lee et al. 2006, Mayor et al. 2004; Butler et al. reverse b and c
–	–	–	–	–	–	–	Lee et al. 2006, Mayor et al. 2004; Butler et al. reverse b and c
–	–	11211.79	0.69	11211.565	0.025	–	Butler et al. 2006, Mayor et al. 2004.
–	–	13694.8	0.3	–	–	–	Johnson (2006a,b)
–	–	13016.31	0.32	13013.705	0.095	–	Butler et al. 2006, Fischer et al. (2005)
–	–	12520	230	12800	260	–	Butler et al. 2006
–	–	11505.33	0.39	11487.03	0.76	–	Butler et al. 2006, Korzennik et al. 2000
–	–	10759.20	2.70	10585.3	2.4	–	Butler et al. 2006, Mayor et al. 2004
–	–	13181.7	3.0	–	–	–	Lovis et al. 2005
1.18	0.07	–	–	13142.5912	0.0003	T	Silva and Cruz 2006; Bouchy et al. 2004; Gillon et al. 2007
1.09	0.09	–	–	–	–	T	Santos et al. 2006; Diaz et al. 2007
0.97	0.06	–	–	12330.44867		T	Santos et al. 2006; Pont et al. 2004; Udalski et al. 2002
0.90	0.09	–	–	12322.55993	–	T, BD?	Fernández 2006; Gallardo et al. 2005
–	–	11171.22	0.69	11168.832	0.031	–	Butler et al. 2006
–	–	10356	34	–	–	–	Butler et al. 2006; Barnes and Greenberg 2006; Naef et al. 2004

(Continued)

Table 16.2. (Continued)

System Designation	Comp	Period (d)	P error (d)	a au	a error (au)	e	e error	M/M (sin i)* (M_J)	M* err (M_J)	i deg	i err deg
47 UMa, HD 95128	c	2594	90	3.79	0.24	0.00	0.12	0.79	0.13	–	–
HD 99109	b	439.3	5.6	1.105	0.065	0.09	0.16	0.502	0.070		
83 Leo B, HD 99492	b	17.0431	0.0047	0.1232	0.0071	0.254	0.092	0.109	0.013		
GJ 436, HD 99492	b	2.643943	0.000084	0.0278	0.0016	0.207	0.052	0.0673	0.0065	–	–
HD 101930	b	70.46	0.18	0.302	0.017	0.110	0.020	0.299	0.043	–	–
HD 102117	b	20.8133	0.0064	0.1532	0.0088	0.121	0.082	0.17	0.02		
HD 102195	b	4.11434	0.00089	0.0491	–	0.060	0.030	0.488	0.015	–	–
HD 104985	b	198.20	0.30	0.779	0.045	0.03	0.02	6.33	0.91		
2MASS WJ 1207334-393254	b	–	–	41	–	–	–	5	2	–	–
HD 106252	b	1516	26	2.60	0.15	0.586	0.065	7.10	0.65	–	–
HD 107148	b	48.056	0.057	0.269	0.016	0.05	0.17	0.210	0.036	–	–
HD 108147	b	10.8985	0.0045	0.1020	0.0059	0.53	0.12	0.26	0.04	–	–
HD 108874	b	395.27	0.92	1.055	0.061	0.068	0.024	1.37	0.12		
HD 108874	c	1599	46	2.68	0.17	0.253	0.042	1.02	0.10	–	–
HD 109749	b	5.23947	0.00056	0.0629	0.0036	0	–	0.277	0.024	–	–
HD 111232	b	1143	14	1.97	0.12	0.200	0.010	6.84	0.98	–	–
HD 114386	b	938	16	1.71	0.10	0.230	0.030	1.34	0.20		
HD 114762	b	83.8881	0.0086	0.363	0.021	0.3359	0.0091	11.68	0.96		
HD 114783	b	496.9	2.3	1.169	0.068	0.085	0.033	1.034	0.089		
HD 114729	b	1114	15	2.11	0.12	0.167	0.055	0.95	0.10		
70 Vir, HD 117176	b	116.6884	0.0044	0.484	0.028	0.4007	0.0035	7.49	0.61	–	–
HD 117207	b	2597	41	3.79	0.22	0.144	0.035	1.88	0.17		
HD 117618	b	25.827	0.019	0.176	0.010	0.42	0.17	0.178	0.021	–	–
HD 118203	b	6.13350	0.00060	0.0703	0.0041	0.309	0.014	2.14	0.31	–	–
tau Boo, HD 120136	b	3.312463	0.000014	0.0481	0.0028	0.023	0.015	4.13	0.34		
HD 121504	b	63.330	0.030	0.329	0.019	0.030	0.010	1.22	0.17		
HD 122430	b	344.95	–	1.02	–	0.68	–	3.71	–	–	–
HD 128311	b	458.6	6.8	1.100	0.065	0.25	0.10	2.19	0.20		
HD 128311	c	928	18	1.76	0.11	0.170	0.090	3.22	0.49		
HD 130322	b	10.70875	0.00094	0.0910	0.0053	0.025	0.032	1.09	0.10		
23 Lib, HD 134987	b	258.31	0.16	0.820	0.047	0.243	0.011	1.62	0.13	–	–
HD 136118	b	1193.1	9.7	2.37	0.14	0.351	0.025	12.0	1.0		
GJ 581, HO Lib	b	5.3660	0.0010	0.0406	0.0023	0	–	0.0521	–		
12 i Dra, HD 137759	b	511.098	0.089	1.275	0.074	0.7124	0.0039	8.82	0.72	–	–
HD 137510	b	804.9	5.0	1.91	0.11	0.359	0.028	22.7	2.4		
GQ Lup	b	–	–	103	37	–	–	21.5	20.5	–	–
HD 330075	b	3.387730	0.000080	0.0392	0.0023	0	–	0.62	0.09	–	–
HD 141937	b	653.2	1.2	1.517	0.088	0.410	0.010	9.7	1.4		
HD 142415	b	386.3	1.6	1.069	0.062	0.500	–	1.69	0.25		
rho Crb, HD 143761	b	39.8449	0.0063	0.229	0.013	0.057	0.028	1.09	0.10		
XO-1, GSC 02041-01657	b	3.941534	0.000027	0.0488	0.0005	0	–	0.90	0.07	87.7	1.2
HD 142022A	b	1928	46	2.93	0.18	0.53	0.20	4.5	3.4	–	–

R (R_J)	R err (R_J)	T_0 (mod. JDN)	T_0 err (d)	T_t (mod. JDN)	T_t err (d)	Comment/ method*	Comments/references
–	–	11360	500	–	–	–	Butler et al. 2006; Barnes and Greenberg 2006; Naef et al. 2004
–	–	11310	80	–	–	–	Butler et al. 2006
–	–	10468.7	1.4	10463.78	0.80	–	Butler et al. 2006
–	–	11551.69	0.11	11549.557	0.068	–	Butler et al. 2006
–	–	13145	2	–	–	–	Lovis et al. 2005
–	–	10942.2	2.6	10931.1	1.0	–	Butler et al. 2006, Lovis et al. 2005
–	–	13731.7	0.5	–	–	–	Ge et al. 2006; Butler et al. 2006
–	–	11990	20	–	–	–	Sato et al. 2003
1.50	–	–	–	–	–	I	Chauvin et al. 2004; 2M1207 b, sp: L5-L9.5
–	–	10385	27	9244	51	BD?	Butler et al. 2006, Perrier et al. 2003, Fischer et al. 2002b
–	–	−31	12	−29	12	–	Butler et al. 2006
–	–	10828.86	0.71	10820.7	1.5	–	Butler et al. 2006, Pepe et al. 2002
–	–	9739	38	–	–	–	Vogt et al. 2005
–	–	9590	110	–	–	–	Vogt et al. 2005
–	–	13014.91	0.85	–	–	–	Fischer et al. 2006
–	–	11230	20	–	–	–	Mayor et al. 2004
–	–	10454	43	–	–	–	Mayor et al. 2004
–	–	9805.36	0.34	9788.23	0.29	BD?	Butler et al. 2006; Latham et al. 1989
–	–	10840	37	10836.0	7.6	–	Butler et al. 2006; Vogt et al. 2002
–	–	10520	67	10515	21	–	Butler et al. 2006
–	–	7239.82	0.21	7138.27	0.21	–	Butler et al. 2006, Naef et al. 2004
–	–	10630	120	10723	41	–	Butler et al. 2006
–	–	10832.2	1.8	10821.8	2.4	–	Butler et al. 2006
–	–	13394.23	0.03	–	–	–	Da Silva 2006
–	–	6957.81	0.54	6956.916	0.028	–	Butler et al. 2006
–	–	11450.0	2.0	–	–	–	Mayor et al. 2004
–	–	–	–	–	–	–	Setiawan et al. 2004
–	–	10211	76	–	–	–	Vogt et al. 2005
–	–	10010	400	–	–	–	Vogt et al. 2005
–	–	212.9	2.1	–	–	–	Butler et al. 2006, Udry et al. 2000
–	–	10331.7	2.2	10119.4	1.8	–	Butler et al. 2006
–	–	10598	13	9734	24	BD?	Butler et al. 2006; Fischer et al. 2002b
–	–	11004.30	0.06	–	–	–	Bonfils et al. (2005)
–	–	12014.59	0.30	12014.32	0.19	–	Butler et al. 2006; Frink et al. 2002
–	–	11762	27	11851.3	9.4	BD?	Butler et al. 2006, Endl et al. 2004
–	–	–	–	–	–	I, BD?	Guenther et al. (2005); star is a T Tau variable
–	–	12878.815	0.003	–	–	–	Pepe et al. 2004
–	–	11847	2	–	–	BD?	Udry et al. 2002; Butler et al. 2006
–	–	11519	4	–	–	–	Mayor et al. 2004
–	–	10563.2	4.1	10539.580	0.350	–	Butler et al. 2006, Noyes et al. 1997
1.30	0.11	–	–	13808.9170	0.0011	T	McCullough et al. 2006, 'Holman et al. 2006 get $R = 1.18R_J$, $1 = 89.3$
–	–	10941	75	–	–	–	Eggenberger et al. 2006

(Continued)

Table 16.2. (Continued)

System Designation	Comp	Period (d)	P error (d)	a au	a error (au)	e	e error	M/M $(\sin i)^*$ (M_J)	M^* err (M_J)	i deg	i err deg
14 Her, HD 145675	b	1754.0	3.2	2.85	0.16	0.3872	0.0094	4.98	0.41	–	–
HD 147506, HAT-P-2	b	5.63341	0.00013	0.0677	0.0014	0.517	0.001	8.64	0.37	90.	0.8
HD 147513	b	528.4	6.3	1.310	0.077	0.260	0.050	1.18	0.19	–	–
HD 149026	b	2.87598	0.00014	0.0432	0.0025	0	–	0.36	0.04	85.8	1.5
HD 150706	b	264.9	5.8	0.802	0.048	0.38	0.12	0.95	0.22	–	–
HD 149143	b	4.07	0.70	0.0531	0.0081	0	–	1.33	0.11	–	–
HD 154345	b	10900	2800	9.21	–	0.474	0.097	2.03	–	–	–
HD 154857	b	398.5	9.0	1.132	0.069	0.510	0.060	1.85	0.16	–	–
mu Arae, HD 160691	b	630.0	6.2	1.510	0.088	0.271	0.040	1.67	0.17	–	–
mu Arae, HD 160691	c	2490	100	3.78	0.25	0.463	0.053	1.18	0.12	–	–
mu Arae, HD 160691	d	9.550	0.030	0.0924	0.0053	0.00	0.02	0.0471	–	–	–
mu Arae, HD 160691	e	310.55	0.83	0.921	–	0.067	0.012	0.522	–	–	–
OGLE 2005-BLG 071	b	2900	–	1.8	–	–	–	0.9	–	–	–
HD 162020	b	8.428198	0.000056	0.0751	0.0043	0.277	0.002	15.1	2.1	–	–
OGLE-TR 10	b	3.101269	0.000040	0.04169	0.00347	0	–	0.54	0.14	86.5	+0.6
TrES-3	b	1.30619	0.00001	0.0226	0.0013	0	–	1.92	0.23	82.2	0.2
TrES-4	b	3.553945	0.000075	0.0488	0.0022	0	–	0.84	-.10	82.8	0.3
OGLE 2005-BLG 390L	b	3500	–	2.6	+1.5, −0.6	–	–	0.017	+0.02, −0.01	–	–
OGLE-TR-56	b	1.21192	–	0.0225	–	0	–	1.18	0.13	82	1
ADS 11003A, HD 164922	b	1155	23	2.11	0.13	0.05	0.14	0.360	0.046	–	–
MOA 2003-BLG 53 = OGLE-BLG-235	b	–	–	4.3	+2.5, −0.8	–	–	2.60	0.70	–	–
EWS 2005-BLG 169	b	3300	–	2.8	–	–	–	0.04	–	–	–
HD 168443	b	58.11055	0.00086	0.300	.017	0.5296	0.0032	8.01	0.65	–	–
HD 168443	c	1764.3	2.4	2.92	0.17	0.2175	0.0015	18.10	1.50	–	–
HD 168746	b	6.4040	0.0014	0.0659	0.0038	0.107	0.080	0.248	0.023	–	–
HD 169830	b	225.62	0.22	0.817	0.047	0.310	0.010	2.9	1.3	–	–
HD 169830	c	2100	260	3.62	0.43	0.33	0.02	4.1	1.6	–	–
SCR J 1845-6537	b	–	–	>4.5	–	–	–	>8.5	–	–	–
TrES 1	b	3.030065	0.000008	0.0393	0.0007	0	–	0.75	0.07	88.2	1.8
HD 177830	b	410.1	2.2	1.227	0.071	0.096	0.048	1.53	0.13	–	–
TrES-2	b	2.47063	0.00001	0.0367	0.012	0	–	1.28	0.09	83.9	0.2
HD 178911B	b	71.511	0.011	0.345	0.020	0.139	0.014	7.35	0.60	–	–
HD 179949	b	3.092514	0.000032	0.0443	0.0026	0.022	0.015	0.916	0.076	–	–
HD 183263	b	635.4	3.9	1.525	0.088	0.363	0.021	3.82	0.34	–	–

R (R_J)	R err (R_J)	T_0 (mod. JDN)	T_0 err (d)	T_t (mod. JDN)	T_t err (d)	Comment/ method[*]	Comments/references
–	–	11368.4	5.9	11530.0	4.9	–	Butler et al. 2006, Naef et al. 2004
1.41	0.04	14236.007	0.004	14212.8599	0.0007	T	Bakos et al. (2007), Loeillet et al. (2007)
–	–	11123	20	–	–	–	Mayor et al. 2004
0.73	0.06	13526.3685	0.009	13527.08746	0.00088	T	Charbonneau et al. 2006, Sato et al. 2005
–	–	11580	26	–	–	–	Udry et al. 2003a,b, Mayor et al. 2003
–	–	13483.9	1.2	–	–	–	Fischer et al. 2006, Da Silva 2007
–	–	12380	850	13300	1100	–	Wright et al. 2007
–	–	11963	10	–	–	–	McCarthy et al. 2004
–	–	10881	28	10596	18	–	Butler et al. 2006, Santos et al. 2004b
–	–	11030	110	10750	110	–	Butler et al. 2006, Santos et al. 2004b
–	–	13168.940	0.050	–	–	–	Butler et al. 2006, Santos et al. 2004b
–	–	12708.7	8.3	–	–	–	Pepe et al. 2007
–	–	–	–	–	–	G	Udalski et al. 2005; q = 0.0071
–	–	11990.677	0.005	–	–	BD?	Butler et al. 2006, Udry et al. 2002
1.16	0.05	–	–	12761.808	0.001	T	Holman et al. (2006, 2007); Bouchy et al. (2005a)
1.30	0.08	–	–	14185.9101	0.0003	T	O'Donovan et al. (2007)
1.67	0.09	–	–	14230.9053	0.0005	T	Mandushev et al. (2007)
–	–	–	–	–	–	G	Beaulieu et al. 2006; $M_{star} = 0.22 M_{sun}$; q = 0.000076
1.25	0.09	–	–	12075.1046	–	T	Bouchy et al. 2005a find q = 0.114; Santos et al. 2006
–	–	11100	280	10780	68	–	Butler et al. 2006
–	–	–	–	–	–	G	Bennett et al. 2006; Bond et al. (2004)
–	–	–	–	–	–	G	Gould et al. 2006; q = 0.00008; $M_{star} = 0.49 M_{sun}$
–	–	10047.454	0.034	10042.919	0.043	–	Butler et al. 2006, Udry et al. 2002
–	–	10255.8	4.6	10335.6	2.7	BD	Butler et al. 2006, Udry et al. 2002
–	–	11757.83	0.47	11758.92	0.19	–	Butler et al. 2006, Pepe et al. 2002
–	–	11923.0	1.0	–	–	–	Mayor et al. 2004
–	–	12516	25	–	–	–	Mayor et al. 2004
–	–	–	–	–	–	BD?	Biller et al. 2006
1.08	0.11	–	–	13186.8060	0.0002	T	Alonso et al. 2004
–	–	10254	42	10154.4	9.1	–	Butler et al. 2006; ESP Cat. has e = 0.43
1.22	0.04	–	–	13957.6358	0.0010	T	O'Donovan et al. (2006); Sozzetti et al. (2007)
–	–	11378.23	0.83	11364.97	0.33	–	Butler et al. 2006, Zucler et al. 2002
–	–	11002.36	0.44	11001.510	0.020	–	Butler et al. 2006
–	–	12103.0	7.5	11910	11	–	Butler et al. 2006

(Continued)

Table 16.2. (Continued)

System Designation	Comp	Period (d)	P error (d)	a au	a error (au)	e	e error	M/M $(\sin i)^*$ (M_J)	M^* err (M_J)	i deg	i err deg
HD 185269	b	6.838	0.001	0.0766	–	0.28	0.04	0.909	–	–	
16 Cyg B, HD 186427	b	798.5	1.0	1.681	0.097	0.681	0.017	1.68	0.15	–	–
HD 187123	b	3.096598	0.000027	0.0426	0.0025	0.023	0.015	0.528	0.044	–	–
HD 187085	b	1147.0	4.0	2.26	0.13	0.75	0.10	0.98	0.43	–	–
HD 188015	b	461.2	1.7	1.203	0.070	0.137	0.026	1.50	0.13	–	–
HD 188753A	b	3.3481	0.0009	0.0446	0.0010	0	–	1.14	0.10	–	
HD 189733	b	2.218573	0.000020	0.0313	0.0004	0	–	1.15	0.04	85.79	0.24
HD 190228	b	1146	16	2.25	0.13	0.499	0.030	4.49	0.70	–	–
GJ 777 A	b	2891	85	3.99	0.25	0.36	0.03	1.55	0.14	–	–
GJ 777 A	c	17.100	0.015	0.1303	0.0075	0.01	0.10	0.0587	0.0078	–	–
HD 192263	b	24.3556	0.0046	0.1532	0.0088	0.055	0.039	0.641	0.061	–	–
HD 195019	b	18.20132	0.00039	0.1388	0.0080	0.0138	0.0044	3.69	0.30	–	–
WASP-2	b	2.152226	0.000004	0.0307	0.0011	0	–	0.88	0.07	84.74	0.39
HD 196050	b	1378	21	2.54	0.15	0.228	0.038	2.90	0.26	–	–
HD 196885A, IDS 20351+1054A	b	1348	11	2.63	–	0.46	0.03	2.96	–	–	–
HD 202206	b	255.87	0.06	0.823	0.048	0.435	0.001	17.3	2.4	–	–
HD 202206	c	1383.4	18.4	2.52	0.15	0.267	0.021	2.40	0.35	–	–
HD 208487	b	130.08	0.51	0.524	0.030	0.24	0.16	0.520	0.082	–	–
HD 209458	b	3.52474554	0.00000018	0.0474	0.0027	0	–	0.689	0.057	86.54	0.02
HD 210277	b	442.19	0.50	1.138	0.066	0.476	0.017	1.29	0.11	–	–
HD 212301	b	2.24572	0.00028	0.0341	0.0020	0	–	0.396	–	–	–
HD 213240	b	882.7	7.6	1.92	0.11	0.421	0.015	4.72	0.40	–	–
GJ 876, IL Aqr	b	60.940	0.013	0.208	0.012	0.0249	0.0026	1.93	0.27	50.2	3.1
GJ 876, IL Aqr	c	30.340	0.013	0.1303	0.0075	0.2243	0.0013	0.619	0.088	50.2	3.1
GJ 876, IL Aqr	d	1.93776	0.00007	0.020807	0.0000005	0	–	0.0185	0.0031	50.2	3.1
tau[1] Gru, HD 216435	b	1311	49	2.56	0.17	0.070	0.078	1.26	0.13	–	–
rho Ind, HD 216437	b	1353	25	2.54	0.15	0.319	0.025	2.26	0.19	–	–
HD 216770	b	118.45	0.55	0.456	0.026	0.37	0.06	0.65	0.11	–	–
51 Peg, HD 217014	b	4.230785	0.000036	0.0527	0.0030	0.013	0.012	0.472	0.039	–	–
HAT-P-1	b	4.46529	0.00009	0.0551	0.0015	0	–	0.53	0.04	85.9	0.8
HD 217107	b	7.12690	0.00022	0.0748	0.0043	0.13	0.02	1.41	0.12	–	–
HD 217107	c	3200	1000	4.3	1.2	0.55	0.20	2.21	0.66	–	–
91 psi[1] Aqr	b	182	–	0.3	–	–	–	2.9	–	–	–
gamma Cep, HD 222404	b	905.0	3.1	2.14	0.12	0.12	0.05	1.77	0.28	–	–
HD 222528	b	572.38	0.61	1.347	0.078	0.725	0.012	7.75	0.65	–	–
HD 224693	b	26.73	0.02	0.192	–	0.05	0.03	0.71	–	–	–

[*] Columns 9 and 10 contain projected masses, i.e., M sin i, unless column 19 indicates that the planet has been detected through transit (T), gravitational lensing (G) or direct imaging (I) observations.

R (R_J)	R err (R_J)	T_0 (mod. JDN)	T_0 err (d)	T_t (mod. JDN)	T_t err (d)	Comment/ method*	Comments/references
—	—	13153.80	0.04	—	—		Johnson 2006b
—	—	6549.1	6.6	6546.3	6.4	—	Butler et al. 2006
—	—	10806.75	0.39	10807.363	0.016	—	Butler et al. 2006, Naef et al. 2004
—	—	10910	110	—	—	—	Jones et al. 2006; ESP Cat. has $e = 0.47$
—	—	11787	17	11634.4	5.7	—	Butler et al. 2006
—	—	13146.81	0.05	—	—	—	Konacki et al. 2005
1.15	0.03	—	—	13629.3942	0.0002	T	Bouchy et al. 2005b; Bakos et al. 2006; Butler et al:$P = 2.2190(5)$
—	—	11236	25	—	—	—	Perrier et al. 2003
—	—	10630	100	—	—	—	Vogt et al. 2005, Naef et al. 2003
—	—	10000.07	0.90	—	—	—	Vogt et al. 2005, Naef et al. 2003
—	—	10994.3	3.9	10987.22	0.39	—	Butler et al. 2006, Santos et al. 2003; but see Henry et al. 2002
—	—	11015.5	1.1	11008.449	0.040	—	Butler et al. 2006
1.038	1.050	—	—	14008.73205	0.00028	T	Collier Cameron (2007), Charbonneau et al. (2007)
—	—	10843	56	10573	42	—	Butler et al. 2006, Mayor et al. 2004
—	—	51236	18	—	—	—	Chauvin et al. (2006); Correia et al. (2008)
—	—	12175.6	0.2	11206.40	29.90	—	Correia et al. (2005)
—	—	—	—	—	—	—	Correia et al. (2005)
—	—	10999	15	10994	10	—	Butler et al. 2006
—	—	12853.94426	0.00014	12854.82545	0.00014	—	Wittenmyer et al. 2004; Naef et al. 2004; Williams 2000.
—	—	10104.3	2.6	10092.8	2.1	—	Butler et al. 2006, Naef et al. 2001b
—	—	13549.1950	0.0040	—	—	—	LoCurto et al. (2006)
—	—	11499	12	11347.7	9.4	—	Butler et al. 2006, Santos et al. 2001
—	—	—	—	—	—	—	Rivera et al. 2005, Delfosse et al. 1998; i from boot-strap fitting
—	—	12517.633	0.051	—	—	—	Rivera et al. 2005; i from boot-strap coplanar fittings of all planets
—	—	12490.756	0.027	—	—	—	Rivera et al. 2005; i from boot-strap coplanar fittings of all planets
—	—	10870	210	10837	53	—	Butler et al. 2006
—	—	10605	29	10647	24	—	Butler et al. 2006, Mayor et al. 2004
—	—	12672.0	3.5	—	—	—	Mayor et al. 2004
—	—	10001.51	0.61	10001.881	0.018	—	Butler et al. 2006, Naef et al. 2004
1.36	0.10	—	—	13984.397	0.009	T	Bakos et al. (2007)
—	—	1.58	0.17	—	—	—	Vogt et al. 2005, Naef et al. 2001b
—	—	11030	300	—	—	—	Vogt et al. 2005, Naef et al. 2001b
—	—	—	—	—	—	—	Mitchell et al. (2003), Raghavan et al. (2006)
—	—	—	—	—	—	—	Hatzes et al. 2003
—	—	10706.7	2.8	10199.8	3.8	—	Butler et al. 2006
—	—	13193.9	3.0	—	—	—	Johnson (2006a,b)

Despite the question of their true masses, the objects in Tables 16.1 and 16.2 cannot *all* be brown dwarfs, because a plot of the distribution of projected masses with semi-major axis indicates a dearth of more massive objects; an

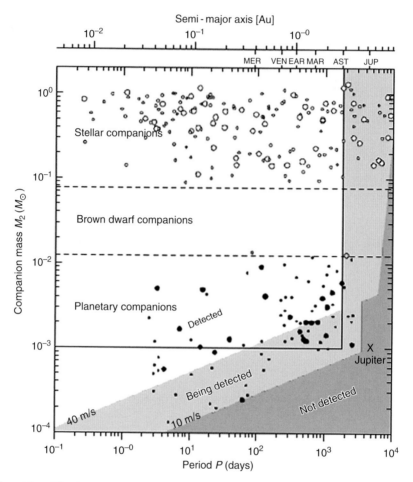

Fig. 16.1. The brown dwarf desert as illustrated in the survey of objects within 25 pc (*large symbols*) and 50 pc (*small symbols*) of the Sun. The figure, taken from Grether and Lineweaver (2006) and reproduced here by permission of the AAS, plots the masses of the companions to objects of stellar mass against the period of the binary systems. The *dashed lines* indicate the approximate dividing lines between stars and brown dwarfs ($80M_J = 0.076M_\odot$), and between brown dwarfs and planets ($13M_J = 0.0124M_\odot$). The *solid line* indicates the regime where masses are greater than 10^{-3} $M_\odot$ and periods less than 5 years. The *dark gray region* in the lower right indicates the regime where objects are not yet being detected by current techniques. Note the dearth of companions in the brown dwarf companion regime. Courtesy D. Grether and C. H. Lineweaver.

Table 16.3. Brown dwarfs—Isolates or companions

Sub-stellar object	Sp.type	a (au)/ $a \sin i$	P(d)	e	$M_{pl}(M_J)$/ $M_{pl} \sin i$	Comments
Ap 326*	M7.5	–	–	–	60–70	In α Per; Li
BD-04°782b*	K5V	0.7	240.92	0.28	21	
Cha Hα 1*	M7.5	–	–	–	20–40	In Cha cluster; CI
Cha Hα 8a	M5.75–M6.5V	–	–	–	100	In Cha I cluster
Cha Hα 8b	–	0.21	4.35 y	0.49	16–20	Found by RV curve
Cha 1109–7734	>M9	–	–	–	8 / 3	Planet?; possesses CS disk
ε Indi Ba	T1	2.65	–	–	47	imaged binary, age, μ
ε Indi Bb	T6	2.65	–	–	28	imaged binary, age, μ
GD 165b	L4*	–	–	–	>65	imaged binary; WD companion; CI
GG Tau Bb*	M8	–	–	–	50–60	T Tau quad. sys.; Li
Gliese 196-3b*	L2	300	–	–	20–25	M2.5V companion; imaged bin.; Li
Gliese 229b*	T*	40	200 y	–	35–50	M1V companion; imaged bin.; CI
Gliese 570d*	T	–	–	–	30–70	imaged binary; quad. sys.; CH$_4$
Gliese D1400	L6.5	–	–	–	–	IR; sp. binary with WD
GY 11*	M6.5	–	–	–	30–50	In ρ Oph; CI; CO, H$_2$O
HD 18445b		0.9	554.67	0.54	39	K2V companion
HD 29587b		2.5	3.17 y	0.	20–60	G2V companion; HIPP. π, μ
HD 38529b		3.38	5.95 y	0.36	12.8	G4IV companion
HD 89707b		–	198.25	0.95	54	G1V companion
HD 98230b		0.06	3.98	0.00	37	F8.5V companion
HD 110833b		0.8	270.04	0.69	17	K3V companion
HD 112758b		0.35	103.22	0.16	35	K0V companion
HD 114762b		–	–	–	>11	F9 companion; RVs; planet?
HD 127506b		–	–	–	20–60	K3 companion; HIPP. π, μ
HD 137510b		1.85	798.2	0.40	26	G0 IV companion
HD 140913b		0.54	147.94	0.61	46	G0V companion
HD 162020b		0.07	8.428	0.28	14.4	K2V: companion
HD 168443c		0.46	4.76 y	0.23	16.9	G5 companion
HD 202206b		0.83	256.00	0.43	17.5	G6V companion
HD 217580b		1	454.66	0.52	60	K4V companion
HD 283750b		0.04	1.79	0.02	50	K2 companion
J0205-1159*	2 × L7V	9.2:	–	–	65–80	imaged binary; DENIS-P
J0255-4700*	L6	–	–	–	>60	DENIS-P; CI
J0535-0546	≥M6.5	0.04	9.78	0.32	56.7, 35.6	EB, SB2; 2MASS
J0700+3157A*	L3.5	2.1:	–	–	75–85	imaged binary; trig. parallax

Table 16.3. (Continued)

Sub-stellar object	Sp.type	a (au)/ $a \sin i$	P(d)	e	$M_{\rm pl}(M_J)$/ $M_{\rm pl} \sin i$	Comments
J0700+3157B*	L6	2.1:	–	–	65–90	DENIS-P; br.dw. binary; Li
J1228-1547*	L4.5	5	–	–	<60	2MASSW; Li
J1553+2109*	L5.5	–	–	–	<60	2MASSW; CH$_4$
J1225-2739*	T	–	–	–	<70	2MASSW; CH$_4$
J1632+1904*	L6	–	–	–	>60	2MASSW; CI
J2151-4853*	T4.5	–	–	–	–	2MASS, CH$_4$
Kelu 1A*	L2	5.4	37 y	–	63(10)	imaged binary (AO); PM; Li
Kelu 1B*	L3.5	5.4	37 y	–	58(10)	binary motion detected
LHS 102b*	L4	–	–	–	>65	imaged bin., M3 companion; CI
LP 944-20*	M9	–	–	–	<60	PM; Li
PIZ 1*	M9	–	–	–	45–50	In Pleiades; CI
PPL 15*	M6.5	–	–	–	60–70	In Pleiades; Li
ρ Oph BD 1*	M8.5	–	–	–	20–40	In ρ Oph; Li
Roque 25*	L0	–	–	–	35–40	In Pleiades; CI
S Ori 47*	L1	–	–	–	10–20	In σ Ori; Li
SDSS 1624+00*	T	–	–	–	<70	CH$_4$
SIMP J013656.5093347	T2.5	–	–	–	–	bright T dwarf
J1212+0136*	L7	0.061	–	0	~20	Binary with WD; $i = 90°$?
Teide 1*	M8	–	–	–	55–60	In Pleiades

This table is characteristic only; no attempt has been made to make it complete.

Among the non-RV studied brown dwarfs (tagged with*), the comments indicate the cluster or other details of combination, and the basis for the brown dwarf identification (CI = color index, i.e., in CMD; Li = presence of lithium or CH$_4$ = methane in spectra); bin = binary; AO = adaptive optics; PM = detected in proper motion survey; IR = infrared observations; EB = eclipsing binary; SB2 = double-lined spectroscopic binary; CS = circumstellar; WD = white dwarf. Other surveys identified by acronyms: 2MASS, DENIS-P, HIPP. = HIPPARCOS. A colon suffix indicates uncertainty.

In many cases, we have truncated the name (composed of the right ascension in h,m,s; and the declination in °,'') to save space. Where the object is in a binary system, some estimated orbital properties are given.

The HD 168443 system also harbors a possible planet with $M \sin i = 7.7$ (Udry et al. 2002).

The uncertainties are such that objects recorded here as brown dwarfs near the planetary cut-off limit may actually be planets.

The eclipsing 2MASS system J0535-0546 also provides radii for this pair of brown dwarfs: 0.67 and 0.51$\Re_\odot$, for components a and b, respectively; in addition, the ratio of temperatures is deduced to be $T_2/T_1 = 1.054$, so that the lower mass object is hotter (Stassun et al. 2006).

References to many of the 2MASS and DENIS stars may be found in Reid et al. (2006), Basri (2000), and references therein. The Kelu reference is Gelino et al. (2006). ϵ Indi data are from McCaughrean et al. (2004). Cha Hα 8 data are from Joergens & Müller (2007).

Table 16.4. Millisecond-pulsar planets

Pulsar system	a (au)	P (d/y)	e	i	$M[\sin i]$ (M_{J} or $M_{\oplus}$)
PSR B1257+12 b	0.19	25.262 d	0.0	–	0.0200
c	0.36	66.5419 d	0.0186	53°	4.3010 (M_{E})
d	0.46	98.2114 d	0.0252	47°	3.9007 (M_{E})
e	~40	~ 170 y	–	–	~100 (M_{E}) (Comet?!)
PSR B1620-26 b	23	100 y	–	55°	2.5 M_{J}

Discovery of a planet around PSR 1257+12 was reported by Wolszczan and Frail (1992), arguably the first valid detection of an extra-solar planetary system. The most recent study is by Konacki and Wolszczan (2003), from which the data in this table have been updated. Note that the ratio of the periods of planets b and c is ~2:3, so the planets are locked in a mean motion ratio of ~3:2. An alternate model has $i = 127°$ and $133°$ for orbits of c and d, respectively. Significant gravitational perturbations are observed. The 'Comet?!' object has been detected at a specific radio wavelength (430 MHz) but not at others (Wolszczan et al. 2000), providing some evidence that a large coma is being observed. This very long period object needs further confirmation.

The pulsar PSR B1620-26 system is located near the edge of the core of a globular cluster, M4. It is a binary star consisting of the pulsar and a white dwarf. The inclination of the orbit is 55°. Fregau et al. (2006) conclude that the system is most likely the result of an exchange encounter of a pulsar-evolving star binary with a planet-bearing main-sequence star; the encounter left the planet around the binary and the main sequence star was ejected. Many such exchange systems have been found in globular clusters.

increase in numbers at this end of the scale would be expected if they were brown dwarfs (Mayor et al. 1998).

Brown dwarfs have been said to be absent in binary combinations with stars, a circumstance referred to in the literature as the "brown dwarf desert." The 2MASS, DENIS, and Sloan near-infrared surveys show that they are present in the field and in combination with other low-mass, low-luminosity objects. An eclipsing binary with both components identified as brown dwarfs has been detected (Stassun et al., 2006), providing firm radii as well as masses. It appeared for a while that the deficiency in brown dwarf companions of normal stars in binary star systems might be, as Basri (2000) put it, merely a brown-dwarf "desert island," partly attributable to the faintness and extreme redness (due to their low temperatures) of the objects. However, the deficit continues to be apparent as more and more planets, including objects strongly suspected of being "free-floating planets," continue to be discovered, whereas brown dwarf numbers in binary systems are being discovered at much lower rates. The "brown dwarf desert" is well illustrated in Figure 16.1, taken from Grether and Lineweaver (2006). To the present, extra-solar planets have been detected primarily through low-amplitude radial velocity surveys, conducted in the visual region of the spectrum, whereas brown dwarfs have been uncovered through long-wavelength photometric surveys and through direct visual and infrared imaging. Therefore, there may yet be room for systematic effects leading the statistics in these two cases. However, a recent direct infrared imaging search for widely-separated companions of late-type, low-luminosity stars failed to find any (Allen et al. 2007). In Section 16.3 we discuss the boundaries separating stars, brown dwarfs, and planets, and the origins of the brown dwarfs and planets in Section 16.5.

Individual properties of the brown dwarfs are mentioned briefly in Table 16.3, where we list some of the objects identified as brown dwarfs in the exoplanet websites and in the literature. There has been no attempt to ensure that the list is complete, even at current writing (the topic is just beyond our mandate, here, after all!), but should be taken as illustrative of the properties of these objects, as currently understood. Allen et al. (2007) state that more than 350 L dwarfs have been identified.

Table 16.4 lists the properties of the known pulsars that have low-mass objects orbiting them. Their nature is highly uncertain, but they appear not to be artifacts of the observing process. We discuss the technique by which they were detected in Section 16.2.6.

16.2 Methods to Find "Small"-Mass Companions

There are more or less six direct and a few indirect methods to detect the presence of a brown dwarf or of a planet near a star:

- Radial velocity variations
- Transit eclipses
- Astrometric variations
- Direct imaging/spectroscopy
- Gravitational lensing
- Time delays (for pulsar planets)
- Indirect effects

We now discuss each of these methods in turn.

16.2.1 Radial Velocity Variations of the Visible Component

Periodic variations in the Doppler shift of the star as seen in its spectrum are a dead giveaway for something pulling the star around. Because masses of planets are much smaller than those of stars, the orbital motion of the star around the common center of mass is correspondingly smaller.

Therefore, high accuracy and precision are required. The technique usually involves a comparison with a high-resolution spectrum of a source, such as the absorption spectrum of an iodine vapor cell placed in the beam from the telescope. The absorption spectrum is superimposed on the observed stellar spectrum at the same high dispersion and resolution. The technique is well described by Butler et al. (1996), who discuss the use of an echelle spectrograph (see the solar spectrum produced by an echelle spectrograph in Figure 4.9, Milone & Wilson, 2008, Chapter 4.3) and cross-disperser to separate out many highly resolved spectral orders. The European exoplanet teams use the ELODIE and CORALIE spectrographs, also equipped with echelle gratings and cross-dispersing elements, and a thorium lamp or other calibrating device.

Detection of planets through stellar radial velocity variations has been the major method of detection thus far. The situation is comparable to that of a single-lined spectroscopic binary where the spectrum of the secondary component cannot be seen because the spectral radiance of the primary component overwhelms that of the less luminous companion. This is the case for all planets detected by this technique, except those that can also be detected by transit eclipses (such as HD 2909458b; see Charbonneau et al. 2000) or imaged directly (and therefore capable of revealing its own orbital motions). Technical improvements in spectrograph stability, in spectral comparison techniques, and in analysis methods have lowered the uncertainties to $\sim$3 m/s. In addition to detection improvement, long-term averaging of data is beginning to yield long-period, low-amplitude effects in the radial velocity signatures of the parent stars—the effects of planets several astronomical units or more from the star—in other words, the searches have begun to probe the region occupied by giant planets in our own solar system, and may improve even further as described in the comprehensive review given by Perryman (2000).

Achievement of further improvements to $\sim$1 m/s or greater means that investigators must enter a realm dominated by noise effects in the atmospheres of the stars. Solar-like activity may generate localized velocity variations that will be modulated by both stellar rotation and magnetic activity cycle intervals. The separation of these effects from the effects of multiple low-amplitude planetary periodicities will become a major problem, but it is not insurmountable, given enough time and dedication. For instance, the rotation period of the star may be obtained through the modeling of line profiles produced by high-resolution spectroscopy. The stellar activity (or starspot) cycle on the star is not strictly periodic, but rather cyclic, as in the Sun, in the sense that the amplitudes of the variations fluctuate in a more stochastic way rather than with strictly periodic variations due to planetary gravitational influences. This having been said, it may take decades rather than years to find the effects of low-mass planets. In the case of planets that are very distant from their stars, the orbital motion may be negligible over intervals of only a few years, even if their masses approach the brown dwarf boundary, which is to discussed in Section 16.3.

The method of finding the mass of a sub-stellar object from the radial velocity variation of the star is analogous to the determination of the mass of the unseen component in a single-line spectroscopic binary. The method is as follows.

Because only one set of spectral lines is visible in the spectrum, the mass of each component is not directly obtainable. Instead one finds a function of the masses. From Kepler's third law (Milone & Wilson, 2008, Equation 3.34),

$$\kappa \left(M_1 + M_2\right) P^2 = a^3 \qquad (16.1)$$

where P is the period of the orbit, M_1 and M_2 are the masses of the star and the (invisible) second component and the constant $\kappa = G/4\pi^2$ disappears if the units for M, P, and a are solar masses, years, and au, respectively.

However, radial velocity measurements provide only the foreshortened quantity: $a \sin i$. Rewriting (1) with the appropriate units,

$$\left(M_1 + M_2\right) P^2 \sin^3 i = (a \sin i)^3 \qquad (16.2)$$

The masses and semi-major axes are related through the definition of the center of mass by

$$M_1/M_2 = a_2/a_1 \qquad (16.3)$$

where a_1 and a_2 are the (average) separations of M_1 and M_2 from the center of mass, and $a = a_1 + a_2$. Now note that

$$a^3 = a_1{}^3 \left(1 + a_2/a_1\right)^3 = a_1{}^3 (1 + M_1/M_2)^3 \qquad (16.4)$$

so that (2) may be written as

$$\left(M_1 + M_2\right) P^2 \sin^3 i = a_1{}^3 \left(1 + M_1/M_2\right)^3 \sin^3 i$$
$$= a_1{}^3 \left(M_2 + M_1\right)^3 \sin^3 i/M_2{}^3 \qquad (16.5)$$

from which we get the *mass function* for a single-line spectroscopic system:

$$f(M) = \left(M_2 \sin i\right)^3 / (M_2 + M_1)^2 = a_1{}^3 \sin^3 i/P^2 \qquad (16.6)$$

However, the RHS can be written more conveniently. From the radial velocity curve, supposing that the orbits are circular,

$$v_1 = 2\pi a_1/P \quad \text{and} \quad v_2 = 2\pi a_2/P \qquad (16.7)$$

where $v_{1,2}$ are expressed in units of au/y. Substituting for a_1 from (16.7) in (16.6), we obtain:

$$f(M) = (M_2 \sin i)^3/(M_2 + M_1)^2 = Pv_1{}^3 \sin^3 i/(2\pi)^3 \qquad (16.8)$$

To convert to v in km/s, substitute

$$v\,(\text{au/y}) = v\,(\text{km/s})\,(1/1.496 \times 10^8)\,(\text{au/km}) \times (3.156 \times 10^7\,\text{s/y})$$
$$= 0.2110\,v\,(\text{km/s})$$

and divide by $(2\pi)^3 = 248.050$. Then we arrive at an expression for the mass function in terms of the known quantities P and $v_1 \sin i$, even though neither v_1 nor i is known separately:

$$f(M) = M_2^3 \sin^3 i/(M_2 + M_1)^2 = 3.785 \times 10^{-5}\,Pv_1{}^3 \sin^3 i \qquad (16.8a)$$

so $f(M)$ or, as it is sometimes written, $f(M_1)$, is computable from $v_1 \sin i$. From (16.8a),

$$M_2 = [f(M_1)\,(M_1 + M_2)^2]^{1/3}/\sin i \qquad (16.9)$$

where P is in years, v_1 is in km/s, and the masses are in solar masses. Therefore, if M_1 is known, the projected mass of the other component, $M_2 \sin i$, can be determined directly if $M_2 \ll M_1$, because $M_1 + M_2 \approx M_1$. This is typically the case if M_2 is a planet and M_1 is a star.

More generally (for elliptical orbits) the amplitude of the radial velocity curve, K_1, is used:

$$K_1 = (v_1 \sin i)_{\max} \qquad (16.10)$$

(we actually observe only the projected velocity in any case). The mass function for a single-line spectroscopic binary (SB1) becomes:

$$f(M) = (M_2 \sin i)^3/(M_1 + M_2)^2 = 1.036 \times 10^{-7}K_1^3\,P(1 - e^2)^{1.5} \qquad (16.11)$$

where K_1 is in km/s, P is in days, e is the eccentricity, and M is in solar masses. The eccentricity is determinable from a highly precise radial velocity curve, so all quantities on the far RHS of (16.11) are known.

Under special circumstances, M_2 may be found even if it is not small compared to M_1. In the case of β Lyrae, for example, even though the companion appears to be obscured by a thick disk of material, eclipses are seen, so the inclination must be close to $90°$, and M_2 can be found by iteration.

When both components are observable, in a double-lined spectroscopic binary (SB2), we have:

$$M_1 \sin^3 i = 1.036 \times 10^{-7} \, (K_1 + K_2)^2 \, K_2 \, P \, (1 - e^2)^{1.5} \qquad (16.12)$$

and

$$M_2 \sin^3 i = 1.036 \times 10^{-7} \, (K_1 + K_2)^2 \, K_1 \, P \, (1 - e^2)^{1.5}$$

Note that without the inclination from a visual or an eclipsing binary analysis, only *lower limits* on the masses are obtainable. Further discussion is beyond our mandate here, but for a fuller discussion of the benefits and limitations of eclipsing and spectroscopic binary analyses, see, for example, Kallrath and Milone (1999). For planetary work, (16.9)–(16.11) are the operative equations.

The expressions used by Butler et al. (2006) to define the projected mass and semi-major axis which they tabulate [their equations (1) and (2)] are:

$$M_{\mathrm{planet}} \sin i = K_{\mathrm{star}} \, (1 - e^2)^{\frac{1}{2}} \, [P \, (M_{\mathrm{star}} + M_{\mathrm{planet}} \sin i)^2 / 2\pi G]^{1/3} \qquad (1)$$

$$(a/au)^3 = [(M_{\mathrm{star}} + M_{\mathrm{planet}} \sin i)/M_\odot](P/y)^2 \qquad (2)$$

where we have added the subscripts to the masses in the numerator for clarity. They are written this way instead of what is expected from (16.4) because $\sin i$ is not known and because the stellar mass is not a measured quantity; the quantities K_{star} and a_{star} are small compared to K_{planet} and a_{planet}, but the end result is that both a and $M_{\mathrm{planet}} \sin i$ will be underestimated, by the factors $r = (M_{\mathrm{star}} + M_{\mathrm{planet}} \sin i)/(M_{\mathrm{star}} + M_{\mathrm{planet}})$ for expression (2) and $r^{2/3}$ for (1). We prefer not to redefine well-established quantities by approximations to those quantities, no matter how close the approximation. Nevertheless, we reproduce the values quoted by Butler et al. (2006), because the uncertainties are probably greater than the error introduced by the approximations.

16.2.2 Transit eclipses

In 1999, radial velocity variations detected with the 1.5-m telescope at the Harvard College Observatory revealed a planetary candidate around the field star HD 209458. Charbonneau monitored the star for photometric eclipses and, with the help of observers at Texas and Hawaii, succeeded in observing one (Charbonneau et al. 2000; Henry et al. 2000). A transit was subsequently observed with the HST (Brown et al. 2001). The integrated spectral data were so precise that limits were established on perturbations due to moons and rings (none have been seen). Subsequent investigation revealed transits in

HIPPARCOS satellite data also. A careful analysis of the HST data set revealed the spectral signature of sodium from the absorption of the starlight as it passed through the planet's atmosphere (Charbonneau et al. 2001). This is the first such identification of the composition of an extra-solar planet. Subsequently, Vidal-Madjar et al. (2004) showed that the planet is outgassing at an enormous rate, and may well appear as a giant comet. See Figures 16.2 and 16.3 for the modeling of the HST-integrated light photometry and the radial velocity light curves, from Williams's (2001) MSc thesis, probably containing the best analysis, to that date and somewhat later, of the HST and other transit light curves. Figure 16.4 depicts the expected L light curve of the occultation of the planet.

The unprecedented precision of the spectrally integrated HST light allowed an extraordinary degree of precision in the parameter modeling; both stellar and planetary radii were extracted, but the single-lined radial velocity curve, shown in Figure 16.3, did not permit unique determinations of the mass of both star and planet.

The detections of the secondary eclipses in TrES-1, HD 209458, and HD 189733 have been reported by Charbonneau et al. (2005) and Deming et al.

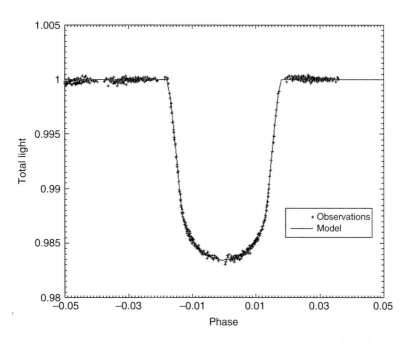

Fig. 16.2. HD 209458 HST observations from Charboneau et al. (2000) modeled with the University of Calgary version of the Wilson–Devinney program, wd95k93h, as modified by Williams (2001) to handle the small grid sizes needed to model planetary transits. Courtesy, Michael D. Williams

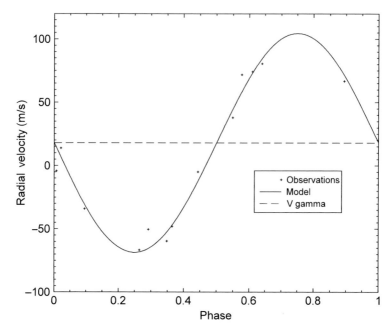

Fig. 16.3. The preliminary HD 209458 single-lined radial velocity curve subsequently published forwarded by D. Charbonneau and later greatly enhanced by Mazeh et al. (2000) with the global solution obtained with the wd98k93h light curve modeling code by Williams (2001). The derived system parameters are listed in Section 16.4.1. Courtesy, Michael D. Williams

(2005, 2006), who used the infrared spectrometer on the Spitzer Space Telescope. In TrES-1, Charbonneau et al. (2005) found eclipse depths of 0.00066 ± 0.00013 at 4.5 µm and 0.00225 ± 0.00036 at 8 µm, from which they infer a blackbody temperature of 1060 K ± 50 K for the planet and a Bond albdeo A = 0.31 ± 0.14. For HD 209458, Deming et al. (2005) found a depth of 0.0026 ± 0.0005 at 24 µm, and a planetary flux of 55 ± 10 µJy, from which they obtain a brightness temperature of 1130 ± 150K. For HD 189733, they found a depth of 0.0055±0.0003 at 16 µm, from which they derived a heated face brightness temperature of 1117±42 K.

The OGLE lensing survey has revealed an increasing number of potential transit-like events of very low depth, that are repeating, suggesting planetary transits. It is not always clear if these are caused by transits by planets or grazing eclipses by stars. Several of the discoveries have been followed up with radial velocities studies on large telescopes. Thus far at least three have proven to be planetary transits: TR-56b, TR-113b, and TR-132b (see Konacki et al. 2004). The planets in these systems are even closer to their stars than the previously found "hot Jupiters." These objects are very faint and require

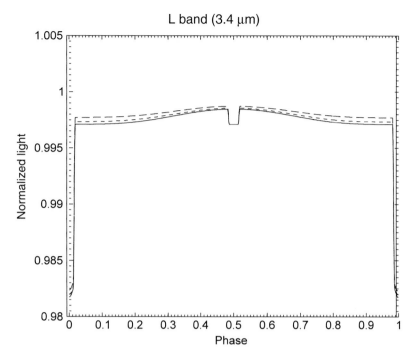

Fig. 16.4. The expected L passband light curves of HD 209458 for several different effective temperatures of the planet as modeled by Williams (2001) with wd98k93h. The occultation eclipse is shown centered, at phase 0.5. The only way to find the temperature difference between star and planet is to observe both eclipses, and this is possible only in the thermal infrared, where the planet has detectable brightness. Nevertheless, the secondary, occultation eclipse is extremely shallow. Courtesy, Michael D. Williams

very large telescopes to investigate further. Silva and Cruz (2006) suggested that an additional 28 candidates of a recent sample of 177 observed eclipse cases are likely to be planetary transits; both suspected and confirmed lists continue to increase.

More recently, large-field surveys of brighter stars have begun to reveal transits. The first such detection, of TrES-1, was announced by Alonso et al. (2004). It was found to have an orbit similar to that of HD 209458b, and similar mass, but smaller radius ($\sim 1.08 R_{\rm J}$).

The method to investigate the properties of the planet from a transit eclipse are well known from eclipsing binary studies (see Kallrath and Milone 1999, for detailed discussions of the methods). The analysis of an optical transit light curve is simpler because the secondary, occultation eclipse of the planet produces very little measurable decrease in light of the system. In a high-precision light curve obtained in the intermediate infrared (say in a modern variation of the L, M or N passbands), however, it is possible and has now been detected in several cases.

Both the depth of the eclipse and the shape can reveal the relative radii of the star and planet. The ratio of the radii is given by $k = r_\mathrm{s}/r_\mathrm{g}$, where r is the radius and subscripts s and g refer to the smaller and greater objects, respectively, in the notation of Henry Norris Russell (for the latest full description of this method see Russell and Merrill 1952). In the simplest case, for central eclipses, the light of the system during transit is:

$$\ell = L_\mathrm{g}(1 - f) + L_\mathrm{s}, \qquad (16.13)$$

where L_g is the light of the star, L_s that of the planet, and $f = k^2$ is the light lost during the transit. For optical wavelengths, the planet is expected to contribute almost nothing to the light of the system, so $L_\mathrm{s} \approx 0$. Modifications are required in the more realistic case that the star is limb-darkened (see Milone & Wilson 2008 Section 4.5.2), and a contribution for the light of the planet may be required for infrared light curves. If the planet is not small compared to the star itself (as in a massive Jovian planet around a red or brown dwarf), the contribution of the planet to the light is important, and the shape of the shoulders of the minimum must be modeled. For non-central eclipses, $a \sin i$ is present and i must be optimized. For further modeling details, see Kallrath and Milone (1999).

The old treatment of Russell and Merrill (1952) is adequate for the case of a purely spherical star and planet, but the modern light curve codes can provide improved fitting, as well as allow for the appropriate limb-darkened effects of the star's atmosphere. Similar comments have been made by Giménez (2006), but his assertion that programs such as that of Wilson-Devinney program are unsuitable for these analyses is wrong, as the work done by Williams (2001) has demonstrated. The correct physics for the intensity distribution of the light across the disk of the star is provided in the most modern codes; it was not in the coding of the purely geometrical models of Russell–Merrill and successor programs that failed to consider the correct radiative and, for interactive stars, even the correct geometric properties of the stars. The distinctions are addressed in Kallrath and Milone (1999). Most of the work done on the evaluation of planetary transits in the literature appears to have been carried out by people who have had little previous experience analyzing such light curves, and who do not discuss the details of their methods in doing so. Therefore, we report their results here as they reported them, but in most cases further work will have to be performed to verify them and perhaps to optimize the determinations. The future of photometric searches is very bright, especially with the COROT, GAIA, and Kepler missions, either in operation (COROT) or well-advanced in preparation.

16.2.3 Astrometric Variations

Proper motions are secular angular motions in the plane of the sky, usually expressed in units of arc-sec/y in coordinates such as right ascension and declination. They indicate slight changes in the direction of the star as seen from the Sun. The size of this motion depends on the objects, distance as well as the object's linear motion across the line of sight. Periodic variation in the proper motion of a star is a sign of binarity. Astrometric binaries are binary star systems where only one star is visible, yet sinusoidal variation in proper motions are seen. The amplitude of the variation depends on the ratio of the masses of the stars, which is inversely proportional to the separation of the two stars from the center of mass of the system. If the mass of the visible star is known, for example, through a well-calibrated mass–luminosity relation (determined through binary star studies), the mass of the second star can be determined. One such procedure is demonstrated here:

Begin with Kepler's third law, (Milone & Wilson, 2008, Equation 3.34),

$$G(M_1 + M_2)P^2 = 4\pi^2 \, (a_1 + a_2)^3 \tag{16.14}$$

where P is the period of the orbit, M_1 and M_2 are the masses of the visible and non-visible component, and a_1 and a_2 are the (average) separations of M_1 and M_2 from the center of mass. Their sum is the semi-major axis of the relative orbit: $a = a_1 + a_2$. From the definition of the center of mass, setting the origin to the center of mass,

$$M_1 \, a_1 - M_2 \, a_2 = 0 \tag{16.15}$$

From the latter, we get:

$$M_2 = M_1 \, (a_1/a_2) \tag{16.16}$$

and

$$a_2 = a_1 \, (M_1/M_2) \tag{16.17}$$

We assume that P, M_1, and a_1 are known (i.e., that the absolute visual orbit of the visible star is known and the orbital elements have been derived). We begin with a guess for M_2. From (16.17), we find a_2. Then, we solve for M_2 from (16.14), and return to (16.17) to find a_2. Iteration continues until successive iteration values of M_2 are in agreement.

The method is applicable to any low-mass companions, and so to brown dwarfs and planets as well as stars.

Advanced astrometry space missions, following up the HIPPARCOS–Tycho mission, may be capable of finding some variations due to precise and frequent astrometric measurements. These missions include NASA's SIM (Space Interferometry Mission) and ESA's GAIA, both to be launched, according to current schedules, near the beginning of the second decade of the twenty-first century. These missions are planned to achieve several microarc-sec of positional precision. Both proper motions and parallaxes will be determined, providing distance and kinematic detail to any discoveries. Although each will have high-precision astrometry capabilities, SIM is a pointed mission, while GAIA will be a survey instrument. In addition, GAIA will be equipped with a radial velocity spectrometer as well as photometric detectors. Although the GAIA spectrometer's resolution ($\sim$11,200) will not be high enough to detect variations due to planets in stellar spectra, astrometric and photometric effects most likely will be seen (see below).

The systems Lalande 21185 and ϵ Eri were studied by Gatewood (1996); however, the claimed planetary discoveries in these systems require confirmation; this seems to be provided for ϵ Eri (see Section 16.4 below), but not, thus far, for Lalande 21185.

Benedict et al. (2002) have made astrometric measurements of a star perturbed by an orbiting planet previously detected from radial velocities of stellar reflective motions, Gliese 876, an M4 dwarf, also known as Ross 780. The measurements were made using the Fine Guidance Sensor on the Hubble Space Telescope. They yielded an unprojected mass of $1.89 \pm 0.34 M_J$ for the planet, Gliese 876b. Further radial velocity work has yielded three planets in this system; this system, too, is discussed further, in Section 16.4.

16.2.4 Gravitational Lensing

Einstein predicted that the gravitational field of a star could cause light from more distant objects to be bent. Thus the star acts as a lens. The passage of a single star (lens) in front of a more distant one causes varying brightness resulting in two detectable sources. Usually, the brightness of the more refracted ray is too low to be measurable due to scattering effects. An exact alignment will produce the "Einstein ring" phenomemon. Most often, however, only one varying peak is seen. From the first detection in 1993 to the present, hundreds of events have been seen. The sources that discuss the geometry and measurable parameters are cited by Perryman (2000), whose discussion is summarized briefly here.

In this lensing context, the effective radius of the lensing object is known as the *Schwarzschild* or *gravitational radius*,

$$R_g = 2GM_L/c^2 \tag{16.18}$$

where M_L is the mass of the lens and c is the speed of light. In gravitational theory the radius of an object is 2M, and G/c^2 is the conversion factor from mass to length units (Misner et al. 1973, pp. 35ff). The source object undergoes distortion as the lens moves in front of it, resulting in a perfect ring when the objects are both in the line of sight. The radius of this ring is the *Einstein radius*,

$$R_E = (2R_g \, \Delta r \, d)^{\frac{1}{2}}, \tag{16.19}$$

where $\Delta r = (r_S - r_L)$ is the difference in distance to the source and lens respectively, and $d = r_L/r_S$, is the distance ratio. The apparent or angular Einstein radius is then given by

$$\Theta_E = R_E/r_L \tag{16.20}$$

which depends implicitly on the square root of M_L, inversely on the square root of r_L, and on the square root of the factor $(1 - d)$.

Usually there is no consensus of the distance of the star that is acting as the lens, but when the lens turns out to be a binary star, the additional lensing action of the second star and an assumption about the motion of the lensing system in the plane of the sky, permit a distance estimate to be made. The time scale is given by

$$t_E = R_E/v_L, \tag{16.21}$$

where v_L is the transverse speed. This is the time of crossing by the *caustic* (named after a surface of maximum brightness created by spherical aberration in a spherical lens/mirror) and it may last for days for a massive object such as a star or cluster, or hours for a planet. If there is a planet in the lensing system, a sharp spike will be seen, in addition to the star's broader amplification effect (see Figure 16.5).

The ground-based OGLE (for Optical Gravitational Lensing Experiment) survey of the galactic bulge region of the galaxy has detected many planetary

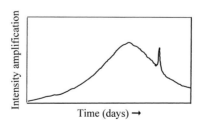

Fig. 16.5. The amplification of signal of a distant source due to microlensing by a star (*broad variation*) and planet (*spike*) as the lensing system passes in front of the source. The example shown is very similar to the lensing object OGLE-2005-BLG-390Lb (Beaulieu et al. 2006)

candidates through light curves, but not directly through lensing. The confirmed or strongly suspected cases (at present writing) are included in Table 16.4.

Three planetary lensing cases have been confirmed to present writing: that of OGLE-2003-BLG-235/MOA-2003-BLG-53; OGLE-2005-BLG-071L; and OGLE-2005-BLG-390LB. In addition, OGLE-2005-BLG-05-169L is a strong candidate (Gaudi et al. 2005; Gould et al. 2006) and also OGLE-TR-109, although the uncertainty in the object's mass in too large to determine on which side of the planet-brown dwarf boundary it falls. For these five systems, the estimated planet masses are $\sim$2.6, 0.9, 0.04, 0.017, and $14M_J$, respectively.

It is difficult to determine from photometry alone whether an isolated planetary gravitational lensing event (without the stellar event in Figure 16.5) is caused by a free-floating planet or by a bound planet widely-separated from its star. Han (2006) notes that the astrometric gravitational lensing event lasts much longer than the photometric event, and shows that follow-up observations with the Space Interferometry Mission should detect the centroid shift induced by the primary (if present) to V $\approx$19 and projected separations to $\sim$100 au, thus allowing separation of the two possibilities.

Gould et al. (2006) argue that of all the techniques for finding planets, gravitational microlensing from space will be the only one capable of detecting objects significantly smaller than an Earth mass.

16.2.5 Direct Imaging and Spectroscopy

In both optical and infrared spectral regions, one can look for faint companions to nearby stars, but true planets (at least the "hot Jupiters," located far closer to the star than Mercury is to the Sun in our system) are not likely to be luminous enough to be seen directly. It would be even more difficult to obtain any kind of spectral resolution to discern identifying features in the spectrum of any such candidate objects. Because of the overwhelming light of the parent star, it has not been possible to separate the flux of the known planets from their stars, for these types of planets, thus far, but coronographic (Lyot 1933) and diffraction techniques are quite promising, and claims have already been made for somewhat more distant planetary companions. New generation space telescopes are an obvious answer to this need, but high-resolution techniques on existing telescopes may permit detection sooner.

As noted above, candidate planets are being detected, relatively far from their parents. There are several candidates thus far, but at least two may be brown dwarfs. The strongest candidate is a low-mass object orbiting the star 2MASSW J1207334-393254, with mass $5 \pm 2M_J$ at 41 au from a brown dwarf with spectral type M8, and a mass of only $0.025M_\odot$ ($25M_J$). The object itself has a spectral type in the range L5–L9.5. A second candidate orbits the variable star GQ Lupi. The companion is much fainter but appears to have the same proper motion as the star on the plane of the sky. The planet is located at 103 ± 37 au from a faint, red, variable star with spectral type K7e V. Therefore its orbital motion over the few years it has been observed is very small, and so far, unobservable. The companion to the K2 V star AB Pic is another candidate with mass $13.5 \pm 0.5 M_J$, but this is right at the boundary between planets and brown dwarfs. Finally, there is an object with mass $> 8.5 M_J$, orbiting the object SCR J1845-6357, with spectral type M8.5 V. The companion also may be a brown dwarf.

Infrared surveys, and imaging in methane and ammonia bands are turning up very red and faint objects, and a number of these have been confirmed to be brown dwarfs through subsequent spectroscopy. Some very red objects in clusters are also turning out to be sub-stellar objects (see Basri 2000 for an early but still useful summary). One object that we mention briefly here appears to be a single object surrounded by a disk. Discovered by Luhman et al. (2005), the object is located in the direction of the Chamaeleon I star-forming region, which is $\sim$165 pc from the Sun. It is designated Cha 110913-773444 and is estimated to have a mass of 8^{+7}_{-3} M_J as determined from a mass-luminosity relation derived for similar objects. There are several stages in this determination that require further scrutiny, but the possibility of the formation of an isolated object of approximately planetary mass is indeed intriguing (see the discussion of proplyds in Chapter 15 for the possibility of the formation of even lower mass objects in star-forming regions).

16.2.6 Pulsar Timings

Whereas the radial velocity technique has been carried out in the visual part of the spectrum, and the photometry, direct imagery, and lensing monitoring in visual and infrared light, pulsar timings have been carried out at radio frequencies. The work on the rapidly rotating (millisecond!) pulsar PSR 1215+12 by Wolszczan and collaborators (e.g., Wolszczan and Frail 1992; Wolszczan 1996; Wolszczan et al. 2000a,b; Konacki and Wolszczan 2003) has provided convincing evidence of the existence of planets around this object. Unless there was a complicated history of migrations, one would expect the three reasonably certain components (b, c, and d, which Wolszczan and colleagues refer to as planets A, B, and C) to have been *inside* the super-giant star that preceded the supernova of which the pulsar is the remnant core. Presumably, therefore, the existing planets accreted from the debris disk produced in the eruption, perhaps from the interaction with a companion star. The latter may have contributed to the tremendous spin-up of the pulsar.

The detection involves periodic shifts in the pulse beams emitted by the pulsar, which is thought to be a dual beamed, rotating neutron star. As the sharply collimated beams rotate into the line of sight, the system briefly brightens and quickly fades. The time delay in the pulses indicates the reflex motion of the pulsar around the center of mass in the same way that the motion of a star reveals the reflex motion of a planet in other systems. The variation in arrival time of the pulses is periodic due to the increased and decreased distance to the observer of the pulsar.

Radio frequency observations of the pulsar PSR 1257+12 were used by Wolszczan and Frail (1992) to determine the existence of the planets described in Table 16.4 and in the notes to that table. Millisecond pulsars are often found in globular clusters, although PSR 1257+12 is a field object. Pulsar planets are clearly rare but many have been found in binary systems (probably arising through collisional interactions), as in the white dwarf, binary pulsar, and brown dwarf/planet objects in PSR B1620-26 in the M4 cluster. In at least three binary pulsar cases, both components are pulsars.

16.2.7 Indirect Effects

These include O–C curves, and various effects on stellar disks: warps, gaps, and clumps.

A team headed by E. F. Guinan (Villanova University) claimed detection of one or more planets in the CM Draconis system, an eclipsing M-dwarf binary. The Villanova group claims only one photometric event (which another group disputes), but has studied the timing of the mutual eclipses and has compiled an O–C (for *Observed – Computed* instants of mid-eclipse) curve of the eclipsing system, which, Guinan feels, furnishes evidence of the gravitational effect on the orbits of the two stars. At present, a planet in this system remains unconfirmed.

Proto-planetary disks have been seen around several stars, including β Pictoris and Vega, and remnants of disks have been seen around older stars. Gaps and warping have been attributed to the presence of planets or protoplanets in some of these systems.

Gorkavyi et al. (2004) summarize the case for a $10 M_J$ planet orbiting β Pictoris, the first star discovered to have a disk around it. There are now other known cases of candidate planets around stars with remnant disks, such as ϵ Eridani and GQ Lupi. Clumpings of material in the disks of ϵ Eri and Vega have been interpreted as evidence of mean motion resonances (see Milone & Wilson, 2008, Chapter 3 for a discussion of commensurabilities) involving planets, to be discussed in later sections for these two stars. However, few details about the planets can be obtained through this method, should it prove to be reliable.

Having now discussed the methods by which planets may be discovered, we turn to a thorny but essential issue: the definition of a planet.

16.3 Definitions of Planets and Brown Dwarfs

It is a great irony that although the planets have been studied as long as they have borne their identifying characteristic of wandering among the stars, up to now, planetary scientists have failed to agree as to exactly what it is that makes a planet a planet.

Until now naming controversies have raged only at the low mass and diameter end of the scale, where the distinction among moon-sized bodies, asteroids, and planets has become more difficult with increasing discoveries. The various factors underlying the controversy are reviewed by Basri and Brown (2006). As far as we are concerned, the term "minor planet" could well have been used to distinguish objects that are smaller and less massive than any existing planetary moon, whether or not they happen to be currently locked in an orbital resonance. It is up to the International Astronomical Union (IAU) to adjudicate matters of astronomical nomenclature, and a definition was adopted at the IAU General Assembly meeting in Prague in 2006. The definition of a planet approved by vote at that meeting, is that (a) the body must be in orbit about the Sun (presumably "star" in other systems); (b) the body must be massive enough to have sufficiently strong gravitational acceleration to result in a hydrostatic equilibrium shape (that is, nearly round, unlike, say the asteroid Eros described in Chapter 15); and (c) it must be the dominant object in its region of space in order to have cleared the region of other objects (hence the Moon or Charon could not be considered planets). Part (c) of the definition must be considered flexibly so that satellites or objects caught in resonances are not considered as uncleared objects. With these definitions, there are now eight planets, because Pluto allegedly fails the "clearing" test. Pluto is relegated to a class known as *dwarf planets*, to which the main belt asteroid Ceres, the Edgeworth–Kuiper Belt object Eris initially designated (2003-UB313), and other round bodies belong. Pluto is, however, the prototype of the trans-Neptunian objects (TNOs, a group to be given another name at some future date—recently *Plutoids* has been suggested as a name for TNO dwarf planets), and the eponymous prototype of a subgroup of the TNOs, the *Plutinos*, which are icy bodies locked in a 3:2 mean motion resonance with Neptune. Interestingly, the barycenter of Pluto and its principal moon Charon is outside Pluto's radius. This makes the Pluto–Charon system a double dwarf planet, with two small moons orbiting in nearly the same plane—see Chapter 13. Most asteroids, such as the shape-challenged Eros, clearly do not qualify as dwarf planets; thus, no asteroid (or planetoid or minor planet, as these objects have been called) and no

Kuiper Belt object (KBO), can be reassigned to dwarf planet status unless its shape can be determined to be round. All objects not in the planet or dwarf planet categories are simply called "small solar-system bodies," with disputed cases to be handled in some way by the IAU. The definitions are likely to be challenged, and probably will be further refined, in practice, if not by writ, but in any case, no one can now claim that astronomers have not provided some leadership in offering a definition of a planet!

At the high mass end of planetary definitions, though, there is somewhat less disagreement, probably because we do not have enough clear-cut examples of objects in the transition region! Theoretical work done so far suggests that deuterium burning begins at a mass $\sim 13 M_J$, providing a convenient dividing line between planets and low mass stellar objects (Saumon et al. 1996). Sub-stellar objects above this mass are assumed to be brown dwarfs. Objects at the brown dwarf–star interface are considered to be stars if their masses are greater than $\sim 75 M_J$ (or $0.072 M_\odot$), if they have solar composition. For stars with no metals, this limit increases to $90 M_J$.

As a practical matter, one may be able to distinguish among these three objects, namely, a planet from a brown dwarf, and a brown dwarf from a low-mass star, through observational means: brightness, color and/or spectral characteristics. One possible criterion is the presence of lithium (Li), which is easily destroyed in stars through large-scale convection. According to Basri (2000, p. 494), any object with spectral class later than M7, and having lithium, must be sub-stellar. According to Chabrier and Baraffe (2000), the characteristic spectral features for ranges of temperature at the cool end of the spectral sequence can be summarized as follows:

$\lesssim 4000$ K: M dwarfs. Most of the hydrogen is in the form of H_2 and most of the carbon in CO. O is bound mostly in TiO, VO, and H_2O, some in OH and in monatomic O, and metal oxides. Metal hydrides (e.g., CaH, FeH, MgH) are also present. In optical spectra, TiO and VO dominate; in the IR, H_2O and CO features are seen.

$\lesssim 2800$ K: O-rich compounds condense in the atmosphere and may go into perovskite ($CaTiO_3$).

$\lesssim 2000$ K: The realm of the *L dwarfs* (example: GD 165B). Some TiO remains, but metal oxides and hydrides disappear from the spectra. Alkali metals are present in atomic form. Some methane may be seen.

$\lesssim 1800$ K: Refractory elements (e.g., Al, Ca, Ti, Fe, V) condense into grains. Corundum (Al_2O_3), perovskite condense. Depending on the pressure, rock-forming elements such as Mg, Si, Fe may condense as metallic iron, forsterite (Mg_2SiO_4), or enstatite ($MgSiO_4$).

1700–1000 K: Region of cross-over to *methane dwarfs* or *T dwarfs*
(example: Gliese 229B). Methane absorption is strong
in the infrared broad passbands H (1.7 μm), K
(2.4 μm), and L (3.3 μm), giving rise to a steep
spectrum at shorter wavelengths, with J–K $\lesssim$ 0, but
with I–J $\gtrsim$ 5.

A planetary temperature is typically less than ~1200 K, but this depends
on the surface temperature of the star, the proximity to it, and the planet's
albedo. Moreover, as we shall note later, this assumes that the planet is in
thermal equilibrium. There are substantial reasons to doubt the validity of
this assumption for at least some of the lower mass planets in very close
proximity to their stars, for example in HD 209458b, which is detectably
outgassing (see the discussion below).

As with stellar characteristics, the observed properties change with time.
Figure 16.6 demonstrates the changes in absolute K passband magnitude
vs. temperature expected for objects of various masses between one and five

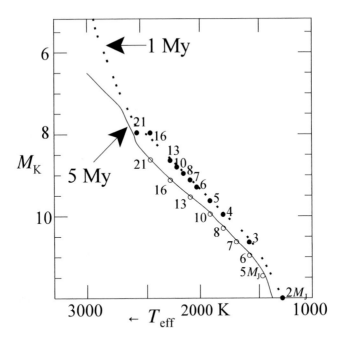

Fig. 16.6. A form of a Hertzsprung–Russell diagram for planets and brown dwarfs.
The isochrones (lines of constants ages) for objects of 1 and 5 My ages; the masses
(in M_J) of the objects along the isochrones are indicated. Diagram adapted from
that in Jayawardhana and Ivanov (2006) from models of Baraffe et al. (2002) and
Chabrier et al. (2000). Reproduced and modified by permission of the authors and
the American Astronomical Society.

million years. Note the changes in position of lower mass objects in the course of time between the isochrones.

All simulations suggest that the lower temperature brown dwarfs would have clouds and other atmospheric features usually associated with gaseous planets. Recent planetary modeling of hot Jupiters indicates equilibrium temperatures to be, very roughly, $\sim$1000 K. Williams (2001) estimated the temperature of HD 209458b to be $\sim$915 K, assuming no significant internal energy sources. Charbonneau et al. (2005) found from Spitzer observations that the temperature of HD 209458b is 1350 K. This is hotter than some of the brown dwarfs that have been observed, as deduced from their infrared colors. Presumably as more of these objects are studied, criteria will become definitive. See Reid et al. (2006) for a catalog of L dwarfs and Burgasser et al. (2006) for a list of T dwarfs and a discussion of objects at the L/T boundary.

16.4 Extra-Solar Planets Detected or Strongly Suspected

At present writing, well over 200 confirmed/suspected extra-solar planets have been discovered. Table 16.1 is a position-ordered list of stars which are known or strongly suspected to contain planets. The sub-stellar entries on each line are arranged in increasing distance from the star. These are mainly objects with $M \sin i$ values less than $13M_J$ but there are exceptions: the second low-mass companion to HD 168443 is included in Table 16.1, just to indicate that it is in the same system as a likely planet object. Because the mass is a lower limit for all but a few cases (one of these is HD 209458b), some of these may turn out to exceed this critical mass which has been used to divide brown dwarfs from planets.

Note the large number of "hot Jupiters," large planets with small distances from their stars. Although many of these have low eccentricities, as expected because circularization effects operate in close binary star systems over time scales of millions or tens of millions of years, some do not. The latter can only be interpreted as the result of recent migration into these orbits. Thus the presence of "hot Jupiters," especially with sizable eccentricities, may imply dynamically active systems.

The metallicities of stars with identified planetary candidates appear to be greater than for the Sun. This also points to a relatively young stellar environment. This conclusion is further enhanced by a failure to see any evidence of planetary transits in more than 8 days of continual monitoring of more than 50,000 stars in core and peripheral HST frames of the globular cluster 47 Tucanae (Gilliland et al. 2000). In this case, however, the latter deficiency also may be explained by intra-system dynamical and extra-system tidal effects to which the planets would have been subjected over the great age of the globular cluster (∼13 Gy). A ground-based study by Weldrake et al. (2005) which concentrated on the outlying fields of 47 Tuc, confirmed the apparent absence of planets in that cluster. However, searches for planets in other clusters, such as the relatively metal-rich and star-rich cluster, NGC 6791, have also failed to find any planetary transits. Even in sparser clusters orbiting planets are rare:. One that has been found is in the Hyades, a relatively young and nearby open cluster; it is orbiting the red giant star ϵ Tauri (Sato et al., 2007). This suggests that stellar number density, as well as metallicity, are key environmental factors in determining the existence of giant, short-period planets (surveys tend to be biased toward this group, but as the time-line increases, the bias against long-period, and lower mass objects should be reduced). The tidal radius of a star becomes a critical matter in planetary systems in dense clusters. If planets tend to be formed further out than 5 au, and if the mean free path through the dense portion of a cluster is small enough, the planets may be stripped from their parent stars and left to be free-ranging. The discovery of such objects probably requires the very large mid-infrared space telescopes planned for the future, and gravitational lensing.

To complement Table 16.2, Table 16.3 lists some of the newly found brown dwarfs. See Basri (2000) for a review of the observational properties and Chabrier and Baraffe (2000) for the theory of these objects. In addition to these low mass objects, the planetary objects around two pulsars are listed in Table 16.4. These are among the lowest mass sub-stellar objects detected. Wang et al. (2006) have identified the infrared signature of a debris disk around one of these objects, 4U 0142+61, in which planet formation may occur, so the mystery of what these objects are and how they arose may have been solved.

There is an increasing number of systems being found where two or more low-mass objects are orbiting a star. As of Nov., 2007, twenty five systems were known to have at least two confirmed planets, and eight systems (υ And, HD 37124, HD 69830, HD 74156, 55 Cnc, μ Arae, Gl 581, and GJ 876), had at least three planets each. The 55 Cnc system had five, and the μ Arae system, four confirmed planets. Two pulsars provide other examples of systems of multiple planets, if the current interpretation of the data is correct. Although the orbital eccentricities are often not determined with high precision, they are secure enough to demonstrate that the solar system appears to be unusual, in its relatively well-behaved, low-eccentricity planetary orbits.

Another growing group of planets are those detected through planetary transits. These include TrES-1b, XO-1b, and a steadily increasing number found in the course of the gravitational lens survey of the galactic bulge: OGLE-TR-10b, -56b, -109b (though probably a brown dwarf), -111b, -113b, and -132b. More can be expected. These are important because they provide independent values for the radius of the planets, and, in combination with radial velocity data, the true masses, because careful photometric analysis can yield orbital inclinations. Other planets have been found to transit after they were discovered through radial velocity variations. These include HD 209458b, HD 149026b, and HD 189733b.

If only a transit eclipse is observed, the relative radii (the radius of each object divided by the semi-major axis) are obtainable, and unless the star is not very large compared with the planet and/or the eclipse is central, so that all four contact points can be observed, the planetary radius determination depends on the assumed radius for the star. In a transit eclipse, however, the darkened limb of the star results in a gradual decrease in light during the beginning of the transit, though the limb-darkening coefficients must be determined or known for a star with similar characteristics. When a radial velocity curve becomes available, even though single-lined, the absolute radii are obtainable, because the transit can be modeled to obtain an optimum inclination (which, in any case, will likely not be too different from $90°$). The planetary mass depends on the assumed mass of the star, as noted in Section 16.2.1, because basically only the mass function is known from analyses of a single-lined RV curve. Therefore, in discussing what is known about extra-solar planets, the stellar radii and masses are often included in full discussions. Consequently, we include them in our descriptions of some of the more interesting systems of Table 16.2 in the sections to follow. References for mass, radius, and chemical composition of the other stars of Table 16.1 can be found in Butler et al. (2006) and from the online extra-solar planets encyclopedia and exoplanets websites mentioned above, as well as from the Simbadwebsite: http://simbad.u-strasbg.fr/Simbad.

16.4.1 HD 209458b

This is the first extra-solar planet to have been observed transiting its star and consequently to have its radius measured and the first to have had a constituent of its atmosphere detected. The HST observations, in particular, have been so precise that both extensive rings and satellites of this object can be ruled out. The HST and previous light curves were analyzed with new light curve analysis programs in use at the University of Calgary. The best fit parameters to both light curve and RV curves of that study, by Williams (2001), are compared to published results by Charbonneau et al. (2000, 2005) and summarized in Table 16.5.

Thanks to the transit observations, the radius of this planet is determined to be $R_p = 1.37 \pm 0.01 R_J$, so that the density can be found: 330 ± 10 kg/m^3. The epoch of mid-eclipse is $t_t = 2451254.587 \pm 0.002$. The orbit appears to be circular to high precision, with $a = 10.06 R_\odot$ (assumed). The spectroscopic properties of the star indicate it to have $\log g = 4.382$, where g is the

Table 16.5. The HD 209458 planetary system

Planet	Period	T_{peri} (JD- 2450000)	e	ω	K (m/s)	M_p/M_J	i	a (au)
b	$3^d52478(5)$	2.093	0.0000(1)	73°	70.2	0.69(1)	86°54(2)	0.048

Star	v sin i	Sp. type	M ($M_\odot$)	$\Re/\Re_\odot$	V	r (pc)	[Fe/H]	T
a	4.49 m/s	F8V	1.09(1)	1.145(3)	7.65	12.47	0.14	6099 K

gravitational acceleration in CGS units. (A single star provides us with no direct information about its radius or mass, but, with log g, in combination with an estimate of the luminosity, both can be estimated.) The movement of the planet across the limbs of the star during the transit produces a slight distortion in the profile of a spectral line which leads to a distortion in the derived radial velocity curve (commonly known as the *Rossiter* or sometimes, the *Rossiter–McLaughlin effect*). From an analysis of the effect in HD 209458, Winn et al. (2005) derived a v sin i value of 4.7 km/s, similar to that found in line profile analyses.

The above values may be compared to the solutions based on new radial velocity measurements by Naef et al. (2004):

$$a_a \sin i = 2.76 \pm 2.76 \times 10^{-5} \text{ au}, \ a = 0.048 \text{ au}; \ M_p \sin i = 0.699 \pm 0.007 \, M_J;$$
$$R_p = 1.27 \pm 0.02 \, R_J \ ; \ P = 3.5246 \pm 0.0001 \text{ d}; \ T_{periapse} = 2452765.790 \pm 0.021; \ T_t = 2452618.66774 \pm 0.007.$$

The projected mass that they derive is based on a mass of the star of $1.15 M_\odot$. With the inclination given in Table 16.5, this yields a mass of precisely $0.700 \, M_J$.

The radius of the planet is larger than one would expect for a planet less massive than Jupiter, especially when compared to the radii of other "hot Jupiters." A larger size for such a planet (relative to *our* Jupiter) is expected because of its proximity to its star, which would increase the equilibrium temperature and therefore increase the pressure scale height of the atmosphere, but the radius of HD 209458b exceeds predictions of models that take this into account. With a mean density of $330 \pm 10 \text{ kg/m}^3$ (Williams 2001), less than half that of Saturn, the result is not in doubt: the atmosphere is distended!

Subsequent observations by Vidal-Madjar et al. (2004) have shown evidence for a trailing cloud of hydrogen, carbon, and oxygen, indicating hydrodynamic loss at atmosphere from this planet. The group refers to the planet as Osiris, after the Egyptian god who was killed and dismembered by Seth (see Kelley and Milone, 2005, Ch. 8.1.2). Vidal-Madjar et al. suggest that planets older and closer to their parent stars than HD 209458b may have been deprived of their atmospheric envelopes and become a new class of planets (*chthonian*).

The confirmed OGLE planets do constitute a closer and therefore even hotter class of hot Jupiters, but their sizes appear to be smaller than HD 209458b. Work on the evolution of hot Jupiters by Baraffe et al. (2004) suggests that for objects lower than a given mass at a particular distance from the star, the initial hydrogen envelope evaporates, after which the icy and rocky components will expand and evaporate also. The time scale depends greatly on the XUV radiation from the star. These authors investigated the varying fluxes of solar-type stars at different ages for the orbital radii of HD 209458b (0.046 au) and OGLE-TR-56b (0.023 au) and found that the planets in the mass range of 0.5–$5M_J$ have evaporation rates varying from $10^{-8}\,M_J$/y at young ages, to $\sim 10^{-12}$ for ages greater than $5\,$Gy.

Finally, from Spitzer Space Telescope data, infrared observations obtained during secondary eclipse imply a temperature for the planet on its heated face of $T = 1130\pm150\,$K (Deming et al. 2005). This result is very close to that found for the transiting system HD 189733 by Deming et al. (2006), namely $T = 1117\pm17\,$K, observed in a more sensitive mode and to higher precision than previous planetary occultation observations. Therefore, in the lower density planets, such as HD 209458b, we may be witnessing a more evolved state of hot Jupiters.

16.4.2 The Multi-Planet System of v Andromedae

This was the first multi-planet system discovered. Data provided by Marcy's California & Carnegie Planet Search Team, as of September 24, 2002 are summarized in Table 16.6.

One, two, and finally, three-planet fittings to the radial velocity data show successively smaller residuals. Note the 'hot Jupiter', v And b, with a low eccentricity orbit, is close to the star while v And c, although still fairly close to v And a, has a much higher eccentricity. Modeling studies of the

Table 16.6. v Andromedae planetary system

Planet	Period	T_{peri} (JD-2450000)	e	ω	K (m/s)	$M_J \sin i$	a (au)
v And b	$4\overset{d}{.}61710$	2.093	0.012	73°	70.2	0.69	0.059
v And c	$241\overset{d}{.}5$	160.5	0.28	250°	53.9	1.89	0.829
v And d	1284^{d}	64	0.27	260°	61.1	3.75	2.53

Star	P (rotn)	Sp. type	Mass ($M_\odot$)	V	r (pc)	[Fe/H]
v And a	$10\overset{d}{.}2$	F8V	1.32	4.09	12.47	0.15

orbital behavior of hot Jupiters in multiple planet systems reveal that the interactions among planets can induce oscillations in such orbital parameters as the eccentricity, semi-major axis, and inclination, just as has been found for asteroids and for irregular moons of the gas giants in our solar system (Section 13.3.5).

16.4.3 The Multi-Planet System of 55 Cancri

At present writing, this is the only system of Table 16.2 that has as many as five detected planets (McArthur et al. 2004; Butler et al. 2006, Fischer et al. 2007). This is typical of the results of extended monitoring as the velocity residuals over longer and longer intervals of time reveal additional periodicities. The innermost planet, e, has as high an eccentricity as the orbit of the most massive and outermost planet found thus far, d. Internally consistent elements from a dynamical fit to all data from Fischer et al. (2007, Table 4) are included in Table 16.7. See Naef et al. (2004) for an independent orbital analysis of planets b and d.

16.4.4 The Multi-Planet System of HD 37124

This is another multi-component system. Vogt et al. (2005) summarize the findings and model fittings to this system. Their results are shown in Table 16.8. The authors present two three-planet models for the system. The properties of the planets according to the first model are included in the table; it has the property that the eccentricity of component d is fixed at 0.2. In the second model, the eccentricity of component d is taken as a free parameter. The derived parameters between the models are greatly different, especially the eccentricities, which become 0.25, 0.15, and 0.16, and the projected masses, 0.66, 0.73, and 0.17, for components b, c, and d, respectively. In the second model, the period of the third planet would be only 29.3

Table 16.7. The 55 Cancri planetary system

Planet	Period	T_{peri} (JD-2450000)	e	ω	K (m/s)	$M_J \sin i$	a (au)
b	$14\overset{d}{.}651126$	17572.0307	0.0159	$164\overset{\circ}{.}001$	71.84	0.8358	0.115
c	$44\overset{d}{.}378710$	17547.5250	0.0530	$57\overset{\circ}{.}405$	20.06	0.1691	0.241
d	$5372\overset{d}{.}8207$	16862.3081	0.0633	$162\overset{\circ}{.}658$	47.20	3.9231	5.901
e	$2\overset{d}{.}796744$	17578.2159	0.2637	$256\overset{\circ}{.}500$	3.73	0.0241	0.038
f	$260\overset{d}{.}6694$	17488.0149	0.0002	$205\overset{\circ}{.}566$	4.75	0.1444	0.785

Star	P (rotn)	Sp. type	$M/M_\odot$	V	r (pc)	[Fe/H]
a	40–45^{d}	G8V	0.91	5.95	12.53	0.32

Table 16.8. The HD 37124 planetary system

Planet	Period	T_{peri} (JD-2449000)	e	ω	K (m/s)	$M_J \sin i$	a (au)
b	154^{d}46	1000.11	0.055	140.5°	27.5	0.61	0.53
c	843^{d}6	409.4	0.14	61°	15.4	0.60	1.64
d	2295^{d}	606	0.2	201°	12.2	0.66	3.19

Star	P (rotn)	Sp. type	$M/M_\odot$	V	r (pc)	[Fe/H]	
a	25^{d}	G4 IV-V	0.83	7.68	33.24	−0.44	

days. The second model has slightly higher rms, however, so we show only the first model results in Table 16.8. With a larger baseline of data, which model turns out to best represent the system should become clear.

16.4.5 The Multi-Planet System of HD 69830

This is an interesting triple planet system, analyzed by Lovis et al. (2006). The properties of the system are shown in Table 16.9. Unless the inclinations of these planets turn out to be significantly different from 90°, these objects are in the Uranus (0.046 M_J)–Neptune (0.054 M_J) class of planets. Note, however, even though they may be lesser giants, all three orbits are crammed within an astronomical unit of their star.

The system has been found to emit strong infrared radiation attributed to silicate grains at a temperature ~400 K, possibly due to an *asteroid mill* (replenishment of an asteroid belt by collisions), a source of grains in the solar system (Lovis et al. 2005).

Perhaps one of their most interesting findings is that the outer planet is in the *habitable zone*, a region where an object with sufficiently great surface gravitational acceleration can retain water in liquid phase on its surface.

Table 16.9. The HD 69830 planetary system

Planet	Period	T_{peri} (JD-245000)	e	ω	K (m/s)	$M_J \sin i$	a (au)
b	8^{d}667	3496.80	0.10	340°	3.5	0.032	0.079
c	31^{d}56	3469.6	0.13	221°	2.7	0.037	0.19
d	197^{d}0	3358.	0.07	224°	2.2	0.057	0.633

Star	P (rotn)	Sp. type	$M/M_\odot$	V	r (pc)	[Fe/H]	
a	25^{d}	G8/K0 V	0.86	5.95	12.58	−0.05	

16.4.6 The Multi-Planet System of Gliese J 876

This three-planet system (Rivera et al. 2005) is the closest known planetary system to Earth. There appears to be very little RV jitter in this star, a faint, red dwarf variable star, IL Aquarii. It is one of a small but growing number of M dwarfs that have been discovered to harbor planets (the others are: GJ 436, GJ 674, and GJ 581 with three planets). The properties of the system are summarized in Table 16.10. If the planets are coplanar, then Rivera finds an optimized inclination of $50°2$ and masses for planets b, c, and d of 2.53, 0.79, and $0.0237 M_J (= 7.53 M_\oplus)$, respectively. Planet d is among the lowest mass objects found through the radial velocity technique. No evidence has been found of transit eclipses of any of the three objects in this system (Rivera et al. 2005; Shankland et al. 2006), in contradiction to the astrometric conclusion of Benedict et al. (2002) that the outer planet has an inclination of ~86°.

The large masses of two of these planets are unexpected around such a low-mass star, because the time scale of formation of such planets exceeds that of the disk from which planets are expected to form. This provides difficulties for both a core-accretion theory as well as clumping in the disk due to gravitational instabilities. Moreover, the estimate of metals in this star is relatively low. If accurate, this compounds the problem of how these planets were formed, because stars with metallicities higher than that of the Sun are found to be more likely to have planets, indicating that the dust in the planetary disk plays a key role in the formation of planets.

16.4.7 The ϵ Eridani System

This is an interesting system for a number of reasons. First, the star is of a late spectral type, and it demonstrates the large velocity jitter due to strong photospheric effects that are expected in late-type stars. It is also

Table 16.10. The Gliese J 876 planetary system

Planet	Period	T_{peri} (JD-245000)	e	ω	K (m/s)	$M_J \sin i$	a (au)
b	60^d94	–	0.034	186°	212.8	1.93	0.208
c	30^d46	–	0.263	197°	87.1	0.61	0.131
d	1^d93774	–	0	–	6.32	0.018	0.021

Star	Sp. type	$M/M_\odot$	$\Re/\Re_\odot$	V	r (pc)	[Fe/H]
a	M4 V	0.32	0.3	10.17	4.70	0.02

Table 16.11. The ϵ Eridani planetary system

Planet	Period	$T_{\rm peri}$ (JD-2450000)	e	ω	K (m/s)	$M_{\rm J}$ $\sin i$	a (au)
b	2500$^{\rm d}$	8940.	0.25	6°	18.6	1.06	3.38
{b	154$^{\rm d}$46	1000.11	0.055	140°5	27.5	0.61	0.53}
{c	843$^{\rm d}$6	409.4	0.14	61°	15.4	0.60	1.64}

Star	$v \sin i$	Sp. type	$M/M_\odot$	$\Re/\Re_\odot$	V	r (pc)	[Fe/H]
a	1.7 km/s	K2 V	0.82	0.76	3.73	3.22	–

known to possess spotted regions (see, for example, Frey et al. (1991)). Nevertheless, radial velocity variations have been reported as due to a planet (Hatzes et al. 2000). Table 16.11 summarizes the properties of this interesting system. Butler et al. (2006) list only one planet, with period 2500 ± 350 d, $T_{\rm peri} = 8940\pm520$, $e = 0.25\pm0.23$, $\omega = 6°$, $K = 18.6\pm2.9$ m/s, $M \sin i = 1.06\pm0.16 M_{\rm J}$, and $a = 3.38\pm0.43$ au. Little supporting evidence for a pair of planets, deduced from structures in the disk, has been found, so parameters for the latter interpretation have been placed in brackets. Clumps of material peaking in brightness at $\sim$18 arc-sec from the star ($\sim$60 au) and extending to 35 arc-sec from the star, have been interpreted as remnants of a debris disk, seen nearly pole-on (Greaves et al. 1998). Other studies of the disk and attempts to observe planets through direct imaging are summarized by Marengo et al. (2006), who report three possible objects seen in their Spitzer Space Telescope data at 3.6 μm, but no evidence of objects with the expected colors and fluxes appropriate for masses as low as $\sim$2$M_{\rm J}$ within the disc, or 1$M_{\rm J}$ outside it.

With inclinations of the planetary orbits as low as that suggested for the disk ($i \approx 30\pm15°$), the masses could be 2–4$\times$ greater, but they orbit much closer to the star than the debris disk, and could not be the cause of resonant clumpings within the disk (a hypothesis advanced by Quillen and Thorndike 2002). Nevertheless, spot modeling of long series of observations made with the MOST satellite suggests an inclination of $30\pm3°$ (Croll et al. 2006) of the rotation axis to the line of sight. This work also confirmed differential rotation from the tracking of two spot regions and, based on an assumed radius of 0.76$R_\odot$, produced an equatorial rotation period of 11$^{\rm d}$20 and an equatorial rotation rate of 3.42 km/s (from which the entry $v \sin i = 1.7$ km/s is derived).

16.4.8 The TrES-1 System

This is one of the transit-discovered systems with a small, wide-field camera, and subsequently studied spectroscopically to determine the mass. The

Table 16.12. The TrES-1 planetary system

Planet	Period	T_{tr} (JD-2450000)	e	ω	R (R_J)	$M_{p/J}$	i	a (au)
b	$3^d_.030065(8)$	3186.806	0	–	1.08(18)	0.61(6)	88(2)°	0.039(1)

Star	$v \sin i$	Sp. type	M ($M_\odot$)	$\Re/\Re_\odot$	V	r (pc)	[Fe/H]	T
a	10.359 m/s	K0 V	0.89(5)	0.83(5)	11.79	157(6)	0.00(9)	5250 K

system is described by Alonso et al. (2004). The properties of the system are listed in Table 16.12. The most recent discussions of the host star are given by Sozzetti et al. (2004, 2006b). The object is both smaller than but of similar mass to HD 209458b, making it similar in density to Jupiter. It would appear from these objects as well as the OGLE transit cases (56b, 111b, 113b, 132b, which have densities in the range 600–1500 kg/m^3), that HD 209458b is quite anomalous; however, note the results for planet XO-1b, discussed in Section 16.4.9, and OGLE-TR-10, discussed in Section 16.4.10. Charbonneau et al. (2005) succeeded in observing the secondary eclipse (occultation of TrES-1b), enabling the flux with and without the planet to be measured in the infrared with the Spitzer Space Telescope. Assuming thermal equilibrium, they derived $T = 1060 \pm 50$ K and a Bond Albedo, $A = 0.31 \pm 0.14$, the first such determination for an extra-solar planet.

16.4.9 The XO-1 System

This system is another case of a transiting planet discovered in a photometric survey, and subsequently studied spectroscopically. The details can be found in McCullough et al. (2006). The most recent photometry has been carried out and analyzed by Holman et al. (2006, 2007) and by Wilson et al. (2006). Note that from the data of Table 16.13, the density of this object is only 40% that of Jupiter or ~510 kg/m^3, which, if correct, would make the density of HD 209458, although still lower, less anomalous. The radius determined for the planet depends on the radius adopted for the star and, as well, the planetary mass depends on the mass of the star, as discussed in Section 16.2. All the sources cited by Holman et al. (2006) derived $R_{star} = 0.938 R_\odot$ from their light curve whereas McCullough et al. (2006) and Wilson et al. (2006) use $1.0 R_\odot$. Holman et al. (2006) derive $R_p = 1.184\,R_J$ to obtain a density of ~670 kg/m^3; Wilson et al. (2006) derived $R_p = 1.34 \pm 0.12\,R_J$ which implies a density of ~370 kg/m^3. The uncertainties are large, but it appears that XO-1b is, in fact, less dense than TrES-1b and the OGLE transit cases discussed in Section 16.4.8.

Table 16.13. The XO-1 planetary system

Planet	Period	T_{tr} (JD-2450000)	e	ω	R (R_3)	$M_{p/J}$	i	a (au)
b	$3\overset{d}{.}94153(3)$	3808.917(1)	0	–	1.34(12)	0.90(7)	$89\overset{\circ}{.}3(5)$	0.049

Star	$v \sin i$	Sp. type	M ($M_\odot$)	$\Re/\Re_\odot$	V	r (pc)	[Fe/H]	T
a	1.11 km/s	G1 V	1.00(3)	0.93(2)	11.30	200(20)	0.15	5750 K

The size of a "hot Jupiter" may depend on its initial mass and age because of the effects of evaporation, as argued by Baraffe et al. (2004). We comment further on this in Section 16.4.10.

16.4.10 The OGLE-TR-10 System

This object was discovered through the transit technique by Udalski et al. (2002) and subsequently studied spectroscopically by Bouchy et al. (2005a) and by Konacki et al. (2005), whose values we cite in Table 16.14.

The system is of interest, because it, together with HD 209458b, has a significantly smaller mass and larger radius (and therefore lower density) than the other half dozen systems studied both photometrically and spectroscopically to date. Konacki et al. (2005) obtain a radius $R = 1.24\pm0.09R_J$, and a density of 380 ± 100 kg/m. OGLE-TR-10b and HD 209458b may well be expanding, as suggested by Baraffe et al. (2004), in the course of the evaporation process. Konacki et al. (2005) notes that among the "hot Jupiters" and "very hot Jupiters," the shorter period (and thus closer and hotter) planets tend to have larger masses, a finding that may be related to the survival of the planets. HD 209458b and OGLE-TR-10b are among the "hot Jupiters" with anomalously large radii and low densities. Baraffe et al. (2004) noted that the more massive a planet, the more it appeared to be resistant to evaporation. Even so, it appears that these two planets apparently are not in imminent danger of disappearing anytime soon.

Table 16.14. The OGLE-TR-10 system

Planet	Period	T_{tr} (JD-2450000)	e	ω	K (m/s)	$M_{p/J}$	i	a (au)
b	$3\overset{d}{.}10139(3)$	70.222(3)	0	–	80(17)	0.57(12)	$89(2)^\circ$	0.0416(7)

Star	$v \sin i$	Sp. type	M ($M_\odot$)	$\Re/\Re_\odot$	V	r (pc)	[Fe/H]	T
a	3(2) m/s	G	1.00(5):	1.0(1):	15.8	–	0.0(2)	5740 K

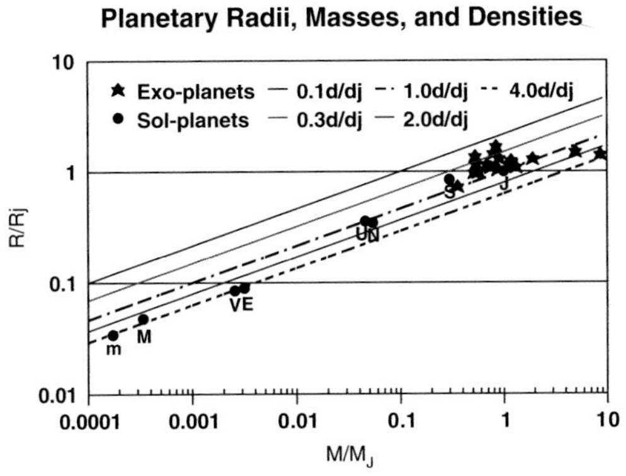

Fig. 16.7. The log radius vs. log mass relation of transit planets, amid lines of constant density relative to Jupiter. The solar system's major planets are also shown for comparison (m = Mercury, M = Mars, etc.)

More recently, planets have been discovered that are both more dense and less dense than those discussed here. At present writing, HAT P-1b and WASP-1b appear to be similar low-density planets and HD 149026 the most dense found to date. See Figure 16.7 for a graphical summary.

This concludes our brief examination of some individual planetary systems. We now discuss briefly the origins of the planets and their more massive cousins, the brown dwarfs.

16.5 Origins of Brown Dwarfs and Planets

The nature of the planetary discoveries since 1995 has demonstrated the variety of planetary systems that can exist around stars, both near to us, and quite distant. As we have noted often, the masses of the great majority of the planets discovered through radial velocity variations of their stars are underestimated; therefore at least a few of the objects in Table 16.2 are brown dwarfs. This is also the case for some of the objects discovered through direct imaging, if the 13 Jovian mass limit for planets turns out to be correct generally (for example, for all chemical compositions). Unless and until the existence of "free-floating" field planets can be verified, one must suppose that they are all born within a disk surrounding a proto-star, or, perhaps, a proto-brown dwarf, as we discussed at the end of the Chapter 15. Work by Muench et al. (2001) points quite clearly to an origin for brown dwarfs similar to that of stars (and therefore different from that of planets), and this seems to be generally accepted.

The implications of planets at small distances from the star, yet possessing discernible eccentricities, raises interesting questions about how long such planets could have remained in such configurations. The results of much modeling in the multi-planet systems have shown that the dynamic interactions among planets are such as to render the orbits non-Keplerian and in some cases chaotic. The result is variation in orbital elements. The likelihood is that planets migrate from their initial birthplaces in response to the viscous environment from which they formed, but that some circumstances, such as fortuitous locations of low viscosity "dead zones" (Matsumura and Pudritz 2003; Matsumura et al. (2007)) may enable some of them to escape tumbling into the star. Outward migrations of some planets, induced by perturbations of large gas giants, may also occur, as has been suggested for Uranus and Neptune. One of the more interesting extra-solar cases is that of HD 188753Ab. The star is in a triple system, with a short-period spectroscopic binary. The planet orbits the single star, but it is located only at 0.046 au from it. Clearly, hierarchical systems exist. Planets have also been found around M dwarfs and around K giant stars. Moreover, it is quite likely that planets exist around hotter non-solar type stars, such as Vega; at least, the hypothesis has been put forward that clumpy features in intermediate and the far-infrared are due to the trapping of disk material in mean motion resonances of a Neptune-mass planet in an orbit 65 au from the star (Marsh et al. 2006).

Brown dwarfs, on the other hand, seem to be rarely found in the company of normal stars (see Figure 16.1). Yet, the evidence seems to suggest that they are born like stars, from a fragmenting molecular cloud. If so, why is there a "brown dwarf desert?" In fact, there seem to be "brown dwarf deserts": there is a paucity also of wide substellar binaries (Allen et al. 2007). The "deserts" can be understood if brown dwarfs do form in clusters, as stars do. Preferential ejections (with probability $P \propto M^{-3}$ according to Anosova, 1986) of low-mass objects occur through collisions. Hence, solitary substellar objects may be the rule and pairs of them exceptions. Such a "dynamical bias" was found by Clarke (1996) in simulations, and later used by Reipurth and Clarke (2001) to support a hypothesis that brown dwarfs result through dynamical disruption by collisions of possibly late-forming protostellar embryos before they could accrete sufficient mass to become stars, and so are "ejected stellar embryos."

The origins of at least some types of planets are also somewhat uncertain at present. A birth within a disk around a proto-star appears very likely, but is this due to core-accretion, where accretion through collisions results in objects sufficiently massive to attract large amounts of gas, or is it due to the onset of gravitational instabilities (where the thermal gas pressure is unable to prevent the collapse of material into gravitationally viable clumps), as suggested by Boss (1997)? The latter mechanism may occur in the cooler, outer portions of the disk, and over short time scales, ~ 1000 years or less, while core-accretion may take ~ 8 million years to form a Jupiter-mass planet.

Because the lifetime of proto-planetary disks is believed to be less than a few million years at best, and may be less than a million years in high-mass star forming regions, the rapid time scale promised by gravitational instability seems to be required in at least some systems. Further work by Boss (2002, for example) suggests that the clumps formed in realistic simulations satisfy Jeans mass and length criteria (they are massive enough so that gravity triumphs over thermal dissipation), and that at least some of them can survive to continue to accrete material and grow as the disk dissipates. The most recent work we have seen suggests that core-accretion, which is favored by some observations (Fischer and Valenti 2005) and by some theoretical work (Rafikov 2004, 2005), may occur under certain circumstances, quickly enough, afterall (Thommes et al. 2007). In any case, it appears that terrestrial planets and objects as large as Neptune still need to be formed through accretion. Future studies of disks and disk structures may clarify if the instability does occur.

Modeling of the evolution of brown dwarfs (and planetary objects—see Figure 16.5) suggests that they are initially bright, and dim with age. Similarly, their rotational velocities initially speed up (presumably corresponding to gravitational contraction, in conservation of angular momentum), and then slow down with age. Planets must follow a similar path, although with no internal nuclear energy source.

The detection and study of the orbits of extra-solar planets has led to a closer look at the stability of the solar system. The evidence for migration of Saturn, Uranus and Neptune outward, where they perturbed some of the Edgeworth-Kuiper Belt objects into a scattered population, and the inward migration of Jupiter seems very strong. We appear to be living at the right time and place to be able to contemplate such things, because the solar system is only marginally dynamically stable.

This is where the research clues have led us thus far. But grand mysteries yet remain to be unraveled. Is there life elsewhere in the solar system or beyond it? At present writing, the closest Earth analog detected thus far may be Gl 581c with m sin i $= 5.0 M_\oplus$ at 0.073 au, on the hot end of the star's *habitable zone* (Udry et al. 2007). There are programs to seek terrestrial planets within their stars' habitable zones, and then to look for the signatures of biological activity. The next generation of space missions may provide the answers. For now, we comment on "habitable zones" as follows.

The habitable zone for any star is the range of distances from the star within which life may exist on a planet or large moon. We include here the usual (but perhaps not necessary!) restriction to carbon-based life forms that require liquid water (i.e., similar to life on the Earth), and also to life on the planetary surface, including surface oceans. The latter restriction excludes some regions of a planetary system where life may in fact be possible, such as subsurface oceans on Europa, Ganymede, Callisto and Enceladus, or within the cryo-volcanoes of Titan, all dependent on non-solar sources of heat.

We locate the inner and outer boundaries of the habitable zone at those distances for which the mean equilibrium temperature is, respectively, 373 and 273 K, the boiling and freezing points of water. As defined in Chapter 10.1, the mean equilibrium temperature of a planet is the blackbody temperature the planet must have (in the absence of internal sources of heat) in order to radiate the same total energy per second that it receives from the star. As such, it should be identical to the mean bolometric temperature as observed from space, and is the same (when the mean is taken over the entire planet) for any planet of a given albedo placed at that distance from the star.

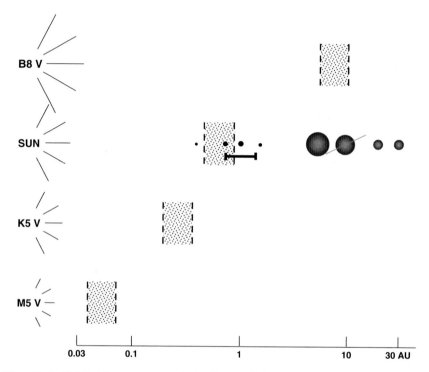

Fig. 16.8. Habitable zones around the Sun and three other main sequence stars: B8V, K5V and M5V. The inner and outer edges of the habitable zones were calculated for airless, quickly-rotating planets of mean equilibrium temperatures 100°C at the inner edge of the zone and 0°C at the outer edge, and an assumed Bond albedo of 0.306 for a "typical" terrestrial planet (the Earth). These criteria place the Earth outside the Sun's habitable zone, because the equilibrium temperature for a planet at 1 au is ~248 K. However, the mean temperature of the Earth is ~288 K, at least in part because of the thermal effects of the atmosphere. With such an atmosphere, the habitable zone is located further from the Sun, as indicated by the bracketed line. Similar effects could be expected of exo-planetary systems also, but the amount of shift for other planets depends critically on the abundance of greenhouse gases (e.g., methane, carbon dioxide, and water vapor) in the planets' atmospheres.

The rapid-rotator case discussed in Milone & Wilson (2008), Chapter 6.3, and Chapter 10.1 in the present volume gives this temperature directly.

Figure 16.8 shows the habitable zones for a planetary albedo equal to that of the Earth, for four stars, including the Sun, characterized by the following effective temperatures and radii, according to data tabulated in Cox (2000): for B8V, 11,400 K, 3.0 $R_\odot$; K5V, 4410 K, 0.72 $R_\odot$; M5V, 3170 K, 0.27 $R_\odot$. For the Sun, we took T_{eff} = 5770 K and $R_\odot$ = 6.955 x 10^8 m. The radial distance to the habitable zone increases as the square root of both the stellar luminosity and the absorptivity, 1-A (where A is the albedo) by Milone & Wilson (2008) equation (6.31) or the equation below (10.6), if we replace the solar luminosity by the stellar luminosity. From Milone & Wilson (2008) Example 6.1, the Earth's equilibrium temperature is 248 K, which, as Figure 16.8 shows, places the Earth outside the Sun's habitable zone by the definitions used here. However, the existence of life does not depend (directly) on the mean bolometric temperature, but on the actual surface temperature of the planet, and the latter can be strongly influenced by atmospheric and oceanic circulations and the greenhouse effect. We retain the definitions used here because the latter effects can vary markedly from planet to planet (note Venus!!) and are difficult to predict. In Fig. 16.8, the bracketed line indicates where the habitable zone would be located if it were centered on the Earth, where the mean surface temperature is 288 K instead of 323 K. One must be cautious in interpreting the boundaries of the zone in this case, because if the Earth were arbitrarily moved to a location elsewhere within this zone, particularly near the boundaries, atmospheric and oceanic feedback mechanisms (e.g., runaway greenhouse effect) might in fact result in the Earth becoming uninhabitable.

References

Allen, P. R., Koerner, D. W., McElwain, M. W., Cruz, K. L., and Reid, I. N. 2007. "A New Brown Dwarf Desert? A Scarcity of Wide Ultracool Binaries," *Astronomical Journal*, **133**, 971–978.

Alonso, R., Brown, T. M., Torres, G., Latham, D. W., Sozzetti, A., Mandushev, G., Belmonte, J. A., Charbonneau, D., Deeg, H. J., Dunham, E. W., O'Donovan, F. T., and Stefanik, R. P. 2004. "TrES-1: The Transiting Planet of a Bright K0 V Star," *Astrophysical Journal*, **613**, L153–L156.

Anosova, J. P. 1986. "Dynamical Evolution of Triple Systems," *Astrophysics and Space Science*, **124**, 217–241.

Baraffe, I., Chabrier, G., Allard, F., and Hauschildt, P. H. 2002, "Evolutionary Models for Low-Mass Stars and Brown Dwarfs: Uncertainties and Limits at Very Young Ages," *Astronomy and Astrophysics*, **382**, 563–572.

Baraffe, I., Selsis, F., Chabrier, G., Barman, T. S., Allard, F., Hauschildt, P. H., and Lammer, H. 2004. "The Effect of Evaporation on the Evolution of Close-In Giant Planets," *Astronomy and Astrophysics*, **419**, L13–L16.

Barnes, R. and Greenberg, R. 2006. "Extrasolar Planets near a Secular Separatix," *Astrophysical Journal*, **638**, 478–487.

Basri, G. 2000. "Observations of Brown Dwarfs" in *Annual Review of Astronomy and Astrophysics*, eds G. Burbidge, A. Sandage, and F. H. Shu, **38**, pp. 485–519.

Basri, G. and Brown, M. E. 2006. "Planetesimals to Brown Dwarfs: What is a Planet?" *Annual Review of Earth and Planetary Sciences*, **34**, 193–216.

Beaulieu, J.-P. et al. (72 co-authors). 2006. "Discovery of a Cool Planet of 5.5 Earth Masses through Gravitational Microlensing," *Nature*, **439**, 437–440.

Benedict, G. F., McArthur, B. E., Forveille, T., Delfosse, X., Nelan, E., Butler, R. P., Spiesman, W., Marcy, G., Goldman, B., Perrier, C., Jeffreys, W. H., and Mayor, M. 2002. "A Mass for the Extrasolar Planet Gliese 876b Determined from *Hubble Space Telescope* Fine Guidance Sensor 3 Astrometry and High-Precision Radial Velocities," *Astrophysical Journal*, **581**, L115–L118.

Bennett, D. P., Anderson, J., Bond, I. A., Udalski, A., and Gould, A. 2006. "Identification of the OGLE-2003-BLG-235/MOA-2003-BLG-53 Planetary Host Star," *Astrophysical Journal*, **647**, L171–L174.

Biller, B., Kasper, M., Close, L., Brandner, W., and Kellner, S. 2006. "Discovery of a Very Nearby Brown Dwarf to the Sun: A Methane Rich Brown Dwarf Companion to the Low Mass Star SCR 1845-6357," *Astrophysical Journal*, **641**, L141–L144.

Bond, I. et al. (31 co-authors). 2004, "OGLE 2003 BLG-235/MOA 2003-BLG-53: A Planetary Microlensing Event," *Astrophysical Journal*, **606**, L155–L158.

Bonfils, X., Forveille, T., Delfosse, X., Udry, S., Mayor, M., Perrier, C., Bouchy, F., Pepe, F., Queloz, D., Bertaux, J.-L. 2005. "The HARPS Search of Southern Extra-solar Planets. VI. A Neptune-Mass Planet Around the Nearby M Dwarf Cl 581," *Astronomy and Astrophysics*, **443**, L15–L18.

Boss, A. P. 1997. "Giant Planet Formation by Gravitational Instability," *Science*, **276**, 1836–1839.

Boss, A. P. 2002. "Evolution of the Solar Nebula. V. Disk Instabilities with Varied Thermodynamics," *Astrophysical Journal*, **576**, 462–472.

Bouchy, F., Pont, F., Santos, N. C., Melo, C., Mayor, M., Queloz, D., and Udry, S. 2004. 'Two New "Very Hot Jupiters" among the OGLE Transiting Candidates," *Astronomy and Astrophysics*, **421**, L13-L16.

Bouchy, F., Pont, F., Melo, C., Santos, N. C., Mayor, M., Queloz, D., Udry, S. 2005a. "Doppler Follow-up of OGLE Transiting Companions in the Galactic Bulge," *Astronomy and Astrophysics*, **443**, 1105–1121.

Bouchy, F., Udry, S., Mayor, M., Moutou, C., Pont, F., Iribarne, N., da Silva, R., Ilovaisky, S., Queloz, D., Santos, N. C., Ségransan, D., and Zucker, S. 2005b. "ELODIE Metallicity-biased Search for Transiting Hot Jupiters: II. A Very Hot Jupiter Transiting the Bright Star HD 189733," *Astronomy and Astrophysics*, **444**, L15–L19.

Brown, T. M., Charbonneau, D., Gilliland, R. L., Noyes, R. W., and Burrows, A. 2001. "Hubble Space Telescope Time-Series Photometry of the Transiting Planet of HD 209458," *Astrophysical Journal*, **552**, 699–709.

Burgasser, A. J., Kirkpatrick, J. D., Cruz, K. L., Reid, I. N., Leggett, S. K., Liebert, J., Burrows, A., and Brown, M. E. 2006. "Hubble Space Telescope

NICMOS Observations of T Dwarfs: Brown Dwarf Multiplicity and New Probes of the L/T Transition," *Astrophysical Journal*, **166**, 585–612.

Burke, C. J., et al. (17 co-authors) 2007. "XO-2b: Transiting Hot Jupiter in a Metal-Rich Common Proper Motion Binary," *Astrophysical Journal*, in press.

Butler, R. P., Marcy, G. W., Williams, E., McCarthy C., Dosanjh, P., and Vogt, S. S. 1996. "Attaining Doppler Precision of 3 m s^{-1}". *Publications of the Astronomical Society of the Pacific*, **108**, 500–509.

Butler, R. P., Wright, J. T., Marcy, G. W., Fischer, D. A., Vogt, S. S., Tinney, C. G., Jones, H. R. A., Carter, B. D., Johnson, J. A., McCarthy, C., and Penny, A. J. 2006. "Catalog of Nearby Exoplanets," *Astrophysical Journal*, **646**, 505–522.

Campbell, B., and Walker, G. A. H. 1979. "Precision Radial Velocities with an Absorption Cell," *Publications of the Astronomical Society of the Pacific*, **91**, 540–545.

Campbell, B., Walker, G., and Yang, S. 1988, "A Search for Substellar Companions to Solar-type Stars," *Astrophysical Journal*, **331**, 902–921.

Chabrier, G. and Baraffe, I. 2000. "Theory of Low-Mass Stars and Substellar Objects", in *Annual Review of Astronomy and Astrophysics*, eds G. Burbidge, A. Sandage, and F. H. Shu, **38**, pp. 337–377.

Chabrier, G., Baraffe, I., Allard, F., and Hauschildt, P. 2000. "Evolutionary Models for Very Low-Mass Stars and Brown Dwarfs with Dusty Atmospheres," *Astrophysical Journal*, **542**, 464–472.

Charbonneau, D., Brown, T., Noyes, R., and Gilliland, R. 2001. "Detection of an Extrasolar Planet Atmosphere," *Astrophysical Journal*, **568**, 377–384.

Charbonneau, D., Brown, T., Latham, D., Mayor, M., and Mazeh, T. 2000. "Detection of Planetary Transits Across a Sun-like Star," *Astrophysical Journal*, **529**, L45–L48.

Charbonneau, D., Winn, J. N., Everett, M. E., Latham, D. W., Holman, M. J., Esquerdo, G. A., and O'Donovan, F. T. 2007. "Precise Radius Estimates for the Exoplanets WASP1-b and WASP-2b," *Astrophysical Journal*, 6**58**, 1322–1327.

Charbonneau, D., Allen, L. E., Megeath, S. T., Torres, G., Alonso, R., Brown, T. M., Gilliland, R. L., Latham, D. W., Mandushev, G., O'Donovan, F. T., and Sozzetti, A. 2005. "Detection of Thermal Emission from an Extrasolar Planet," *Astrophysical Journal*, **626**, 523–529.

Charbonneau, D., Wynn, J. N., Latham, D. W., Bakos, G., Falco, E. E., Holman, M. J., Noyes, R. W., Csák, B., Esquerdo, G. A., Everett, M. E., and O'Donovan, F. T. 2006. "Transit Photometry of the Core-Dominated Planet HD149026b," *Astrophysical Journal*, **636**, 445–453.

Chauvin, G., Lagrange, A. M., Dumas, C., Zuckerman, B., Mouillet, D., Song, I., Beuzit, J.-L., and Lowrance, P. 2004. "A Giant Planet Candidate near a Young Brown Dwarf: Direct VLT/NACO Observations Using IR Wavefront Sensing," *Astronomy and Astrophysics*, **425**, L29–L32.

Chauvin, G., Lagrange, A. M., Zuckerman, B., Dumas, C., Mouillet, D., Song, I., Beuzit, J.-L., Lowrance, P., and Bessell, M. S. 2005. "A Companion to AB Pic at the Planet/Brown Dwarf Boundary," *Astronomy and Astrophysics*, **438**, L29–L32.

Chauvin, G., Lagrange, A. M., Udry, S., Fusco, T., Galland, F., Naef, D., Beuzit, J.-L., and Mayor, M. 2006. "Probing Long-Period Companions to Planetary

Hosts. VLT and CFHT Near Infrared Coronographic Imaging Surveys," *Astronomy and Astrophysics,* **456**, 1165–1172.

Clarke, C. J. 1996, "The Formation of Binaries in Small N Clusters," in *The Origin, Evolution, and Destinies of Binary Stars in Clusters,* eds. E. F. Milone and J.-C. Mermilliod, ASP Conference Series, **90**, pp. 242–251.

Cochran, W. D. et al. (15 co-authors). 2004. "The First Hobby-Eberly Telescope Planet: a Companion to HD 37605," *Astrophysical Journal,* **611**, L133–L136.

Collier Cameron, A. et al. (38 co-authors) 2007. "WASP-1b and WASP-2b: Two New Transiting Exoplanets Detected with SuperWASP and SOPHIE," *Monthly Notices of the RAS,* **375**, 951–957.

Correia, A. C. M., Udry, S., Mayor, M., Laskar, J., Naef, D., Pepe, F., Queloz, D., and Santos, N. C. 2005. "The CORALIE Survey for Southern Extra-Solar Planets. XIII. A Pair of Planets around HD 202206 or a Circumbinary Planet?" *Astronomy and Astrophysics,* **440**, 751–758.

Correia, A. C. M., Udry, S., Mayor, M., Eggenberger, A., Naef, D., Beuzit, J.-L., Perrier, C., Queloz, D., Sivan, J.-P., Pepe, F., Santos, N. C., and Segransan, D. 2008. "The ELODIE Survey for Northern Extra-solar Planets. IV. HD 196885, a Close Binary Star with a 3.7-year Planet," *Astronomy and Astrophysics,* **479**, 271–275.

Cox, A. N., ed. (2000). *Allen's Astrophysical Quantities.* 4th ed. (New York: Springer-Verlag).

Croll, B., Walker, G. A. H., Kuschnig, R., Matthews, J. M., Rowe, J. F., Walker, A., Rucinski, S. M., Hatzes, A. P., Cochran, W. D., Robb, R. M., Guenther, D. B., Moffat, A. F. J., Sasselov, D., and Weiss, W. W. 2006. "Differential Rotation of ϵ Eridani Detected by MOST," *Astrophysical Journal,* **648**, 607–613.

Da Silva, R., Udry, S., Bouchy, F., Mayor, M., Moutou, C., Pont, F., Queloz, D., Santos, N. C., Ségransan, D., and Zucker, S. 2006. "Elodie Metallicity-Biased Search for Transiting Hot Jupiters. I. Two Hot Jupiters Orbiting the Slightly Evolved Stars HD 118203 and HD 149143," *Astronomy and Astrophysics,* **446**, 717–722.

Delfosse, X., Forveille, T., Mayor, M., Perrier, C., Naef, D., and Queloz, D. 1998. "The Closest Extrasolar Planet: A Giant Planet Around the M4 Dwarf Gl876," *Astronomy and Astrophysics,* **338**, L67–L70.

Deming, D., Seager, S., Richardson, L. J., and Harrington, J. 2005. "Infrared Radiation from an Extrasolar Planet," *Nature,* **434**, 740–742.

Deming, D., Harrington, J., Seager, S., and Richardson, L. J. 2006. "Strong Infrared Emission from the Extrasolar Planet HD 189733," *Astrophysical Journal,* **644**, 560–564.

Díaz, R. F., Ramirez, S., Fernández, J. M., Gallardo, J., Gieren, W., Ivanov, V. D., Mauas, P., Minniti, D., Pietrzynski, G., Pérez, F., Ruíz, M. T., Udalski, A., Zoccali, M. 2007. "Millimagnitude Photometry of OGLE-TR-113-b in the Optical and Near-IR," *Astrophysical Journal,* **660**, 850–857.

Eggenberger, A., Mayor, M., Naef, D., Pepe, F., Santos, N. C., Udry, S., and Lovis, C. 2006. "The CORALIE Survey for Southern Extrasolar Planets. XIV. HD 1420222 b: A Long-Period Planetary Companion in a Wide Binary," *Astronomy and Astrophysics,* **447**, 1159–1163.

Endl, M., Hatzes, A. P., Cochran, W. D., McArthur, B., Allende Prieto, C., Paulson, D. B., Guenther, E., and Bedalov, A. 2004. "HD 137510: An Oasis in the Brown Dwarf Desert," *Astrophysical Journal,* **611**, 1121–1124.

Endl, M., Cochran, W. D., Wittenmyer, R. A., and Hatzes, A. P. 2006. "Determination of the Orbits of the Planetary Companion to the Metal-Rich Star HD 45350," *Astronomical Journal*, **131**, 3131–3134.

Fernández, J. M., Minniti, D., Pietrzynski, G., Gieren, W., Ruiz, M. T., Zoccali, M., Udalski, A., and Szeifert, T. 2006. "Millimagnitude Optical Photometry for the Transiting Planetary Candidate OGLE-TR-109," *Astrophysical Journal*, **647**, 587–593.

Fischer, D. A. and Valenti, J. 2005. "The Planet–Metallicity Correlation," *Astrophysical Journal*, **622**, 1102–1117.

Fischer, D. A., Marcy, G. A., Butler, R. P., Laughlin, G., and Vogt, S. S. 2002a. "A Second Planet Orbiting 47 Ursae Majoris," *Astrophysical Journal*, **564**, 1028–1034.

Fischer, D. A., Marcy, G. A., Butler, R. P., Vogt, S. S., Walp, B., and Apps, K. 2002b. "Planetary Companions to HD 136118, HD 50554, and HD 106252," *Publications of the Astronomical Society of the Pacific*, **114**, 529–535.

Fischer, D. A. et al. (23 co-authors). 2005. "The N2k Consortium. I. a Hot Saturn Planet Orbiting HD 88133," *Astrophysical Journal*, **620**, 481–486.

Fischer, D. A., Marcy, G. W., Butler, R. P., Vogt, S. S., Laughlin, G., Henry, G. W., Abouav, D., Peek, K. M. G., Wright, J. T., Johnson, J. A., McCarthy, C., Isaacson, H. 2007. "Five Planets Orbiting 55 Cancri," *Astrophysical Journal*, in press.

Fischer, D. A. et al. (22 co-authors). 2006. "The N2K Consortium. III. Short-Period Planets Orbiting HD 149143 and HD 109749," *Astrophysical Journal*, **637**, 1094–1101.

Fregau, L., Chatterjee, S., and Rasio, F. 2006. "Dynamical Interactions of Planetary Systems in Dense Stellar Environments," *Astrophysical Journal*, **640**, 1086–1098.

Frey, G. J., Grim, B., Hall, D. S., Mattingly, P., Robb, S., and Zeigler, K. 1991. "The Rotation Period of ϵ Eri from Photometry of Starspots," *Astronomical Journal*, **102**, 1813–1815.

Frink, S., Mirchell, D. S., Quirrenbach, A., Fischer, D. A., Marcy, G. W., and Butler, R. P. 2002. "Discovery of a Substellar Companion to the K2 III Giant ι Draconis," *Astrophysical Journal*, **576**, 478–484.

Galland, F., LaGrange, A. M., Udry, S., Chelli, A., Pepe, F., Beuzit, J.-L., and Mayor, M. 2005. "Extrasolar Planets and Brown Dwarfs Around A-F Type Stars. II. A Planet Found with ELODIE Around the F6V Star HD 33564," *Astronomy and Astrophysics*, **444**, L21–L24.

Gallardo, J., Minniti, D., Valls-Gabaud, D., and Rejkuba, M. 2005. "Characterisation of Extrasolar Planetary Candidates," *Astronomy and Astrophysics*, **431**, 707–720.

Gatewood, G. 1996. "Lalande 21185," *Bulletin, American Astronomical Society*, **28**, 885.

Gaudi, B. S., Gould, A., Udalski, A., An, D., Bennett, D., Zhou, A., Dong, S., Rattenbury, N. J., Yock, P. C., Bond, I. A., Christie, G. W., Horne, K., Anderson, J., Stanek, K. Z., MicroFUN Collaboration, OGLE Collaboration, RobotNet Collaboration 2005 "Microlens OGLE-2005-BLG-169 Implies Cool Neptune-Like Planets are Common," *Bulletin, American Astronomical Society*, **38**, 88.

Ge, J., van Eyken, J., Mahadevan, S., DeWitt, C., Kane, S. R., Cohen, R., Vanden Heuvel, A., Fleming, S. W., Guo, P., Henry, G. W., Schneider, D. P., Ramsey, L. W., Wittenmyer, R. A., Endl, M., Cochran, W. D., Ford, E. B., Martín, E. L., Israelian, G., Valenti, J., and Montes, D. 2006. "The First Extrasolar Planet Discovered with a New-Generation High-Throughput Doppler Instrument," *Astrophysical Journal*, **648**, 683–695.

Gelino, C. R., Kulkarni, S. R., and Stephens, D. C. 2006. "Evidence of Orbital Motion in the Binary Brown Dwarf Kelu-1AB," *Publications, Astronomical Society of the Pacific*, **118**, 611–616.

Gilliland, R. L., Brown, T. M., Guhathakurta, P., Sarajedini, A., Milone, E. F., Albrow, M. D., Baliber, N. R., Bruntt, H., Burrows, A., Charbonneau, D., Choi, P., Cochran, W. D., Edmunds, P. D., Frandsen, S., Howell, J. H., Lin, D. N. C., Marcy, G. W., Mayor, M., Naef, D., Sigurdsson, S., Stagg, C. R., Vandenberg, D. A., Vogt, S. S., and Williams, M. D. (2000) "A Lack of Planets in 47 Tucanae from a Hubble Space Telescope Search," *Astrophysical Journal*, **545**, L47–L51.

Gillon, M., Pont, F., Moutou, C., Santos, N. C., Bouchy, F., Hartman, J. D., Mayor, M., Melo, C., Queloz, D., Udry, S., and Magain, P. 2007. "The Transiting Planet OGLE-TR-132b Revisited with New Spectroscopy and Deconvolution Photometry," *Astronomy and Astrophysics*, **466**, 743–748.

Gimenez, A. 2006, "Equations for the Analysis of the Light Curves of Extra-Solar Planetary Transits," *Astronomy and Astrophysics*, **450**, 1231–1237.

Gorkavyi, N., Heap, S., Ozernoy, L., Taidakova, T., and Mather, J. 2004. "Indicator of Exo-Solar Planet(s) in the Circumstellar Disk Around β Pictoris" in *Planetary Systems in the Universe — Observation, Formation, and Evolution*, eds A. J. Penny, P. Artymowicz, A. M. Lagrange, and S. S. Russell, Procs., IAU Symposium No. 202 (San Francisco, CA: ASP), pp. 331–334.

Gould, A. et al. (35 co-authors). 2006. "Microlens OGLE-2005-BLG-169 Implies that Cool Neptune-like Planets are Common," *Astrophysical Journal*, **644**, L37–L40.

Greaves, J. S., Holland, W. S., Moriarty-Schieven, G., Jenness, T., Dent, W. R. F., Zuckerman, B., McCarthy, C., Webb, R. A., Butner, H. M., Gear, W. K., and Walker, H. J. 1998. "A Dust Ring around ϵ Eridani: Analog to the Young Solar System," *Astrophysical Journal*, **506**, L133–L137.

Grether, D. and Lineweaver, C. H. 2006. "How Dry is the Brown Dwarf Desert: Quantifying the Relative Nimber of Planets, Brown Dwarfs, and Stellar Companions around Nearby Sun-Like Stars," *Astrophysical Journal*, **640**, 1051–1062.

Guenther, E. W., Neuhäuser, R., Wuchterl, G., Mugrauer, M., Bedalov, A., and Hauschildt, P. H. 2005. "The Low-Mass Companion of GQ Lupi," *Astronomische Nachrichten*, **326**, 958–963.

Han, C. 2006. "Secure Identification of Free-Floating Planets," *Astrophysical Journal*, **644**, 1232–1236.

Hatzes, A. P., Cochran, W. D., McArthur, B., Baliunas, S. L., Walker, G. A. H., Campbell, B., Irwin, A. W., Yang, S., Kürster, M., Endl, M., Els, S., Butler, R. P., and Marcy, G. W. 2000. "Evidence of a Long-Period Planet Orbiting ϵ Eridani," *Astrophysical Journal*, **544**, L145–L148.

Hatzes, A. P., Cochran, W. D., Endl, M., McArthur, B., Paulson, D., Walker, G. A. H., Campbell, B., and Yang, S. 2003. "A Planetary Companion to γ Cephei A," *Astrophysical Journal*, **599**, 1383–1394.

Hatzes, A. P.,Guenther, E. W., Endl, M., Cochran, W. D., Döllinger, M. P., and Bedalov, A. 2005. "A Giant Planet Around the Massive Giant Star HD 13189," *Astronomy and Astrophysics*, **437**, 743–751.

Hatzes, A. P., Cochran, W. D., Endl, M., Guenther, E. W., Saar, S. H., Walker, G. A. H., Yang, S., Hartmann, M., Esposito, M., Paulson, D. B., and Döllinger, M. P. 2006. "Confirmation of the Planet Hypothesis for the Long-Period Radial Velocity Variations of β Geminorum," *Astronomy and Astrophysics*, **457**, 335–342.

Henry, G., Marcy, G., Butler, P., and Vogt, S., 2000. "A Transiting 51 Peg-Like Planet," *Astrophysical Journal*, **529**, L41–L44.

Henry, G. W., Donahue, R. A., and Baliunas, S. L. 2002. "A False Planet around HD 192263," *Astrophysical Journal*, **577**, L111–L114.

Holman, M. J., Winn, J. N., Latham, D. W., O'Donovan, F. T., Charbonneau, D., Bakos, G. A., Esquerdo, G. A., Hergenrother, C., Everett, M. E., and Pál, A. 2006. "The Transit Light Curve (TLC) Project. I. Four Consecutive Transits of the Exoplanet XO-1b," *Astrophysical Journal*, **652**, 1715–1723.

Holman, M. J., Winn, J. N., Fuentes, C. I., Hartman, J. D., Stanek, K. Z., Torres, G., Sasselov, D. D., Gaudi, B. S., Jones, R. L., and Fraser, W. 2007. "The Transit Light Curve Project. IV. Five Transits of the Exoplanet OGLE-TR-10b," *Astrophysical Journal*, **655**, 1103–1109.

Jayawardhana, R. and Ivanov, V. 2006. "Spectroscopy of Young Planetary Mass Candidates with Disks," *Astrophysical Journal*, **647**, L167–L170.

Joergens, V., and Müller, A. 2007. "16-20 M_{Jup} Radial Velocity Companion Orbiting the Brown Dwarf Candidate Cha Hα 8," *Astrophysical Journal*, **666**, L113–L116.

Johnson, J. A., Marcy, G. W., Fischer, D. A., Laughlin, G., Butler, R. P., Henry, G. W., Valenti, J. A., Ford, E. B., Vogt, S. S., and Wright, J. T. 2006a. "The N2K Consortium. VI. Doppler Shifts Without Templates and Three New Short-Period Planets," *Astrophysical Journal*, **647**, 600–611.

Johnson, J. A., Marcy, G. W., Fischer, D. A., Henry, G. W., Wright, J. T., Isaacon, H., and McCarthy, C. 2006b. "An Eccentric Hot Jupiter Orbiting the Subgiant HD 185269," *Astrophysical Journal*, **652,** 1724–1728.

Jones, H. R. A., Butler, R. P., Tinney, C. G., Marcy, G. W., Penney, A. J., McCarthy, C., Carter, B. D., and Pourbaix, D. 2002. "A Probable Planetary Companion to HD 39091 from the Anglo-Australian Planet Search ," *Monthly Notices, Royal Astronomical Society*, **333**, 871–875.

Jones, H. R. A., Butler, R. P., Tinney, C. G., Marcy, G. W., Carter, B. D., Penny, A. J., McCarthy, C., and Bailey, J. 2006. "High-Eccentricity Planets from the Anglo-Australian Planet Search," *Monthly Notices, Royal Astronomical Society*, **369**, 249–256.

Kallrath, J. and Milone, E. F. 1999. *Eclipsing Binary Stars: Modeling and Analysis* (New York: Springer-Verlag).

Kelley, D. H., and Milone, E. F. 2005. *Exploring Ancient Skies: An Encyclopedic Survey of Ancient Astronomy* (New York: Springer).

Konacki, M. 2005. "An Extrasolar Giant Planet in a Close Triple-Star System," *Nature*, **436**, 230–233.

Konacki, M. and Wolszczan, A. 2003. "Masses and Orbital Inclinations of Planets of the PSR B 1257+12 System," *Astrophysical Journal*, **591**, L147–L150.

Konacki, M., Torres, G., Sasselov, D. D., Pietrzynski, G., Udalski, A., Jha, S., Ruiz, M. T., Gieren, W., and Minniti, D. 2004. "The Transiting Extrasolar Giant Planet Around the Star OGLE-TR-113," *Astrophysical Journal*, **609**, L37–L40.

Konacki, M., Torres, G., Sasselov, D., and Jha, A. 2005. "A Transiting Extrasolar Planet around the Star OGLE-TR-10," *Astrophysical Journal*, **624**, 372–377.

Korzennik, S., Brown, T. M., Fischer, D. A., Nisenson, P., and Noyes, R. W. 2000. "A High-Eccentricity Low-Mass Companion to HD 89744," *Astrophysical Journal*, **533**, L147–L150.

Latham, D. W., Mazeh, T., Stefanik, R. P., Mayor, M., and Burki, G. 1989. "The Unseen Companion of HD 114762, A Probable Brown Dwarf," *Nature*, **339**, 38–40.

Lee, M. H., Butler, R. P., Fischer, D. A., Marcy, G. W., and Vogt, S. S. 2006. "On the 2:1 Orbital Resonance in the HD 82943 Planetary System," *Astrophysical Journal*, **641**, 1178–1187.

Lo Curto, G, Mayor, M., Clausen, J., Benz, W., Bouchy, F., Lovis, C., Moutou, C., Naef, D., Pepe, F., Queloz, D., Santos, N., Sivan, J.-P., Udy, S., Bonfils, X., Mordasini, C., Fouque, P., Olsen, E., and Pritchard, J. 2006. "The HARPS Search for Southern Extra-Solar Planets. VII. A Very Hot Jupiter Orbiting HD 212301,"*Astronomy and Astrophysics*, 451, 345–350.

Lovis, C., Mayor, M., Bouchy, F., Pepe, F., Queloz, D., Santos, N. C., Udry, S., Benz, W., Bertaux, J.-L., Mordasini, C., and Sivan, J.-P. 2005. "The HARPS Search for Southern Extra-Solar Planets. III. Three Saturn-Mass Planets Around HD 93083, HD 101930 and HD 102117," *Astronomy and Astrophysics*, **437**, 1121–1126.

Lovis, C., Mayor, M., Pepe, F., Alibert, Y., Benz, W., Bouchy, F., Correia, A. C. M., Laskar, J., Mordasini, C., Queloz, D., Santos, N. C., Udry, S., Bertaux, J.-L., and Sivan, J.-P. 2006. "An Extrasolar Planetary System with Three Neptune-Mass Planets," *Nature*, **441**, 305–309.

Luhman, K. L., Adame, L., d'Alessio, P., Calvet, N., Hartmann, L., Megeath, S. T., and Fazio, G. G. 2005. "Discovery of a Planetary-Mass Brown Dwarf with a Circumstellar Disk," *Astrophysical Journal*, **635**, L93–L96.

Lyot, B. (tr. Marschall, R. K.) 1933. "The Study of the Solar Corona Without an Eclipse," *Journal of Royal Astronomical Society of Canada*, **27**, 225–234; 265–280.

Mandushev, G., O'Donovan, F. T., Charbonneau, D., Torres, G., Latham, D. W., Bakos, G. A., Dunham, E. W., Sozzetti, A., Fernandez, J. M., Esquerdo, G. A., Everett, M. E., Brown, T. M., Rabus, M., Belmonte, J. A., and Hillenbrand, L. A. 2007. "TrES-4: A Transiting Hot Jupiter of Very Low Density," *Astrophysical Journal*, 667, L195–L198.

Marcy, G. W. and Butler, R. P. 1998. "Detection of Extrasolar Giant Planets," *Annual Review of Astronomy and Astrophysics*, **36**, 57–97.

Marengo, M., Megeath, S. T., Fazio, G. G., Stapelfeldt, K. R., Werner, M. W., and Backman, D. E. 2006. "A Spitzer IRAC Search for Substellar Companions of the Debris Disk Star ε Eridani," *Astrophysical Journal*, **647**, 1437–1451.

Marsh, K. A., Dowell, C. D., Velusamy, T., Grogan, K., and Beichman, C. A. 2006. "Images of the Vega Dust Ring at 350 and 450 µm: New Clues to the Trapping of Multiple-Sized Dust Particles in Planetary Resonances," *Astrophysical Journal*, **646**, L77–L80.

Matsumura, S. and Pudritz, R. E. 2003. "The Origin of Jovian Planets in Proto-stellar Disks: The Role of Dead Zones," *Astrophysical Journal*, **598**, 645–656.

Matsumura, S., Pudritz, R. E., and Thomas, E. W. 2007. "Saving Planetary Systems: Dead Zones and Planetary Migration," *Astrophysical Journal*, **660**, 1609–1623.

Mayor, M. and Queloz, D. 1995, "A Jupiter-Mass Companion to a Solar-Type Star," *Nature*, **378**, 355–359.

Mayor, M. and Santos, N. C. 2003. "Extrasolar Planetary Systems," in *Astronomy, Cosmology, and Fundamental Physics*, eds P. A. Shaver, L. DiLella, and A. Giménez (Berlin: Springer), pp. 359–370.

Mayor, M., Queloz, D., and Udry, S. 1998. "Mass Function and Orbital Distri-bution of Substellar Companions," in *Brown Dwarfs and Extrasolar Planets*, eds R. Rebolo, E. L. Martin, and M. R. Zapatero Osorio, ASP Conf. Series, **134**, 140–151.

Mayor, M., Udry, S., Naef, D., Pepe, F., Que loz, D., Santos, N. C., and Burnet, M. 2004. "The CORALIE Survey for Southen Extra-Solar Planets. XII. Orbital Solutions for 16 Extra-Solar Planets Discovered with CORALIE," *Astronomy and Astrophysics*, **415**, 391–402.

Mazeh, T. et al. (19 co-authors). 2000. "The Spectroscopic Orbit of the Planetary Companion Transiting HD 209458," *Astrophysical Journal*, **532**, L55–L58.

McArthur, B. E., Endl, M., Cochran, W. D., Benedict, G. F., Fischer, D. A., Marcy, G. W., Butler, R. P., Naef, D., Mayor, M., Queloz, D., Udry, S., and Harrison, T. E. 2004. "Detection of a Neptune-Mass Planet in the ρ^1 Cancri System Using the Hobby-Eberly Telescope," *Astrophysical Journal*, **614**, L81–L84.

McCarthy, C., Butler, R. P., Tinney, C. G., Jones, H. R. A., Marcy, G. W., Carter, B., Penny, A. J., and Fischer, D. A. 2004. "Multiple Companions to HD 154857 and HD 160691," *Astrophysical Journal*, **617**, 575–579.

McCaughrean, M. J., Close, L. M., Scholz, R.-D., Lenzen, R., Biller, B., Brandner, W., Hartung, M., and Lodieu, N. 2004. "ε Indi Ba, Bb: The Nearest Binary Brown Dwarf," *Astronomy and Astrophysics*, **413**, 1029–1036.

McCullough, P. R., Stys, J. E., Valenti, J. A., Johns-Krull, C. M., Janes, K. A., Heasley, J. N., Bye, B. A., Dodd, C., Fleming, S. W., Pinnick, A., Bissinger, R., Gary, B. L., Howell, P. J., and Vanmunster, T. 2006. "A Transiting Planet of a Sun-Like Star," *Astrophysical Journal*, **648**, 1228–1238.

Milone, E.F., and Wilson, W.J.F. 2008. *Solar System Astrophysics: Background Science and the Inner Solar System* (New York: Springer).

Misner, C. W., Thorne, K. S., Wheeler, J. A. 1973. *Gravitation* (San Francisco: W.H. Freeman and Co.)

Mitchell, D. S., Frink, S., Quirrenbach, A., Fischer, D. A., Marcy, G. W., and Butler, R. P. 2003. "Four Substellar Companions Found Around K Giant Stars," *Bulletin, American Astronomical Society*, **35**, 1234.

Moutou, C., Mayor, M., Bouchy, F., Lovis, C., Pepe, F., Queloz, D., Santos, N. C., Udry, S., Benz, W., Lo Curto, G., Naef, D., Ségransan, D., and Sivan, J.-P. 2005. "The HARPS Search for Southern Extra-Solar Planets, IV. Three Close-in Planets around HD 2638, HD 27894, HD 63454," *Astronomy and Astrophysics*, **439**, 367–373.

Muench, A. A, Alves, J., Lada, C. J., and Lada, E. A. 2001. "Evidence for Circumstellar Disks around Young Brown Dwarfs in the Trapezium Cluster," *Astrophysical Journal*, **558**, L51–L54.

Naef, D., Latham, D. W., Mayor, M., Mazeh, T., Beuzit, J. L., Drukier, G. A., Perrier-Bellet., Queloz, D., Sivan, J. P., Torres, G., Udry, S., Zucker, S. 2001a. "HD 80606 b, A Planet on an Extremely Elongated Orbit," *Astronomy and Astrophysics*, **375**, L27–L30.

Naef, D., Mayor, M., Pepe, F., Queloz, D., Santos, N. C., Udry, S., and Burnet, M. 2001b. "The CORALIE Suvey for Southern Extrasolar Planets. V. 3 New Extrasolar Planets," *Astronomy and Astrophysics*, **375**, 205–218.

Naef, D., Mayor, M., Korzennik, S. G., Queloz, D., Udry, S., Nisenson, P., Noyes, R. W., Brown, T. M., Beuzit, J. L., Perrier, C., and Sivan, J. P. 2003. "The ELODIE Survey for Northern Extra-Solar Planets. II. A Jovian Planet on a Long-period Orbit Around GJ777A," *Astronomy and Astrophysics*, **410**, 1051–1054.

Naef, D., Mayor, M., Beuzit, J. L., Perrier, C., Queloz, D., Sivan, J. P., and Udry, S. 2004. "The ELODIE Survey for Northern Extra-Solar Planets. III. Three Planetary Candidates Detected with ELODIE," *Astronomy & Astrophysics*, **414**, 351–359.

Neuhäuser, R., Guenther, E. W., Wuchterl, G., Mugrauer, M., Bedalov, A., and Hauschildt, P. H. 2005. "Evidence for a Co-Moving Sub-stellar Component to GQ Lupi," *Astronomy and Astrophysics*, **345**, L13–L16.

Noyes, R. W., Jha, S., Korzennik, S. G., Krockenberger, M., Nisenson, P., Brown, T. M., Kennelly, E. J., and Horner, S. D. 1997. "A Planet Orbiting the Star ρ Coronae Borealis," *Astrophysical Journal*, **483**, L111–L114.

O'Donovan, F. T. et al. (19 coauthors). 2007a. "TrES-2: The First Transiting Planet in the Kepler Field," *Astrophysical Journal*, **651**, L61–L64.

O'Donovan, F. T., et al. (19 co-authors). 2007b. "TrES-3: A Nearby, Massive, Transiting Hot Jupiter in a 31 Hour Orbit," *Astrophysical Journal*, **663**, L37–L40.

Pepe, F., Mayor, M., Galland, F., Naef, D., Queloz, D., Santos, N. C., Udry, S., and Burnet, M. 2002. "The CORALIE Survey for Southern Extrsolar Planets. VII. Two Short-Period Saturnian Companions to HD 108147 and HD 168746," *Astronomy and Astrophysics*, **388**, 632–638.

Pepe, F., Mayor, M., Queloz, D., Benz, W., Bonfils, X., Bouchy, F., Lo Curto, G., Lovis, C., Mégevand, D., Moutou, C., Naef, D., Rupprecht, G., Santos, N. C., Sivan, J.-P., Sosnowska, D., and Udry, S. 2004. *Astronomy and Astrophysics*, **423**, 385–389.

Pepe, F., Correia, A. C. M., Mayor, M., Tamuz, O., Couetdic, J., Benz, W., Bertaux, J. L., Bouchy, F., Laskar, J., Lovis, C., Naef, D., Queloz, D., Santos, N. C., Sivan, J.-P., Sosnowska, D., Udry, S., 2007. "The HARPS Search for Southern Extra-solar Planets. VIII. μ Arae, a System with Four Planets," *Astronomy and Astrophysics*, **462**, 769–776.

Perrier, C., Sivan, J.-P., Naef, D., Beuzit, J. L., Mayor, M., Queloz, D., and Udry, S. 2003. "The ELODIE Survey for Northern Extra-Solar Planets. I. Six New Extra-Solar Planet Candidates," *Astronomy and Astrophysics*, **410**, 1039–1049.

Perryman, M. A. C. 2000. "Extra-Solar Planets," *Reports of Progress in Physics*, **63**, 1209–1272.

Pont, F., Bouchy, F., Queloz, D., Santos, N. C., Melo, C., Mayor, M., and Udry, S. 2004. "The 'Missing Link': A 4-day Period Transiting Exoplanet Around OGLE-TR-111," *Astronomy and Astrophysics*, **426**, L15–L18.

Portegies Zwart, S. F. and McMillan, S. L. W. 2005. "Planets in Triple Star Systems: The Case of HD 188753," *Astrophysical Journal*, **633**, L141–L144.

Queloz, D., Mayor, M., Weber, L., Blecha, A., Burnet, M., Confino, B., Naef, D., Pepe, F., Santos, N., & Udry, S. 2000. "The CORALIE Survey for Southern Extra-Solar Planets. I. A Planet Orbiting the Star Gliese 86," *Astronomy and Astrophysics*, **354**, 99–102.

Quillen, A. C. and Thorndike, S. 2002. "Structure in the ε Eridani Dusty Disk Caused by Mean Motion Resonances with a 0.3 Eccentricity Planet at Periastron," *Astrophysical Journal*, **578**, L149–L152.

Rafikov, R. R. 2004 "Fast Accretion of Small Planetesimals by Protoplanetary Cores," *Astronomical Journal*, **128**, 1348–1363.

Rafikov, R. R. 2005. "Can Giant Planets Form by Direct Gravitational Instability?" *Astrophysical Journal*, **621**, L69–L72.

Raghavan, D., Henry, T. J., Mason, B. D., Subasavage, J. P., Jao, W.-C., Beaulieu, T. D., and Hambly, N. C. 2006. "Two Suns in the Sky: Stellar Multiplicity in Exoplanet Systems," *Astrophysical Journal*, **646**, 523–542.

Reid, I. N., Lewitus, E., Allen, P. R., Cruz, K. L., and Burgasser, A. J. 2006. "A Search for Binary Systems among the Nearest L Dwarfs," *Astronomical Journal*, **132**, 891–901.

Reipurth, B., and Clarke, C. 2001. "The Formation of Brown Dwarfs as Ejected Stellar Embryos," *Astronomical Journal*, **122**, 432–439.

Rivera, E. J., Lissauer, J. L., Butler, R. P., Marcy, G. W., Vogt, S. S., Fischer, D. A., Brown, T. M., Laughlin, G., and Henry, G. W. 2005. "A $\sim$7.5 $M_\oplus$ Planet Orbiting the Nearby Star GJ 876," *Astrophysical Journal*, **634**, 625–640.

Russell, H. N. and Merrill, J. E. 1952. *The Determination of the Elements of Eclipsing Binaries* (Princeton, NJ: Princeton University Observatory). Princeton University Observatory Contributions No. 26.

Santos, N. C., Mayor, M., Naef, D., Pepe, F., Queloz, D., Udry, S., and Burnet, M. 2001. "The CORALIE Survey for Southern Extra-Solar Planets. VI. New Long Period Giant Stars around HD 28185 and HD 213249," *Astronomy and Astrophysics*, **379**, 999–1004.

Santos, N. C., Israelian, G., Mayor, M., Rebolo, R., and Udry, S. 2003. "Statistical Properties of Exoplanets. II. Metallicity, Orbital Parameters, and Space Velocities," *Astronomy and Astrophysics*, **398**, 363–376.

Santos, N. C., Israelian, G., and Mayor, M. 2004a. "Spectroscopic [Fe/H] for 98 Extra-Solar Planet-Host Stars: Exploring the Probability of Planet Formation," *Astronomy and Astrophysics*, **415**, 1153–1166.

Santos, N. C., Boucht, F., Mayor, M., Pepe, F., Queloz, D., Udry, S., Lovis, C., Bazot, M., Benz, W., Bertaux, J.-L., Lo Curto, G., Delfosse, X., Mordasini, C., Naef, D., Sivan, J.-P., and Vauclair, S. 2004b. "The HARPS Survey for Southern Extrasolar Planets. II. A 14 Earth-Masses Exoplanet around μ Arae," *Astronomy and Astrophysics*, **426**, L19–L23.

Santos, N. C., Pont, F., Melo, C., Israelian, G., Bouchy, F., Mayor, M., Moutou, C., Queloz, D., Udry, S., and Guillot, T. 2006. "High Resolution Spectroscopy of Stars with Transiting Planets: The Cases of OGLE-TR-10, 56, 111, 113, and TrES-1," *Astronomy and Astrophysics*, **450**, 825–831.

Sato, B., Izumieura, H., Toyoya, E., Kambe, E., Takeda, Y., Masuda, S., Omiya, M., Murata, D., Itoh, Y., Ando, H., Yoshida, M., Ikoma, M., Kokubo, E., Ida, S. 2007.

"A Planetary Companion to the Hyades Giant ϵ Tauri," *Astrophysical Journal*, **661**, 527–531.

Sato, B., Ando, H., Kambe, E., Takeda, Y., Izumira, H., Masuda, S., Watanabe, E., Noguchi, K., Wada, S., Okada, N., Koyano, H., Maehara, H., Norimoto, Y., Okada, T., Shimizu, Y., Uraguchu, F., Yanagisawa, K., and Yoshida, M., 2003 "A Planetary Companion to the G-Type Giant Star HD 104985," *Astrophysical Journal*, **597** , L157–L160.

Sato, B. et al. (20 co-authors) 2005. "The N2K Consortium II. A Transiting Hot Saturn Around HD 149026 with a Large Dense Core," *Astrophysical Journal*, **633**, 465–473.

Saumon, D., Hubbard, W. B., Burrows, A., Guillot, T., Lunine, J. I., and Chabrier, G. 1996. *Astrophysical Journal*, **460**, 993–1018.

Sempels, H. C., Collier Cameron, A., Hebb, L., Smalley, B., and Frandsen, S. 2007. "WASP-1: A Lithium- and Metal-Rich Star with an Oversized Planet," *Monthly Notices of the RAS*, **379**, 773–778.

Setiawan, J. et al. 2003. "Planets Around Evolved Stars," in *Toward Other Earths: DARWIN/TPF and the Search for Extrasolar Terrestrial Planets*. (ESA SP-539), pp. 595–598.

Setiawan, J., Pasquini, L., da Silva, L., Hatzes, A. P., von der Lühe, O., Girardi, L., de Medeiros, J. R., and Guenther, E. 2004. "Precise Radial Velocity Measurements of G and K Giant Stars. Multiple Systems and Variability Trend along the Red Giant Branch," *Astronomy and Astrophysics*, **421**, 241–254.

Setiawan, J., Rodmann, J., da Silva, L., Hatzes, A. P., Pasquini, L., von der Lühe, O., de Medeiros, J. R., Döllinger, M. P., and Girardi, L. 2005. "A Substellar Companion Around the Intermediate-Mass Giant Star HD 11977," *Astronomy and Astrophysics*, **437**, L31–L34.

Shankland, P. D., Rivera, E. J., Laughlin, G., Blank, D. L., Price, A., Gary, B., Bissinger, R., Ringwald, F., White, G., Henry, G. W., McGee, P., Wolf, A. S., Carter, B., Lee, S., Biggs, J., Monard, B., and Ashley, M. C. B. 2006. "On the Search for Transits of the Planets Orbiting Gl 876," *Astrophysical Journal*, 653, 700–707.

Silva, V. R. and Cruz, P. C. 2006. "Search for Planetary Candidates Within the OGLE Stars," *Astrophysical Journal*, **642**, 488–494.

Sozzetti, A., Yong, D., Torres, G., Charbonneau, D., Latham, D. W., Allende Prieto, C., Brown, T. M., Carney, B. W., and Laird, J. B. 2004. "High-Resolution Spectroscopy of the Transiting Planet Host Star TrES-1," *Astrophysical Journal*, **616**, L167–L170.

Sozzetti, A., Udry, S., Zucker, S., Torres, G., Beuzit, J. L., Latham, D. W., Mayor, M., Mazeh, T., Naef, D., Perrier, C., Queloz, D., and Sivan, J.-P. 2006a. "A Massive Planet to the Young Disc Star HD 81040," *Astronomy and Astrophysics*, **449**, 417–424.

Sozzetti, A., Yong, D., Carney, B. W., Laird, J. B., Latham, D. W., and Torres, G. 2006b. "Chemical Composition of the Planet-Harboring Star TrES-1," *Astronomical Journal*, **131**, 2274–2289.

Sozzetti, A., Torres, G., Charbonneau, D., Latham, D. W., Holman, M. J., Winn, J. N., Laird, J. B., and O'Donovan, F. T. 2007. "Improving Stellar and Planetary Parameters of Transiting Planet Systems: The Case of TrES-2," *Astrophysical Journal*, **664**, 1190–1198.

Stassun, K., Mathieu, R. D., and Valenti, J. A. 2006. "Discovery of Two Young Brown Dwarfs in an Eclipsing Binary System," *Nature*, **441**, 305–309.

Takeda, G., and Rasio, F. 2005. "High Orbital Eccentricities of Extrasolar Planets Induced by the Kozai Mechanism," *Astrophysical Journal*, **627**, 1001–1010.

Thommes, E. W., Nilsson, L., and Murray, N. 2007. "Overcoming Migration During Giant Planet Formation," *Astrophysical Journal*, **616**, L25–L28.

Tinney, C. G., Butler, R. P., Marcy, G. W., Jones, H. R. A., Laughlin, G., Carter, B. D., Bailey, J. A., and O'Toole, S. 2006. "The 2:1 Resonant Exoplanetary System Orbiting HD 73526," *Astrophysical Journal*, **647**, 594–599.

Udalski, A., Paczyński, B., Zebruń, K., Szymańsk, M., Kubiak, M., Soszyński, I., Szewczyk, O., Wyrzykowski, L., and Pietrzyński, G. 2002. "The Optical Gravitational Lensing Experiment. Search for Planetary and Low-Luminosity Object Transits in the Galactic Disk. Results of 2001 Campaign," *Acta Astronomica*, **52**, 1–19 (astro-ph/0202320).

Udalski, A. et al. (32 others). 2005. "A Jovian-Mass Planet in Microlensing Event OGLE-2005-BLG-071," *Astrophysical Journal*, **628**, L109–L112.

Udry, S., Mayor, M., and Queloz, D. 2003a. "Extrasolar Planets: from Individual Detections to Statistical Properties," in *Scientific Frontiers in Research on Extrasolar Planets, ASP Conf. Series*, eds D. Deming, and S. Seager (San Francisco, CA: ASP), **294**, pp. 17–26.

Udry, S., Mayor, M., Naef, D., Queloz, D., Santos, N. C., and Burenet, M. 2002. "The CORALIE Survey for Southern Extra-Solar Planets". VIII. The Very Low-Mass Companions of HD 141937, HD 168443 and HD 202206: Brown Dwarfs or "Superplanets?" *Astronomy and Astrophysics*, **390**, 267–279.

Udry, S., Mayor, M., Naef, D., Pepe, F., Queloz, D., Santos, N. C., Burnet, M., Confino, B., and Melo, C. 2000. "The CORALIE Survey for Southern Extra-Solar Planets. II. The Short-Period Planetary Companions to HD 75289 and HD 130322," *Astronomy and Astrophysics*, **356**, 590–598.

Udry, S., Mayor, M., Benz, W., Bertaux, J.-L., Bouchy, F., Lovis, C., Mordasini, C., Pepe, F., Queloz, D., and Sivan, J.-P. 2006. "The HARPS Search for Southern Extra-Solar Planets. V. 14 Earth-Masses Planet Orbiting HD 4308," *Astronomy and Astrophysics*, **447**, 361–367.

Udry, S., Bonfils, X., Delfosse, X., Forveille, T., Mayor, M., Perrier, C., Bouchy, F., Lovis, C., Pepe, F., Queloz, D., and Beraux, J.-L. 2007. "The HARPS Search for Southern Extra-Solar Planets. XI. Super-Earths (5 and 8 $M_{\oplus}$) in a 3-Planet System," *Astronomy and Astrophysics*, **469**, L43–L47.

Udry, S., Mayor, M., Clausen, J. V., Freyhammer, L. M., Helt, B. E., Lovis, C., Naef, D., Olsen, E. H., Pepe, F., Queloz, D., and Santos, N. C. 2003b. "The CORALIE Survey for Southern Extra-Solar Planets. X. A Hot Jupiter Orbiting HD 73256," *Astronomy and Astrophysics*, **407**, 679–684.

Vidal-Madjar, A., Désert, J.-M., Lecavelier des Étangs, A., Hébrard, G., Ballester, G., Ehrenreich, D., Ferlet, R., McConnell, J. C., Mayor, M., and Parkinson, C. D. 2004. "Detection of Oxygen and Carbon in the Hydrodynamically Escaping Atmosphere of the Extrasolar Planet HD209458b," *Astrophysical Journal*, **604**, L69–L72.

Vogt, S. S., Butler, R. P., Marcy, G. W., Fischer, D. A., Pourbaix, D., Apps, K., and Laughlin, G. 2002. "Ten Low-Mass Companions from the Keck Precision Velocity Survey," *Astrophysical Journal*, **568**, 352–362.

Vogt, S. S., Butler, R. P., Marcy, G. W., Fischer, D. A., Henry, G. W., Laughlin, G., Wright, J. T., and Johnson, J. A. 2005. "Five New Multicomponent Planetary Systems," *Astrophysical Journal*, **632**, 638–658.

Wang, Z., Chakrabarty, D., and Kaplan, D. L. 2006. "A Debris Disk around an Isolated Young Neutron Star," *Nature*, **440**, 772–775.

Weldrake, D. T. F., Sackett, P. D., Bridges, T. J., and Freeman, K. C. 2005. "An Absence of Hot Jupiter Planets in 47 Tucanae: Results of a Wide-Field Transit Search," *Astrophysical Journal*, **620**, 1043–1051.

Williams, M. D. 2001. *In the Shadows of Unseen Companions: Modeling the Transits of Extra-solar Planets*. MSc Thesis, University of Calgary.

Wilson, D. M. et al. (21 co-authors). 2006. "SuperWASP Observations of the Transiting Extrasolar Planet XO-1b," *Publications, Astronomical Society of the Pacific*, **118**, 1245–1248.

Winn, J. N., Noyes, R. W., Holman, M. J., Charbonneau, D., Ohta, Y., Taruya, A., Suto, Y., Narita, N., Turner, E. L., Johnson, J. A., Marcy, G. W., Butler, R. P., and Vogt, S. S. 2005. "Measurement of Spin–Orbit Alignment in an Extrasolar Planetary System," *Astrophysical Journal*, **631**, 1215–1226.

Wolszczan, A. 1996. "Further Observations of the Planets Pulsar," in *Pulsars: Problems and Progress*, eds., Johnston, S., Walker, M. A., and Bailes, M. ASP Conf. Series **105**, 91–94.

Wolszczan, A. 1997. "Searches for Planets around Neutron Stars," in *Visual Double Stars: Formation, Dynamics, and Evolutionary Tracks*, eds. Docobo, J. A., Elipe, A., and McAllister, H. (Dordrecht: Springer), pp. 221–231.

Wolszczan, A. and Frail, D. 1992. "A Planetary System Around the Millisecond Pulsar PSR 1257+12," *Nature*, **355**, 145–147.

Wolszczan, A., Hoffman, I. M., Konacki, M., Anderson, S. B., and Xilouris, K. M. 2000. "A 25.3 Day Periodicity in the Timing of the Pulsar PSR B1257+12: A Planet or a Heliospheric Propagation Effect?" *Astrophysical Journal*, **540**, L41–L44.

Wolszczan, A., Doroshenko, O., Konacki, M., Kramer, M., Jessner, A., Wielebinski, R., Camilo, F., Nice, D. J., and Taylor, J. H. 2000. "Timing Observations of Four Millisecond Pulsars with the Arecibo and Effelsberg Radiotelescopes," *Astrophysical Journal*, **528**, 907–912.

Wright, J. T., Marcy, G. W., Fischer, D. A., Butler, R. P., Vogt, S. S., Tinney, C. G., Jones, H. R. A., Carter, B. D., Johnson, J. A., McCarthy, C., and Apps, K. 2007. "Four New Exoplanets and Hints of Additional Substellar Companions to Exoplanet Host Stars," *Astrophysical Journal*, **657**, 533–545.

Zucker, S., Mazeh, T., Santos, N. C., Udry, S., and Mator, M. 2004. "Multiorder TODCOR: Application to Observations Taken with the CORALIE Echelle Spectrograph. II. A Planet in the System HD 41004," *Astronomy and Astrophysics*, **426**, 695–698

Challenges

[16.1] Plot the eccentricities of the objects in Table 16.2 *vs.* their semi-major axes. What can you conclude about the process of orbital circularization for these systems?

[16.2] Now perform the same test for the objects in Table 16.3. What can you conclude about the process of orbit circularization for *these* systems? Can you account for the differences between the plots?

[16.3] Examine the spectral types associated with each star in Table 16.1 with one or more planets. Can you draw any conclusions about the type(s) of star likely to have planets? Justify your response.

[16.4] Compute the Roche limit and the synchronous orbit radius for the three closest planets in Table 16.2. You may need to look up data and make assumptions to do the computation, so name all sources and state all assumptions. Are these planets stable against tidal disruption at present or in the foreseeable future?

[16.5] Compute the equilibrium temperature of HD 209458b and calculate the expected pressure scale height of its upper atmosphere; compare these values to those of Jupiter. Again, you may need to look up data and make assumptions to do the computation, so name all sources and state all assumptions.

[16.6] For the case of HD 209458b, compute the true and Table 16.2 values of $M \sin i$ and a (see Sections 16.2.1 and 16.4).

Index

Absorptivity, 252
Abundance(s), 11, 46, 48, 65–66, 67,
        109, 126–127, 133, 145, 168,
        180–181, 231–232, 258, 265,
        268, 269, 270, 271, 276, 277,
        279, 283, 286, 370
Acetonitrile (CH_3CN), 180, 231
Acetylene (C_2H_2), 180, 182, 231, 232
Adams, John Couch, 134
Adams–Williamson equation, 136
Adiabatic
  convection, 7–8
  lapse rate, 11, 31–32, 37, 44
  pressure–density relation, 7
  processes, 8
Adoration of Magi, 226, 228
Adrastea (satellite, Jupiter XV), 154,
        158, 192, 193
Airy, George Biddell, 134
Albedo
  bolometric, 2, 120, 295
  bond, 120, 150, 344, 365, 370
  geometric (visual), 120, 184, 200
Aluminum isotopes, 268, 302
Alvarez, Luis, 277
Amalthea (satellite, Jupiter V), 154,
        158, 193
Ambipolar diffusion, 60, 304
American Meteor Society, 248
Amidogen radical (NH_2), 231
Ammonia (NH_3), 50, 122, 124,
        125–126, 128, 178, 180, 231,
        236, 239, 351
Ammonium hydrosulfide (NH_4SH),
        125–126
Anaxagoras of Clazomenae, 214
Angular
  diameter(s), 134, 294
  momentum, 42–43, 93–94, 96, 119,
        207, 224, 302, 315, 369
  resolution, 294

velocity/velocities, 11, 12, 15,
        190, 222
Apollonius of Myndus, 214
Aquinas, Thomas, 214
Ariel (satellite, Uranus I), 151, 153,
        156, 160, 161, 184
  albedo, 184
Aristotle, 214, 215–216
Artemis, 258
"Asteroid mill", 250, 286, 362
Asteroid(s) (general)
  albedo(s), 293, 298
  densities, 297, 298
  dimension(s), 294
  double, 297
  inner solar system plot, 288, 290,
        301, 302
  masses, 297
  nomenclature, 287
  orbital properties
    families, 290, 292
    Kirkwood gaps, 201, 288, 301
  outer solar system plot, 289, 293
  radii, 297, 301, 304, 306
  rotations, 301, 305
  thermal emissions, 54, 138, 197
Asteroids (individual)
  2101 Adonis, 290
  1221 Amor, 290
  3554 Amun, 290
  1943 Anteros, 290
  2061 Anza, 290
  1862 Apollo, 290
  197 Arete, 298
  2062 Aten, 290
  1865 Cerberus, 290
  1 Ceres, 286, 298, 300
  2060 Chiron, 239, 293
  1864 Daedalus, 290
  136199 Eris, 157
  433 Eros, 290, 292, 298, 299

Asteroids (individual) (*Cont.*)
  951 Gaspra, 297
  2340 Hathor, 290
  2212 Hephaistos, 290
  944 Hidalgo, 239, 293
  624 Hektor (Hector), 297
  1566 Icarus, 290
  243 Ida, 297
  3 Juno, 287
  2 Pallas, 287, 295, 298
  3200 Phaethon, 242, 290
  5145 Pholus, 239, 293
  1915 Quetzalcoatl, 290
  5381 Sekhmet, 290
  2608 Seneca, 290
  4 Vesta, 298, 299
  1989 PB, 297
  2004 VB61, 290
  2004 WC1, 290
  2004 WS2, 290
Asteroids (types)
  Amor(s), 290, 298
  Apollo(s), 290, 299
  Aten(s), 290
  Centaur(s), 206, 225, 239, 289, 293
  Cubewano(s), 290
  "excited"/ "hot" objects, 293
  Flora family, 291
  Hilda family, 288, 291, 293
  Koronis family, 290, 292, 297
  Kuiper Belt objects, 293, 353
    "classical"/ "cold", 293
    "excited"/ "hot", 293
  main belt, 253, 288, 290, 291, 293,
    298, 353
  Plutino(s), 186, 206, 289, 293, 353
  scattered population, 238, 240,
    294, 369
  trans-Neptunian object(s), 206, 290,
    293, 353
    eccentricities vs. semi-major
      axes, 294
  Trojan(s) & Greek(s), 288, 290, 292
Atlas (satellite, Saturn XV), 155, 159,
  195, 201
Atmosphere/atmospheric
  air mass(es), 27–28
  circulation(s)
    eddy/eddies, 27–28
    Ferrel cell(s), 25, 26, 43
    global, 22–23
    Hadley cell(s), 24–25

    meridional, 25–26, 39, 43
    parasitic cell(s), 25
    polar cell(s), 25
    thermal, 19–20
    wave motion(s), 27
    zonal, 39–41
  constituent(s)
    comparative, 46
  convection, 7, 32, 47
  diffusion, 45
  diffusive equilibrium, 47
Earth
  circulation, 38
  composition, 46
  cyclones, 42, 44
  DALR (dry adiabatic lapse rate),
    32, 37
  diffusion time, 47–48
  exosphere, 34
  heterosphere, 48
  homosphere, 48
  hurricane(s), 16, 44
  lapse rate, 31–32, 33, 36,
    37, 44, 47
  mesopause, 30, 33, 47
  mesosphere, 30, 32–33, 47
  mixing time, 47–48
  ozone, 31, 32–33, 36, 37
  Rossby number, 38
  SALR (saturated adiabatic lapse
    rate), 32
  stratosphere, 32–33
  superrotation, 41
  thermal structure, 30
  thermosphere, 33, 36
  tropopause, 27, 28, 30
  troposphere, 31–32, 36, 37
  winds, 36, 43, 75
eddy/eddies
  effects on heat transport, 26–27
escape mechanisms, 68
isobars, 18–19, 20–22
isotherm(s), 19, 20, 22, 27, 33, 37, 40
jet stream(s), 27–29, 39
Jupiter
  belts, 122, 123, 124, 125, 143
  brown barges, 123, 124
  Great Red Spot, 123, 124, 125, 128
  helium deficiency, 131, 138
  mean molecular weight, 121
  pressure scale height, 147
  white spot(s), 166

winds, 121
zones, 123, 126
Mars
    averaged temperatures, 40
    circulation, 38
    composition, 46
    dust devils, 36
    ionosphere, 66–67, 113–114
    lapse rate, 36
    mixing rate, 64
    Rossby number, 38
    stratomesosphere, 34, 36
    thermal structure, 34
    thermosphere, 36
    troposphere, 36
meridional cross-section(s), 27, 28
mixing, 11, 45–46, 47, 51, 64
Neptune, 54, 133–136
pressure
    baroclinic instabilities, 26, 27
    gradient(s), 5, 15, 16–17, 18–19,
        20–22, 38, 42
    high/low, 18, 21, 22, 24–25, 27, 42
refractive index, 6
Saturn
    features, 129, 196
    structure, 121, 129, 130
    winds, 131
seeing, 294
structure, 4–11, 30, 34, 36, 121,
    126, 127
subtropical fronts, 28
thermal inertia, 34, 36, 44
Uranus, 131–133, 156, 160, 183
Venus
    circulation, 38
    composition, 46
    cryosphere, 37
    cyclostrophic balance, 41–42
    Rossby number, 38, 41
    stratomesosphere, 37, 44
    superrotation, 41, 43–44
    thermal structure, 30, 36
    thermosphere, 36, 37, 41, 44
    troposphere, 37
Aurora(e), 87, 91, 99, 103, 105, 127,
    130–131
Auroral oval, 101, 103–105, 130
Avogadro's number, 6, 9

Bacon, Roger, 214
Baroclinic instability, 26, 27

Bayeaux tapestry, 226, 228
Benzene (C_6H_6), 182
Bessel, Friedrich Wilhelm, 218
Bielids, 242
Bipolar outflow(s), 302, 305
Birkeland currents, 104, 105
Bode, Johann Elert, 131, 286
Bolide(s), 243, 248, 257, 269
Bolometer, 296
Boltzmann constant, 1, 2, 303
Borelli, Giovanni, 216
Bradley, James, 132
Breccia(s)
    anorthositic, 276
    basaltic, 276
    genomict, 273
    monomict, 264, 273
    polymict, 264, 273, 274
    regolith, 273, 274, 276
Bright Star Catalog, 314
Brown dwarf desert, 334,
    337, 368
Brown dwarf(s), 335ff, 346ff,
    367–368
    definition, 353–356
    detection techniques, 338ff
    evolution, 369
    origins, 367ff
    spectral characteristics, 354
Brownlee particles, 235, 249
Brunt-Väisälä frequency, 44–45

Calcium–aluminum inclusion(s) (CAI),
    267, 269–270
Callisto (satellite, Jupiter IV)
    domed crater(s), 170, 172
    interior, 170–172
    moment of inertia, 171
    surface feature(s)
        ringed plain(s), 168, 170, 171
        Valhalla (ringed plain), 170–171
Campbell, William Wallace, 194, 313
Capture mechanisms, 163
Carbon black, 239
Carbon dioxide (CO_2), 31, 68, 175,
    180, 236, 370
Carbon monosulfide (CS), 231, 236
Carbon monoxide (CO), 69, 180, 231,
    232, 235–237
Cardan, Jerome (Cardano,
    Girolamo), 215

Cassini/Cassini's Division, 119, 121,
        123, 125, 127, 129–131, 143,
        151, 164, 174, 175–177, 179,
        185, 192, 194–197, 201, 202
Cassini, Jean-Dominique, 192
Catalog of Meteorites, 259
Centaurs, 206, 239, 289, 293
Ceres (asteroid dwarf planet), 188, 206,
        286, 287, 294, 298, 300, 353
CH radical, 182, 231
Chaeremon the Stoic, 213
Chalcophilic elements, 267–268
Challis, James, 134
Chaotic variation, 301
Charge-exchange reactions, 112
Charon (satellite, Pluto I)
    atmosphere, 188
    composition, 153, 157, 161, 188
    discovery, 153
    orbit/orbital
        elements, 153, 154, 173, 224, 228,
            239, 291, 347, 368
        properties, 164, 189, 225, 238
    spin–orbit resonance, 175
Chinook, 55
Chiron (asteroid), 206, 225, 239, 293
Chladni, Ernst Florenz Friedrich, 257
Chondrites, see under Meteorites
Chondrules
    composition, 267–268, 270, 271
    origin, 271–273
CHON particles, 232
Circulation
    Ferrel cell(s), 25, 26, 43
    global, 22–23
    Hadley cell(s), 24–25, 39, 43–44
    parasitic cell(s), 25
    polar cell(s), 23, 25
Clairaut, Alexis-Claude, 217
CN radical, 231
Coefficient
    ablation, 245
    drag, 245
Comet(s)/cometary
    attrition, 240
    coma, 230, 231, 335
    composition, 231–237
    demise, 240–241
    dust, 230, 236, 241, 250
    envelope, 230
    Halley group, 238–239
    Jupiter family, 220, 238–240

Kracht group, 221
Kreutz group, 220, 240
"lost", 224–225
Marsden group, 220–221,
        224–225, 229
Meyer group, 220
nucleus, 229–230, 231–232,
        235, 239
orbit computation, 218–219, 221,
        224, 287
orbital elements, 226, 228–229
origin, 163, 237–240
parabolic (or long-period), 216,
        220–221, 237
periodic, 217–219, 220, 225
probability of approach, 238
specific angular momentum, 224
structure, 229–230
surface ice, 232
tail(s)
    dust, 227, 231, 234, 236, 242, 396
    ion, 227, 230, 231, 233–234, 237
"virgin", 226
Comets (individual)
    1P/Halley, 219, 228–229, 230–232,
        235, 242
    2P/Encke, 220, 225, 242
    9P/Tempel 1, 225, 230, 232, 235
    13P/Olbers, 225, 226
    19P/Borrelly, 230
    23P Brorsen–Metcalfe, 234
    26P/Grigg–Skjellerup, 225
    29P/Schwassmann–Wachmann, 225
    55P/Tempel–Tuttle, 226, 242
    73P/Schwassmann–Wachmann
        3, 240
    79P/du Toit–Hartley, 225
    81P/Wild, 230
    95P/Chiron, 225
    107P/Wilson–Harrington, 225
    109P/Swift–Tuttle, 225, 229, 242
    153P (Ikeya-Zhang), 231
    C/1577 V1 (Comet of 1577),
        215–216, 225, 226
    C/1908 R1 (Morehouse), 237
    C/1961 R1 (Humason), 237
    C/1980 E1 (Bowell), 221
    C/1995 O1 (Hale–Bopp), 227
    C/1996 B2 (Hyakutake), 231, 233
    C/1997 P2 (Spacewatch), 221
    C/1999 H1 (Lee), 233
    C/2004 Q2 (Machholz), 236

3D/Biela, 226, 242
D/Shoemaker–Levy 9, 226, 240
Condensation sequence, 269, 302–303
Copernicus, 216
Cordelia (satellite, Uranus VI), 156,
        160, 198
Core accretion, 363, 368–369
Coriolis
    force, 11–12, 13, 15, 16, 18, 19,
        20–22, 25, 38–39, 41–42,
        109, 122
    parameter, 122
Coriolis, Gaspar G., 24
Coronal
    holes, 70, 99
    mass ejection, 99
Cosmogenic nuclei, 283
"Counterglow", 250
Crater(s)
    impact, 164, 166, 168, 172, 176,
        177, 185
    volcanic, 164
Cretaceous–Tertiary interface, 277
Critical
    frequency, 75, 139–140
    orbital radius, 190
Cryo-volcanoes, 177, 369
Curie point, 109, 115
Cyanogon (C_2N_2), 237
Cyclostrophic
    balance, 41–42
    winds, 42
Cyclotron
    frequency, 140
    motion, 80
    radius, 80

Dactyl (243 Ida 1), 260, 297
Daphnis (satellite, Saturn XXXV), 156,
        160, 195, 196
d'Arrest, Heinrich Louis (Ludwig),
        134, 226
Darwin, George, 203
Dead zone(s), 307, 368
Debris disk(s), 305
    4U 0142+61, 357
Deimos (satellite, Mars II), 152, 154,
        158, 162–163, 207, 298
Del values, 284–285
Democritus, 214
Descartes, Rene, 133

Despina (Satellite, Neptune V), 157,
        161, 201
D/H (deuterium/hydrogen ratio), 272,
        286, 301
Deuterium, 250, 272, 286, 354
Diacetylene (C_4H_2), 180
Diamagnetism, 83–84
Diapir(s), 171–172
Diffusion time, 47–48, 138
Diffusive equilibrium, 47
Dione (satellite, Saturn IV), 153, 155,
        159, 173, 177, 194, 196
Disk(s)
    dead zone, 307, 368
    debris/remnant, 301, 305, 307, 313,
        351, 364
Dissociative recombination, 58–59, 61,
        63, 67–69
Doldrums, 23
Doppler ranging, 297
Dörffel, Georg S., 216
Dry adiabatic lapse rate (DALR),
        32, 37
Dunham, David W., 295
Dunham, Joan Bixby, 295
Dust
    composition, 232, 235, 249–250
    destinies, 241, 251–253
Dust devils, 36
Dysnomia (satellite, Eris I), 153,
        157, 161

Earth
    atmosphere/atmospheric
        tides, 90
    aurora(e)/auroral
        oval, 78, 101, 103–105, 130
    cycles
        carbon, 48–49
        nitrogen, 49–50
        oxygen, 48–49
        sulfur, 51–53
    ion
        loss mechanisms, 58–59, 61, 63, 67
        production mechanisms, 58, 60,
            62, 66
    ionosphere/ionospheric
        Birkeland current, 104, 105
        charge separation, 60, 70, 103–104
        D layer, 61
        dynamo, 90
        E layer, 60–61

Earth (*Cont.*)
    electrojet(s), 101, 103, 105
    electron density, 57, 67, 73, 75, 111
    field-aligned current, 84–86,
        103–104, 107
    F layer, 57, 63
    Hall current, 86–87, 105–106
    Joule heating, 103
    Pedersen current, 86–87, 105–106
  lower atmosphere properties, 34, 35
  magnetic
    crustal anomalies, 114–115
    field, 75–77
    storm(s), 99–100
    substorm(s), 102–103, 107
  magnetopause, 76–77, 78, 91, 104
  magnetosheath, 76, 77
  magnetosphere/magnetospheric
    Chapman–Ferraro boundary
        current, 91
    convection, 100–101, 104
    magnetotail(s), 77, 91, 100,
        101–103, 112, 142
    neutral point, 91
    neutral sheet, 77–78, 101–102
    plasma sheet, 78, 101, 102, 107
    polar cusp, 77, 91, 108
    reconnection(s), 103
    ring current, 77, 78, 92, 98,
        99–100, 102, 142
    ring prospect, 203
  Van Allen radiation belts
    mirror points, 95
    pitch angles, 97
    precipitation, 95
    ring current, 98
Edgeworth, Kenneth, 289
Edgeworth–Kuiper Belt/Cloud
  shape and size, 353–354
Effective gravity (gravitational
    acceleration), 14, 41, 121, 353,
    359, 362
Einstein radius, 349
Einstein ring, 348
Enceladus (satellite, Saturn II), 153,
    155, 159, 168, 173–175, 194,
    196, 369
  plumes, 175
  tiger stripes, 175
Encke gap, 174, 194, 195, 196, 201, 203
Encke, Johann Franz, 217
Energy budget, 54

Enstatite ($MgSiO_3$), 270, 274
Ephorus of Cyme, 214, 218
Epimetheus (satellite, Saturn XI), 155,
    159, 177, 196
Equation
  Adams–Williamson, 136
  of hydrostatic equilibrium, 4–5, 136
Equation(s) of state, 5, 9, 136, 138
Equatorial
  bulge, 14
Equilibrium
  hydrostatic, 4, 5, 136, 353
  mechanical, 4
  temperature, 1–2, 31, 54, 133, 138,
    164, 188, 296, 356, 359,
    370–371
Eris (dwarf planet)
  albedo, 206
  comparison with Pluto, 153–154,
    188, 206
  radius, 206
  satellite, 153
  temperature, 206
Escape velocity, 1, 120–121, 146, 221,
    240, 281
Ethane (C_2H_6), 177, 178, 180, 182, 231
Ethanol (C_2H_5OH), 232
Ethylene (C_2H_4), 180, 181–182
Ethyl radical(s) (C_2H_5), 181–182
Ethnyl radical (C_2H), 182
Europa (satellite, Jupiter II)
  cratering density, 152
  magnetic field, 166
  mean density, 169
  surface feature(s)
    chaotic terrain, 167
    cracks, 166, 169
    ice rafts, 166
    lenticulae (dark spots), 167
    linea, 167
    mountains, 166–167
    ringed plains, 167, 168
    wrinkle ridges, 166
Eutectic mixture(s), 180
Exobase, 34, 69, 97, 113
Extra-ordinary ray(s), 140
Extra-solar planets (general)
  chthonian, 360
  detection techniques
    astrometric, 347–348
    direct imaging, 350–351, 364
    gravitational lensing, 348, 350, 357

pulsar timings, 350–351
    radial velocity variations, 313–314,
        338–340, 342, 358, 364, 367
    transit photometry, 350
direct imaging candidates, 351
free-floating, 350, 367
habitable zone, 362, 369–370
host stars, 365
L-band simulated light curve, 345
multi-planet systems, 360–361,
        362, 363
O-C analyzes, 352
occultation observations, 295, 360
outgassing evidence, 355
sodium detection, 343
spectroscopy, 314, 350–351
transit observations, 358–359
Extra-solar planets (individual
        systems)
    Cha 110913–773444, 351
    2MASSW J1207334–393, 254, 350
    GJ 436, 316, 328, 363
    GJ 674, 363
    GJ 876, 322, 332, 357, 363
    Gl 581, 357, 369
    HAT P-1, 367
    HD 37124, 316, 324, 357, 361–362
    HD 69830, 318, 326, 357, 362
    HD 74156, 318, 326, 357
    HD 149026, 320, 330, 358, 367
    HD 188753A, 368
    HD 189733, 320, 332, 343, 344,
        358, 360
    HD 209458, 314, 342–345,
        355–360, 365
        L-band simulated light curve, 345
        O–C analyzes, 352
        occultation observations, 295, 360
        outgassing evidence, 355
        sodium detection, 343
        transit observations, 358–359
    OGLE-2003-BLG-235/MOA-2003-
        BLG–53,
        350
    OGLE-2005-BLG-071L, 350
    OGLE-2005-BLG-390LB, 349–350
    OGLE-TR-10, 357, 365, 366–367
    OGLE-TR-56, 360
    OGLE-TR-109, 350
    OGLE-TR-111, 318, 326
    OGLE-TR-113, 318, 326
    OGLE-TR-132, 326

PSR 1257+12, 313, 337, 352
PSR B1620-26, 337, 352
TrES-1, 343–345, 357, 364–365
WASP-1, 367
XO-1, 320, 328, 357, 365
v Andromedae, 360
IL Aquarii, 363
μ Arae, 320, 330, 357
55 Cancri, 361–362
FAYALITE (Fe_2SiO_4), 262
Ferrel cell(s), 25–26
Field-aligned currents, 84, 86, 103,
        104, 107
Flamsteed, John, 131–133
Föhn, 55
Force(s)
    apparent, 11, 13
    centrifugal, 11–14, 41–42
    centripetal, 80, 89, 190
    Coriolis, 11–16, 18–19, 20, 21–22, 24,
        25, 38–39, 41–42, 109, 122
    disruptive differential, 190
    electric, 80, 101
    fictitious, 11, 13–14
    friction, 17
    gravitational, 14, 79, 190–191
    Lorentz, 79, 92
    pressure gradient, 15, 16–17, 18–19,
        20–22, 38, 41–42, 122
    virtual, 11
Formaldehyde (H_2CO), 236
Forsterite (Mg_2SiO_4), 354
Frame(s) of reference, 11–12
Fugacity, 284

Galatea (Satellite, Neptune VI), 157,
        161, 201
Galilean satellites, 152–153
Galilei, Galileo, 216
Galle, Johann Gottfried, 134
Ganymede (satellite, Jupiter II)
    atmosphere, 168, 170
    cratering densities, 169, 170
    crater(s)
        Achelaous, 170
        Gula, 170
    Galileo Regio, 169
    ice polymorph(s), 169
    magnetic field, 169
    mantle, 169
    mean density, 169
    moment of inertia, 169

Ganymede (satellite, Jupiter II) (*Cont.*)
   surface features
      cratered plains, 169, 170
      grooved terrain, 169
      palimpsest(s), 169
      penepalimpsest(s), 169
      ringed plain(s), 170
      strike–slip faults, 167, 169
      sulcus/sulci, 169
Gas(es)
   chemically active, 10
   monatomic, 10
   polyatomic, 10
Gas giants
   albedos, 138
   densities, 121, 132, 134,
      136–137, 139
   heat fluxes, 127, 138
   ionospheres, 139–140
   pressure(s)
      central, 136–137
      internal, 136–137
   properties
      orbital, 164
      physical, 158
   temperature(s), 127
   winds, 221
Gauss, Karl Friedrich, 287
Gauss's law, 72
Gegenschein, 250
Geminids, 242
Georgium Sidus (Uranus), 131
Geostrophic
   balance, 18–19, 41
   winds, 18–19, 20–22, 43, 122
Germane (GeH$_4$), 125
Gram-molecular weight, 9
Gravitational contraction, 369
Gravitational instability/instabilities,
   139, 363, 368–369
Gravitational radius, 348
Greenhouse
   effect, 54, 135, 371
      runaway, 371
   gases, 370
Guiding center, 82–83, 86, 87, 88, 92
Gulliver's Travels, 161
Gyration frequency, 80, 91, 93
Gyromotion, 80–81, 82, 83–84, 92–94,
   116, 117
Gyroradius, 80, 82, 88, 93

Habitable zone, 362, 369–371
Hadley cell(s), 24–26, 40
Hadley, George, 24
Hagecius, Thaddeus, 215
Half-life/lives, 268, 280, 283
Hall, Asaph, 162
Halley, Edmond, 217
Harding, Karl Ludwig, 287
Heat capacity/capacities, 7–9, 20, 26
Heat of ablation, 245
Helium, 121, 126, 127, 131, 133,
   138, 180
Helium rain, 54, 138
Herschel, Caroline, 226
Herschel, John, 218, 226
Herschel, William, 131
Hertzsprung–Russell Diagram
   for planets and brown dwarfs, 355
Hevel, Johannes, 216
Hill, George W., 204
Hirayama, Kiyotsugu, 290
Hohmann orbital transfer, 189
"Horse latitudes", 23
Horseshoe-shaped orbits, 174
"Hot Jupiters", 307, 344, 350, 356,
   359–360, 361, 366
Hurricane(s)
   Wilma, 16
Huygens, Christiaan, 151, 175–180,
   183, 192
Hydrazine (N$_2$H$_4$), 126
Hydrodynamic outflow, 147
Hydrogen, 3, 68, 99, 100, 121, 126, 127,
   133, 141, 142, 145, 180, 183,
   230, 272, 286, 304, 313, 354,
   359, 360
Hydrogen cyanide (HCN), 180, 181,
   231, 232, 236
Hydrogen sulfide (H$_2$S), 125, 236
Hydrostatic equilibrium, 4–5, 136, 353
Hydroxyl radical (OH), 231, 240
Hyperion (satellite, Saturn XVIII),
   155, 159, 174

Iapetus (satellite, Saturn VIII)
   terrain
      dark, 176
Ice(s)
   polymorph (s), 169
   water
      phase diagram, 171, 173

Interplanetary magnetic field (IMF), 76, 108
Interstellar medium, 232, 235, 272, 286, 304
Inter-tropical convergence zone, 23
Ionosphere(s)/ionospheric
    critical frequency, 139–140
    field-aligned current(s), 84–86, 103, 104–105, 107
    index of refraction, 139–140
    plasma frequency, 70, 139–140
    reflection, 63
    refraction, 63, 73, 74, 139–140
Io (satellite, Jupiter I)
    atmosphere, 146–147
    composition, 154
    cratering density, 166
    interior, 166
    loss mechanisms, 146–148
        hydrodynamic outflow, 147
        Jeans escape, 146
        sputtering, 147–148
        volcanism, 147
    mean density, 121, 169
    orbit(al)
        elements, 154
    physical properties, 158
    plasma torus, 145
    plumes, 147
    pressure scale height, 147
    sodium D-line emission, 146–148
    volatiles, 145–148, 164, 165
    volcano(es)
        Pele, 165
        Pillan Patera, 165
        Prometheus, 165
        Tvashtar Catena, 166
Iron
    Curie point, 109
    meteorites (q. v.), 277–279
Isobaric processes, 8
Isochoric processes, 8

Jacobi's integral, 223
Janus (satellite, Saturn X), 134, 155, 159, 177, 195, 202
Jeans
    escape mechanism, 68, 121, 146
    length, 369
    mass, 369
Jeans, James, 3
Jefferson, Thomas, 257

Jet streams, 27–29, 39
Junge layer, 51
Jupiter
    age, 121
    atmosphere
        belts, 122, 123–124, 125, 143
        blue-grey regions, 124
        brown barges, 123, 124
        Great Red Spot, 123, 124, 125, 128
        mean molecular weight, 121
        photochemical products, 126
        scale height, 121, 145
        structure, 121, 125, 126–127
        UV photolysis, 124
        white ovals, 123, 124
        winds, 122, 125
        zones, 122, 123, 126
    aurorae(e), 130
    circulation, 122
    composition, 121
    flattening, 15, 122
    global oscillations, 138
    helium content, 127, 131
    internal structure, 141
    lightning, 127, 128
    magnetic field
        compared to Earth's, 127
        dynamo, 141
        quasi-dipolar, 141
    magnetodisk, 141, 142, 143–144
    magnetosphere
        inner, 141, 143
        interaction with Io, 143
        outer, 142, 144
    magnetotail, 142
    mean density, 121
    metallic hydrogen, 127, 141
    nuclear reactions, 138
    pressure(s)
        central, 136–137
        internal, 136–137
    radiation
        belts, 143
        decametric (DAM), 140, 145
        decimetric (DIM), 141
        excess, 137–139
        synchrotron, 141
    ring(s)
        Gossamer, 192, 193
        Halo, 192, 193
    Rossby number, 122
    rotation periods, 143

Jupiter (*Cont.*)
  satellites
    direct, 163
    Galilean, 140, 152, 163
    orbits, 154, 155
    physical properties, 158–159
    retrograde, 163
    temperature(s), 121, 127

Kamacite crystals, 277
Keeler gap, 194, 195, 196
Keeler, James Edward, 194
Kepler, Johannes, 162, 216
Kirkwood Gaps, 201, 288, 301
Kozai mechanism, 315
Kreutz, Heinrich Carl Friedrich, 220
Kronos (Saturn), 128
Kuiper Belt, *see* Edgeworth–Kuiper
    Belt/Cloud
Kuiper, Gerard Peter, 180, 184, 289
Kursk magnetic anomaly, 114

Lagrangian points, 119, 293
Lalande, Joseph-Jérôme, 348
de Laplace, Pierre Simon, 133, 192
Lapse rate
  adiabatic
    DALR (dry adiabatic lapse rate),
      32, 37
    SALR (saturated adiabatic lapse
      rate), 32
  environmental, 32
Latent heat, 32, 44
Law(s)
  Gauss's, 72
  Kepler's 3rd, 217, 340, 347
  Newton's 2nd, 72, 80, 81
  Newton's 3rd, 245
  Titius–Bode, 134, 286
L dwarf(s)
  GD 165B, 335, 354
Least squares computational scheme,
    287
Lemonnier, Pierre Charles, 131
Leonids, 242
Le Verrier/Leverrier,
      Urbain-Jean-Joseph, 134
Light curve modeling
  Russell–Merrill model, 346
  Wilson–Devinney program, 343, 346
Lithophilic elements, 268

Mab (satellite, Uranus XXVI)
  water ice absorption, 184
Maestlin (Mästlin), Michael, 215
Magnetic
  bottle(s)
    loss cone, 97
  bow shock, 76–77, 107–108,
      112–113, 143
  field(s)
    boundary current, 91–92, 99,
      102, 111
    curvature drift, 88, 89–90
    diamagnetic currents, 83, 91,
      111–112
    dipolar, 75–76, 141
    dynamo model, 109–110
    E x B drifts, 84–87, 102, 118
    field-aligned current(s), 84–86,
      103, 104, 107
    gradient drift, 88–90
    north/south-seeking poles, 76
    pressure, 111–112
    quasi-dipolar, 141
  mirror(s)
    ratio, 96, 116
  moment, 93–94
  ring current, 92–107
  storms, 99–100, 103
  tail(s), 78, 95
Magnetism
  remanent, 109, 114
Magnetotactic bacteria, 276
Marcy, Geoffrey, 314
Mars
  atmosphere, 38, 64
  core, 114–115
  dust devils, 36
  hydrosphere, 284
  impact basins, 114
  ionosphere, 66–67, 113–114
  iron content, 114
  lower atmosphere properties, 35,
      46, 64
  magnetic field
    crustal anomalies, 114–115
    dynamo, 114–115
    global dipole moment, 114
    strengths, 114
  magnetism
    remanent, 109, 114
  magnetosphere, 113–115
  north–south dichotomy, 114

satellites, 154, 158, 161–163
suprathermal atoms and ions, 69
Marsden, Brian G., 218, 220, 224, 225, 229
Mass function, 340–341, 358
Ma Tuan-Lin, 213
Maxwellian distribution, 3
Maxwell, James Clerk, 194
Maxwell's equations, 70–72
Mayer, Tobias, 131
Mayor, Michel, 313, 337
Mean free path, 3, 34, 357
Medicean satellites (Galilean satellites, q.v.), 140, 142, 152, 153, 163
Mercury
  magnetic field
    dipole moment, 107
    origin, 107
  magnetopause, 107–110
  magnetosphere
    bow shock, 107
  molten core, 109
"Metals", in stellar astrophysics, fn 1, 313
Meteor crater(s)
  Ries, 276
Meteorites (general)
  age(s)
    cosmic ray exposure, 283
    crystallization, 284
    differentiation, 284
    ejection, 276, 284
    formation, 284
    gas retention, 279, 282–283
    mean isochron, 282
    radiogenic, 279–282
    terrestrial, 283
  breccia(s) (q.v.), 273–274, 276
  calcium–aluminum inclusions (CAI), 267, 269, 270
  chondrule(s), 267, 270
  classification, 261
  deuterium/hydrogen ratio, 272, 286
  equilibration, 271
  falls and finds, 259
  isochrones, 281–282
  kamacite bands, 277
  nomenclature, 258, 259
  origin(s), 279–284
  petrographic/petrologic properties, 259–260, 262
  shock stages, 258

strewn field, 269
weathering grade, 258
Meteorites (groups/classes/clans/types)
  acapulcoites, 273
  achondrites, 261, 265, 269, 271, 273, 280, 299
  aerolites (stony), 258
  angrites, 273
  Antarctic, 258, 276
  ataxites, 277
  aubrites, 261, 264, 265, 273, 274, 285, 299
  brachinites, 273
  carbonaceous chondrites
    CH (high Fe), 269
    CI (Ivuna), 269
    CK (Karoonda), 269
    CM (Migei), 269
    CO (Ornans), 261, 269
    CR (Renazzo), 269
    CV (Vigrano), 269
  chassignite(s), 265, 275, 299
  chondrites
    carbonaceous (v.s.), 163, 232, 235, 269–270, 284, 299, 300
    enstatite (v.i.), 261, 270–271
    ordinary (v.i.), 270, 299
    Rumurutiite (R), 271
  differentiated
    origin, 262–265
  differentiated silicate-rich (DSR), 262, 264, 273–276
  diogenites, 261, 264
  enstatite(s), 270–271
    EH (high iron), 261, 270, 274
    EL (low iron), 261, 270, 274
  eucrites, 261, 264, 265, 273–274
  HED, 273
  hexahedrites, 265, 277
  howardites, 261, 264, 265, 273–274
  IAB, 261, 271, 277
  IIAB, 261, 266, 278, 279
  IIIAB, 261, 266, 273, 278, 279
  IIICD, 261, 266, 271, 277
  IVA, 261, 266
  igneous, 261, 273, 276, 299
  iron(s), 277–279
    cooling rates, 277
  lodranites, 273
  lunar, 276

Meteorites (groups/classes/clans/
        types) (*Cont.*)
    Martian (SNC), 265, 273, 274–276,
        281, 284
    mesosiderites, 261, 265, 273–274
    microbreccias, 260
    minichondrule, 261, 269
    nakhlites, 265, 275
    non-differentiated, 260–262
    octahedrites, 265
        plessitic, 277
    ordinary chondrites
        H (high iron), 270
        L (low iron), 270
        LL (low iron, metal), 270
    orthopyroxenite, 275
    pallasites, 261, 264, 265,
        273–274, 299
    plessites, 277
    "primitive achondrites" (Iab clan),
        271
    refractory-rich, 261, 269
    shergottites, 265, 274, 281, 284
    Shergotty–nakhla–chassigny (SNC),
        274
    siderites (irons), 258
    siderolites (stony-irons), 258
    tektite(s)
        moldavite, 276
    ureilites, 264, 265, 274
    very-low-titanium (VLT) basalt, 276
    volatile-rich, 261, 269
    winonaites, 273
Meteorite(s) (individual)
    Abee (EH4), 271
    ALH A77005 (SNC), 274
    ALH 84001(SNC), 275–276
    Allende (CV3, 2), 282
    Bruderheim (L3), 271
    Chassigny (SNC), 265
    EET A79001 (SNC), 274
    Governador Valadares (SNC), 275
    Henbury (IIIAB), 278
    Lafayette (SNC), 275
    MET 01210 (lunar), 276
    Millarville (IVA), 259
    Nakhla (SNC), 274, 275, 281–282
    QUE 94201 (SNC), 284
    Shergotty (SNC), 265, 274
    Sikhote-Alin (IIAB), 278
    Zagami (SNC basaltic shergottite),
        283–284

Meteorite Hills Antarctica, 276
Meteoritical Bulletin, 259
Meteor(s)
    ablation rate, 246
    bolide(s) (fireballs), 243, 248
    brightness distribution, 249
    brightness variation, 248
    fast *vs* slow, 246–247
    fragmentation, 247
    ionization probability, 247
    luminous efficiency, 246
    shower, 242–243, 248
    sporadic, 243, 248, 257
    visibility, 243, 246
Meteor shower(s)
    radiant(s), 242, 243
Meteoroid, 243–245, 125, 126,
        132ff, 175ff
Methane (CH_4)
    clathrate hydrate, 177
Methanol (CH_3OH), 231, 232, 236
Methylacetylene (CH_3C_2H), 182
Methylene (CH_2), 181
Methylidyne (CH), 181
Methyl radical(s) (CH_3), 181
Micrometeorites
    asteroidal/cometary dust ratio,
        231–237
    Brownlee particles
        chondritic porous (CP), 249
        chondritic smooth (CS), 249
    destinies, 251
    origins, 250
Millisecond pulsars, 337
Mimas (Saturn I), 153
Minerals
    anhydrous, 285
    augite, 274–275
    enstatite, 270–271
    feldspar, 263
    forsterite, 354
    hibonite, 268
    kamacite, 263
    magnetite, 274
    maskelynite, 274
    melilite, 268
    olivine, 236, 239, 262–264, 271,
        273–275, 299
    orthopyroxene, 236, 264, 271,
        275, 285
    perovskite, 268
    pigeonite, 274

plagioclase, 274, 275
spinel, 268
taenite, 263
Minor Planet Center, 229, 240,
    288–289, 295
Minor planets, *see* Asteroids
Miranda (satellite, Uranus V), 156,
    180, 183
the "racetrack", 183
scarp(s), 183
scissor fault, 183
Mission(s)
Cassini
    Huygens Probe, 177–178
    Orbiter, 177–179
COROT, 314
Deep Impact, 225
Deep Space 1, 230
Friendship, 34
GAIA, 346, 348
Galileo
    Orbiter, 153, 192
    Probe, 121, 125, 126
    thermal imagers, 164
Giotto, 230
HIPPARCOS, 295, 346
Hubble Space Telescope (HST), 314,
    342, 343, 348, 356–358
Kepler, 346
Long Duration Exposure Facility
    satellite, 249
Magellan, 110
Mariner(s), 107–108
Mars Global Surveyor, 113, 162
Mars Opportunity Rover, 259
Mars Spirit Rover, 259
Messenger, 107
MOST, 364
NEAR Shoemaker, 292, 298
Phobos, 163
Pioneer(s)
    Pioneer 10, 125
    Pioneer 11, 125, 131
    Pioneer Venus, 37, 44, 111–112
Space Interferometry (SIM), 348, 350
Spitzer Space telescope, 240, 344,
    360, 364, 365
Stardust, 230
Vega, 232, 305, 352, 368
Venera, 37, 42, 44
Viking(s)
    Orbiter(s), 163

Vostok, 34
Voyager(s)
    Voyager 1, 125, 129, 131, 171, 202
    Voyager 2, 121, 132–136, 183–188,
        198, 200–201
Mixing
    lines, 285
    ratio, 11, 51, 64
    time, 47–48
Moldavite, tektites, 276
Molecular
    charge transfer, 53
    ions, 58, 61
    photodissociation, 61
    photoexcitation, 61
    thermal speeds, 68
    viscosity, 17
    weight, 1–2, 5, 6, 35, 47, 121
Molecular cloud(s)
    density, 250
    mass, 306
    radius, 251
Monatomic gas(es), 10
The Moon
    orbital evolution, 203
Müller, Johann (Regiomontanus), 214

Naiad (satellite, Neptune III), 200
Natural Satellite Data Center, 153
Neptune
    angular diameter, 134
    atmosphere
        Great Blue/Dark Spot, 134
        the "scooter", 135
    composition, 200
    discovery, 133, 205
    electrolytic sea, 136
    excess radiation, 54
    greenhouse effect, 54, 135
    magnetic field, 136
    magnetosphere, 136
    mean density, 136
    pressure(s)
        central, 136
        internal, 136
    properties, 136–137
    rings
        arcs, 200–203
        clumpiness, 200
    satellites, 184–186, 200, 204–206
        orbits, 157
        physical properties, 161

Neptune (*Cont.*)
  winds, 135
Nereid (satellite, Neptune II)
  orbital elements, 153
  orbital properties, 205
Newton, Isaac, 217
Nitrogen (N, N_2), 180, 185, 186, 205
  sublimation, 185
  triple point, 185
Noise
  improvement, 339
Nüremberg Chronicles, 228

Oberon (satellite, Uranus IV), 151, 153,
    156, 160, 184, 204
Observatories, 134, 162, 187–188, 189,
    229, 313–314, 340
  Harward college, 340
  Paris, 314
  Sproul, 313
Occultation(s), 6, 121, 129, 161, 174,
    198, 200, 201, 206, 239, 294,
    295, 314, 343, 345, 360, 365
Olbers, Heinrich Wilhelm Matthias,
    225–226, 287
Olivine, 236, 239, 261, 262–264, 271,
    273–275, 299
Oort Cloud, 226, 237–240, 302–303
  inner, 238
  outer, 238
  population, 238
  radius, 238
Oort, Jan, 237
Ophelia (satellite, Uranus VII), 156,
    160, 198
Optical depth, 6, 305
Optical Gravitational Lensing
    Experiment (OGLE), 349
Optics
  active, 294
  adaptive, 336
Ordinary ray(s), 140
Origen, 213
Orthopyroxene, 236, 264, 271, 275, 285
Osculating orbits, 228
Osiris, 359
Oxygen, 268, 284
Oxygen ions, 142, 145
Oxygen isotopes, 270ff, 285
Ozone (O_3), 31–34, 36–37, 48
  concentration, 33
  production, 32–33

PAH (polycyclic aromatic compound),
    276
Pan (satellite, Saturn XVIII), 155, 159,
    173, 174, 201, 203
Pandora (satellite, Saturn XVII), 155,
    159, 195, 196, 201
Paul of Tarsus (St. Paul), 258
Perfect gas law, 5, 7, 125
Perseids, 242
Petrography, 259, 275
Petrologic classes, 263
Petrology, 259
Phobos (satellite, Mars I), 152, 154,
    158, 162–163, 207, 298
  decaying orbit, 163
Phoebe (satellite, Saturn IX), 155, 159,
    184, 185, 204
Phosphine(PH_3), 125
Photodissociation, 32, 53, 58, 60, 61,
    65, 181
Photolysis, 52, 58, 65, 66, 124,
    180–181, 233
Piazzi, Giuseppe, 286–287
Planet/planetary
  definition(s), 353–356
  distinction from brown dwarfs,
    367–371
  evolution, 360
  migration, 307, 351, 356, 368
  origins, 367–371
    core-accretion, 368–369
    gravitational instability, 368–369
Plasma frequency, 70–75, 139–140
Plasmoids, 103
Plessite(s), 277
Pliny the Elder, 214
Plutino(s), 186, 206, 289, 290,
    293, 353
Pluto (dwarf planet), 152, 188, 206,
    288, 353
  atmosphere, 188
  comparison with Eris, 153, 188, 206
  comparison with Triton, 188,
    205–207
  discovery, 187–188
  interior, 188
  orbit/orbital, 186
    elements, 157
    properties, 161
    resonance with Neptune, 186–187
  satellites, 157, 161, 204–207
  seasons, 188

spin–orbit coupling, 188
temperature(s), 188
Plutoid, 353
Polarization drift, 87–88
Polarized waves, 72–75
Polyacetylene(s), 182
Polycyclic aromatic
        hydrocarbons(PAH), 276
Portia (satellite, Uranus XII), 157, 160,
        199
Posidonius, 214
Poynting–Robertson drag, 253
Poynting–Robertson effect, 251, 253
Pressure
    density relation, 7
    gradient, 5, 17, 19
        force, 16–17, 18, 19, 20–22, 38, 41,
            42, 122
        high/low, 18, 20, 21, 22, 24–25, 27,
            42, 123, 177
    partial, 45–47, 126, 185
    radiation, 229, 251–253, 301–302
    scale height, 6, 19, 34, 35, 38, 41, 55,
        59, 100, 147, 359
        equation, 6
Prometheus (satellite, Saturn XVI),
        155, 159, 165, 195, 196, 201
Propane (C_3H_8), 182
Propylene (C_3H_6), 182
Proteus (satellite, Neptune VIII), 153,
        157, 161
Proto-planetary disks (proplyds), 285,
        305–307, 352, 369
    dead zone, 307, 368
    detection techniques, 313
        coronographic, 313, 350
    heterogenity, 285
    lifetime, 369
    mass, 357, 368–369
    model, 306
    Orion, 305
Puck (satellite, Uranus XV), 157, 160,
        199
Pulsar timings, 351
Purbach, Georg von, 214, 228

Quadrans muralis, 242

Radiants, 243
Radiation, 31, 33, 36–37, 40, 57, 59,
        61–63, 65, 95, 98, 103, 107,
        140, 145, 151, 202, 232, 246,
        283, 296, 301–302, 360, 362
    excess, 54, 137–139
    pressure, 229, 251–253
    coefficient, 252
Radioactive isotopes, 272, 279
    aluminum 26 (^{26}Al), 268, 305
    argon 39 (^{39}Ar), 283
    rubidium 87 (^{87}Rb), 279–280, 283
    uranium 238 (^{238}U), 280, 282
Radioactivity, 152–153
Radio wave reflection, 63–64
Ratio of specific heats, 7, 10
Red giant wind, 305
Refraction, 6, 63, 72–74, 139–140
Refractory material, 268, 302
Regiomontanus (Johann Müller), 214
Rhea (satellite, Saturn V), 153, 155,
        159, 173, 177, 196
Ries crater, 276
Right-hand rule, 12–15, 79, 83,
        84, 89, 91
Ring/ ring structures, 192–203
    albedoes, 151–207
    Jupiter, 192–193
    Neptune, 200–201, 203
    origin, 201–203
    particle sizes, 192
    resonances, 202
    Rhea, 174, 196
    Saturn, 192–197, 201
    scattering function, 202
    spiral waves, 202
    Uranus, 197–200
    vertical oscillations, 203
Roche limit, 190–191, 203
Rosalind (satellite, Uranus XIII), 157,
        160, 199
Rossby number, 38–39, 122
Rossiter/Rossiter–McLaughlin effect,
        359
Rotational flattening, 15
Rothmann, Christoph, 216
Russell, Henry Norris, 346

Sallamu, 218
SALR (saturated adiabatic lapse rate),
        32
Saturn, 15, 54, 119–121, 125, 127,
        128–131, 196, 206, 216
    atmosphere/atmospheric, 216
    belts, 129

Saturn (*Cont.*)
   storms, 130
   velocities, 129
   zones, 129
  aurora(e), 130
  color index, 314, 346
  electroglow, 131, 133
  excess radiation, 54, 137–139
  helium abundance, 127, 133
  helium "rain", 54, 126–127, 138
  magnetic field, 129
  magnetosphere, 131
  metallic hydrogen, 127
  period(s), 128
   revolution, 175, 180
   rotational, 129
  plasma disk, 129
  poles, 129, 130
  pressure(s)
   central, 137
   internals, 136–137
  ring(s), 192–197
   A ring, 192, 194, 195
   associated satellites, 195
   B ring, 194, 195, 197
   Cassini Division, 192, 194–195
   composition, 195
   C ring, 194, 195
   discovery, 192–194
   Encke gap, 194, 195, 196, 201
   E ring, 194, 195, 196
   F ring, 194, 195, 196
   G ring, 194, 195
   Keeler gap, 194, 195, 196
   kinked, 201
   Maxwell gap, 194
   resonances, 202–203
   spokes, 196, 197, 202
   temperatures, 196, 197
  satellites, 151, 153, 173ff
   co-orbital, 174, 195–196
   orbits, 155–156
   physical properties, 159–160
   propeller-shaped moonlets, 173, 174
  temperature(s), 127
Scattered disk, 238
Schwarzschild radius, 348
Seneca, 214, 290
Shepherding satellites, 198, 200, 201
Siderophilic elements, 267–268
Silica, 245

Silicates, 232, 236, 250, 256, 260ff, 271, 274, 284, 302
Silicate emission, 236
SIMBAD database, 314
Sirocco, 55
Snell's law, 73–74
The "Snow Line", 272
Sodium, 145–146, 164, 165, 343
Sodium chloride (NaCl), 165
Solar apex, 237
Solar constant(s), 31, 35
Solar wind, 69–70, 75–78, 91, 99, 100–103, 107, 108, 110–114, 121, 140, 143, 145, 146, 170, 229–230, 233, 234, 251, 272, 274, 301, 302
Spallation products, 283
Specific angular momentum, 207, 224
Specific heat(s)
  capacity, 9
  per unit mass, 9, 11
Spectrograph(s)/spectrographic
  CORALIE, 339
  echelle, 339
  ELODIE, 339
  precision, 339
Spectroscopic
  binaries, 336, 339–342, 368
Sputtering, 112–113, 147
Stable isotopes, 283
  argon, 283
  oxygen, 284
  strontium, 283–284
Standard mean ocean water (SMOW), 284
Star of Bethlehem, 213, 228
Star clusters, 337, 352, 356–357
  47 Tuc, 356–357
  Hyades, 357
  NGC 6791, 357
Stars (classes)
  brown dwarfs, 315, 334–338, 346, 350ff, 367ff
  L dwarfs, 338, 354
  K-stars, 269, 368
  M-stars, 269, 368
  methane dwarf, 355
  red dwarfs, 363
  red giants, 357
  solar-like, 339
  spectroscopic binaries, 336, 339–342, 368

double-line (SB2), 336, 342
single-line (SB1), 339–341,
    343–344, 358
T dwarfs, 336, 355–356
white dwarfs, 337, 352
Stars (individual), 316–323, 324–333
2MASSW J1207334-393254, 318, 350
Barnard's Star, 313
Cha 110913-773444, 351
Gliese876, 348
HD 168443, 320, 330, 335, 336, 356
HD 189733, 320, 332, 343–344, 358,
    360
HD 209458, 314, 322, 332, 342–345,
    355–356, 358–360
Lalande 21185, 348
PSR 1257+12, 313, 337, 352
PSR B1620-26, 337, 352
Ross 780, 348
SCR J1845-6357, 351
SN 1572, 215
CM Draconis, 352
ε Eridani, 305, 352, 363–364
GQ Lupi, 351–352
α Lyrae (Vega), 305, 352, 368
β Lyrae, 341
51 Pegasi, 313
β Pictoris, 305, 313, 352
AB Pictoris, 313
α Piscium Austrini (Fomalhaut), 305
ε Tauri, 357
34 Tauri(Uranus), 131
Star(s)/stellar (general)
limb-darkening, 358
line profiles, 339, 359
magnetic activity cycles, 339
metallicity/metallicities, 356–357,
    363
rotation, 339
spectral types, 260–265
Sulfur, 52–53
boiling point, 164–165
Sulfur dioxide (SO_2), 35, 46, 51, 146,
    147–148, 165
Sulfur oxide (SO), 165
Sung-shih annals, 218
Sun/solar
apex, 237
corona(e), 151–152
    F-corona, 250
disturbed current(s), 103
flare(s), 62, 64, 70, 90, 99, 272

flux, 30, 31, 35, 92, 99
luminosity, 2, 150, 251, 295, 348
nebula, 267, 271–272, 282, 284–285,
    302, 303–305
photon momentum, 251
quiet current(s), 90
wind, 69–70, 75–78, 91, 99, 100–103,
    107, 108, 110, 111–112
Supernova, 217, 304, 348
Survey(s), 131, 165, 216, 334, 336–337,
    344–345, 348–349, 351,
    355, 365
Optical Gravitational Lensing
    Experiment (OGLE), 349
Swift, Jonathon, 161–162
Szebehely, Victor, 204

Taenite crystal(s), 277
Taylor instability, 124
T dwarf(s), 336, 355, 356
Gliese 229B, 335, 355
Tektite(s), 276
Telescopes
Canada–France–Hawaii (CFHT), 313
Frederick C. Gillette (Gemini
    North), 236
Hubble Space (HST), 34, 130,
    188–189, 348
Keck, 200
Spitzer Space, 240–241, 344, 360,
    364–365
Subaru, 236
Temperature(s)
effective, 120, 137, 315, 345, 371
equilibrium, 1–2, 31, 54, 133, 138,
    164, 188, 296, 356, 359,
    370, 371
Tethys (satellite, Saturn III), 153, 155,
    159, 173, 194, 196
Tetraacetylene (C_8H_2), 182
Thalassa (Satellite,Neptune IV), 200
Thebe (satellite, Jupiter XIV), 154,
    158, 193
Thermal inertia, 34, 36, 39–40, 44
Tholins, 239
Tide(s)/tidal, 90, 190–191, 203, 257
action, 152, 240
bulge, 203
critical density, 166
critical distance, 190–191, 204
disruption criteria, 189–190, 191
friction, 191, 203

Tide(s)/tidal (*Cont.*)
  instability, 190
  -raising force, 190
  Roche limit, 190–191, 203–204
Tisserand, François Félix, 224
Tisserand invariant/parameter, 224,
    238–239
Tisserand's criterion, 224, 239
Titania (satellite, Uranus III), 151, 153,
    156, 160, 184
Titan (satellite, Saturn VI), 133, 151,
    153, 155–156, 159–160, 168,
    175–183, 203, 286, 369
  atmosphere, 176, 178, 180–183
    chemistry, 182
    cloud(s), 176, 178–179
    composition, 180
    polar storms, 177
    superrotation, 180
    thermosphere, 183
  "cat scratches", 177
  crustal material, 177
  hydrogen torus, 183
  lake(s), 176–178
  methane cycle, 177–178, 180–181
  orbit/orbital, 175–177, 183
    elements, 155
    properties, 159
  panoramic view, 178
  polar lakes, 177
  rotation period, 175, 180
  surface images, 175, 176, 179
  temperature(s), 180, 183
  tide(s)/tidal, 175, 181
    heating, 181
  water retention, 183
Titius, Daniel von Wittenberg, 286–287
Toscanelli, Paolo, 214
Triacetylene (C_6H_2), 182
Triton (satellite, Neptune I), 151, 153,
    157, 161, 184, 185–186, 187,
    188, 189, 205, 206–207
  atmosphere/atmospheric, 185–186
    cloud(s), 185–186
    composition, 185–186
    pressure, 185
    thermosphere, 185
  "cantaloupe" terrain, 185, 187
  color, 185
  cryovolcanoe(s), 185, 187
  geyser(s), 185, 187
  impact crater(s), 185–186

  Mazomba, 185–186
  mean density, 185
  nitrogen geyser(s), 185, 187
  orbit/orbital, 154–157
    elements, 157
    properties, 161
  seasons, 188
  surface relief, 185
  synchronous rotation, 186
T Tauri stage, 272, 301
T'ung-k'ao annals, 218
Twain, Mark (Samuel Clemens),
    228, 229
Tycho Brahe, 215–216

Umbriel (satellite, Uranus II), 151, 153,
    156, 160, 184
  albedo, 184
Universal gas constant, 6, 9
Uranus, 131–133, 183–184, 198–200
  angular size, 132
  atmosphere, 132–133
  brightness (opposition magnitude),
    132
  color, 132
  composition, 132
  density profiles, 132
  discovery, 131–132
  electroglow, 133
  He/H ratio, 145
  helium abundance, 133
  internal pressure, 136–137
  lightning, 133
  magnetic field, 133
  magnetosphere, 54
  ring(s), 198–200
    discovery, 198
    smoke, 202
    variation, 198
  satellites, 133, 151, 153, 183–184
    orbits, 156–157
    physical properties, 160–161
  seasons, 133
  superheated water, 133

Van de Kamp, Peter, 313
Vector, 4, 11–13, 15–16, 21, 79, 83, 117
  cross-product, 12
The Venerable Bede, 214
Venus
  atmosphere/atmospheric, 36
    loss, 36

cycles, 66
  sulfur, 66
dominant ions, 112–113
exosphere, 112–113
heat flux, 110
ionopause, 111
ionosphere/ionospheric, 110–111
  diamagnetic current, 111–112
  electron density, 111
  plasma pressure, 111
  processes, 111
  solar wind interaction, 110–113
limit on magnetic dipole moment,
  110
lower atmosphere properties, 112
magnetic field pressure, 111
magnetocavity, 112
rotation, 110
surface cratering, 110
Vesta (asteroid), 188, 287, 298
Virial theorem, 303, 304
Volatile substance, 268
Volcano(es), 147, 164–166, 177,
  179, 232
Von Zach, Baron Franz Xaver (János
  Ferenc), 286–287

Water, 16, 20, 44, 65, 69, 85, 128, 133,
  166, 177, 180, 183, 231, 236,
  260, 269, 272, 284, 369, 370
  ice, 44, 126, 166, 169, 171, 173, 178,
    181, 185, 188, 195, 205, 231,
    232, 239

vapor, 11, 31, 44, 55, 128, 239, 370
Whipple, Fred, 231
Widmanstätten figures, 277
Wildt, Rupert, 132
Winds, 16–17, 18–19, 20–25, 28–29, 36,
  39–40, 41–43, 47, 69–70, 91,
  99, 100–103, 107, 108,
  110–114, 121, 140, 143, 145,
  146, 170, 229–230, 233, 234,
  251, 272, 274, 301, 302
Chinook, 55
cyclostrophic, 41–42
easterlies, 24–25, 29
Föhn, 55
geostrophic, 18–19, 20–22, 43, 122
Sirocco, 55
solar, 69–70, 75–78, 91, 99, 100–103,
  107, 108, 110–114, 121, 140,
  143, 145, 146, 170, 229–230,
  233, 234, 251, 272, 274,
  301, 302
thermal, 20–22, 28, 39, 40
trade, 22–25, 29, 42–43, 122
westerlies, 22–27, 29, 43, 122
zonally averaged distribution, 29

Yarkovsky effect, 301
Yarkovsky force, 301

Zodiacal light, 250